C0-GEG-960

NORTH DAKOTA
STATE UNIVERSITY

DEC 1 1 2000

SERIALS DEPT.
LIBRARY

WITHDRAWN

Selected Titles in This Series

196 **Yu. N. Lin′kov,** Asymptotic statistical methods for stochastic processes, 2001

195 **Minoru Wakimoto,** Infinite-dimensional Lie algebras, 2001

194 **Valery B. Nevzorov,** Records: Mathematical theory, 2001

193 **Toshio Nishino,** Function theory in several complex variables, 2001

192 **Yu. P. Solovyov and E. V. Troitsky,** C^*-algebras and elliptic operators in differential topology, 2001

191 **Shun-ichi Amari and Hiroshi Nagaoka,** Methods of information geometry, 2000

190 **Alexander N. Starkov,** Dynamical systems on homogeneous spaces, 2000

189 **Mitsuru Ikawa,** Hyperbolic partial differential equations and wave phenomena, 2000

188 **V. V. Buldygin and Yu. V. Kozachenko,** Metric characterization of random variables and random processes, 2000

187 **A. V. Fursikov,** Optimal control of distributed systems. Theory and applications, 2000

186 **Kazuya Kato, Nobushige Kurokawa, and Takeshi Saito,** Number theory 1: Fermat's dream, 2000

185 **Kenji Ueno,** Algebraic Geometry 1: From algebraic varieties to schemes, 1999

184 **A. V. Mel′nikov,** Financial markets, 1999

183 **Hajime Sato,** Algebraic topology: an intuitive approach, 1999

182 **I. S. Krasil′shchik and A. M. Vinogradov, Editors,** Symmetries and conservation laws for differential equations of mathematical physics, 1999

181 **Ya. G. Berkovich and E. M. Zhmud′,** Characters of finite groups. Part 2, 1999

180 **A. A. Milyutin and N. P. Osmolovskii,** Calculus of variations and optimal control, 1998

179 **V. E. Voskresenskiĭ,** Algebraic groups and their birational invariants, 1998

178 **Mitsuo Morimoto,** Analytic functionals on the sphere, 1998

177 **Satoru Igari,** Real analysis—with an introduction to wavelet theory, 1998

176 **L. M. Lerman and Ya. L. Umanskiy,** Four-dimensional integrable Hamiltonian systems with simple singular points (topological aspects), 1998

175 **S. K. Godunov,** Modern aspects of linear algebra, 1998

174 **Ya-Zhe Chen and Lan-Cheng Wu,** Second order elliptic equations and elliptic systems, 1998

173 **Yu. A. Davydov, M. A. Lifshits, and N. V. Smorodina,** Local properties of distributions of stochastic functionals, 1998

172 **Ya. G. Berkovich and E. M. Zhmud′,** Characters of finite groups. Part 1, 1998

171 **E. M. Landis,** Second order equations of elliptic and parabolic type, 1998

170 **Viktor Prasolov and Yuri Solovyev,** Elliptic functions and elliptic integrals, 1997

169 **S. K. Godunov,** Ordinary differential equations with constant coefficient, 1997

168 **Junjiro Noguchi,** Introduction to complex analysis, 1998

167 **Masaya Yamaguti, Masayoshi Hata, and Jun Kigami,** Mathematics of fractals, 1997

166 **Kenji Ueno,** An introduction to algebraic geometry, 1997

165 **V. V. Ishkhanov, B. B. Lur′e, and D. K. Faddeev,** The embedding problem in Galois theory, 1997

164 **E. I. Gordon,** Nonstandard methods in commutative harmonic analysis, 1997

163 **A. Ya. Dorogovtsev, D. S. Silvestrov, A. V. Skorokhod, and M. I. Yadrenko,** Probability theory: Collection of problems, 1997

For a complete list of titles in this series, visit the
AMS Bookstore at **www.ams.org/bookstore/**.

Asymptotic Statistical Methods for Stochastic Processes

Translations of
MATHEMATICAL
MONOGRAPHS

Volume 196

Asymptotic Statistical Methods for Stochastic Processes

Yu. N. Lin'kov

American Mathematical Society
Providence, Rhode Island

EDITORIAL COMMITTEE

AMS Subcommittee
Robert D. MacPherson
Grigorii A. Margulis
James D. Stasheff (Chair)
ASL Subcommittee Steffen Lempp (Chair)
IMS Subcommittee Mark I. Freidlin (Chair)

Ю. Н. Линьков

АСИМПТОТИЧЕСКИЕ МЕТОДЫ СТАТИСТИКИ
СЛУЧАЙНЫХ ПРОЦЕССОВ

НАУКОВА ДУМКА, КИЕВ, 1993

Translated from the Russian by Victor Kotov

2000 *Mathematics Subject Classification.* Primary 62Mxx; Secondary 60Gxx.

Abstract. For increasing number of observations, asymptotic methods of statistics of semimartingale type processes are considered. Local densities of probability measures generated by semimartingales are introduced, and their asymptotic properties are investigated for various types of asymptotic distinguishability of the corresponding families of statistical hypotheses. Certain problems of simple hypothesis testing and estimation of an unknown parameter from observation of semimartingales, as well as some asymptotic information-theoretic problems of statistics, are solved.

This book can be used by researchers and graduate students working in statistics of stochastic processes and its applications.

Library of Congress Cataloging-in-Publication Data

Lin′kov, IU. N.
[Asimptoticheskie metody statistiki sluchainykh protsessov. English]
Asymptotic statistical methods for stochastic processes / Yu. N. Lin′kov.
p. cm. — (Translations of mathematical monographs, ISSN 0065-9282 ; v. 196)
Includes bibliographical references and index.
ISBN 0-8218-1183-5 (alk. paper)
1. Mathematical statistics—Asymptotic theory. 2. Semimartingales (Mathematics) I. Title. II. Series.
QA276.L548713 ~~2000~~ 2001
519.5—dc21 00-045349

Copying and reprinting. Individual readers of this publication, and nonprofit libraries acting for them, are permitted to make fair use of the material, such as to copy a chapter for use in teaching or research. Permission is granted to quote brief passages from this publication in reviews, provided the customary acknowledgment of the source is given.

Republication, systematic copying, or multiple reproduction of any material in this publication is permitted only under license from the American Mathematical Society. Requests for such permission should be addressed to the Assistant to the Publisher, American Mathematical Society, P. O. Box 6248, Providence, Rhode Island 02940-6248. Requests can also be made by e-mail to reprint-permission@ams.org.

© 2001 by the American Mathematical Society. All rights reserved.
The American Mathematical Society retains all rights
except those granted to the United States Government.
Printed in the United States of America.

∞ The paper used in this book is acid-free and falls within the guidelines
established to ensure permanence and durability.
Visit the AMS home page at URL: http://www.ams.org/

10 9 8 7 6 5 4 3 2 1 06 05 04 03 02 01

Contents

Preface ix

Basic Notation xiii

Chapter 1. Local Densities of Measures and Limit Theorems for Stochastic Processes 1
- 1.1. Basic notions of the theory of stochastic processes 1
- 1.2. Statistic experiments generated by stochastic processes 9
- 1.3. Limit theorems for semimartingales 21

Chapter 2. Asymptotic Distinguishing between Simple Hypotheses in the Scheme of General Statistical Experiments 33
- 2.1. Statistical hypotheses and tests 33
- 2.2. Types of asymptotic distinguishability between families of hypotheses and their characterization 36
- 2.3. Complete asymptotic distinguishability under the conditions of the law of large numbers 45
- 2.4. Complete asymptotic distinguishability under conditions of weak convergence 52
- 2.5. Contiguous families of hypotheses 58
- 2.6. The case of asymptotic expansion of the likelihood ratio 65
- 2.7. Reduction of the problem of testing hypotheses 70

Chapter 3. Asymptotic Behavior of the Likelihood Ratio in Problems of Distinguishing between Simple Hypotheses for Semimartingales 79
- 3.1. Hellinger integrals and Hellinger processes 79
- 3.2. Limit theorems for the likelihood ratio 83
- 3.3. Asymptotic decomposition of the likelihood ratio in parametric formulation 88
- 3.4. Observations of diffusion-type processes 98
- 3.5. Observations of counting processes 112

Chapter 4. Asymptotic Estimation of Parameters 137
- 4.1. Formulation of the problem 137
- 4.2. Properties of the normalized likelihood ratio for semimartingales 140
- 4.3. Observations of diffusion-type processes 151
- 4.4. Observations of counting processes 156

Chapter 5. Asymptotic Information-Theoretic Problems in Parameter Estimation 161
5.1. Asymptotic behavior of the Shannon information in observations with respect to an unknown parameter 161
5.2. Lower bounds for the information about a statistical estimate of a parameter 174
5.3. Bounds for risk functions of consistent estimates 186
5.4. Observations of semimartingales 194

Bibliographical Notes 197

References 203

Index 213

Preface

The asymptotic properties of the likelihood ratio play an important part in solving problems in statistics for various schemes of observations. Wald [**195, 196**] and Le Cam [**168, 169**] were the first to commence developing asymptotic methods in mathematical statistics based on asymptotic properties of the likelihood ratio. At first, sequences of independent random variables were considered using the central limit theorem for the logarithm of the likelihood ratio. These investigations gave rise to the notion of the local asymptotic normality (LAN) of a family of probability measures generated by observed random variables [**169**]. Later on Hájek and Šidák [**14**], Chibisov [**125**], Roussas [**110**], Ibragimov and Has'minskii [**40**], Dzaparidze [**29**], and others developed a rather general asymptotic theory of parameter estimation and hypothesis testing based on asymptotic properties of the likelihood ratio for sequences of mutually dependent random variables.

The extension of statistical methods to time-continuous stochastic processes attracted the attention of many scientists. Among the first works in this field was the work by Ulf Grenander [**24**], which marked the beginning of active research on developing statistical methods for Gaussian and stationary processes [**2, 32, 37, 40, 41**]. Many asymptotic methods for estimating Gaussian diffusion-type processes based on the LAN property can be found in the book by Ibragimov and Has'minskii [**40**]. The extension of the asymptotic methods of mathematical statistics based on the central limit theorem and the LAN property to non-Gaussian and nonstationary stochastic processes gave rise to new ideas in the theory of stochastic processes.

In recent years convenient formulas for densities of probability measures generated by stochastic processes were obtained and limit theorems for various stochastic processes were proved. Eventually, the asymptotic methods became an important tool in studying diffusion-type [**57, 62, 68, 103**] and counting [**57, 71, 78, 183, 186**] processes, processes with independent increments [**73, 129, 163, 199**], Markov processes [**77, 132, 162**], and semimartingales [**83, 84, 119, 166**].

In this book we describe the asymptotic methods for parameter estimation and hypothesis testing based on asymptotic properties of the likelihood ratio in the case where an observed stochastic process is a semimartingale. Semimartingales form a rather wide class of stochastic processes, which include diffusion-type and counting processes, processes with independent increments, Markov processes, and others. In this book we consider only right-quasicontinuous semimartingales. This limitation allowed us to simplify the presentation of the asymptotic method and to make it accessible to engineers.

As was observed by Chibisov (see the corresponding remark in [**110**]) and Ibragimov and Has'minskii [**40**], the asymptotic method developed by Wald and Le Cam is rather general by its nature. It can be applied to any model of observations for which the likelihood ratio possesses the properties required by this method. Therefore, further development of the method of Wald and Le Cam and its application to particular models of observations reduce to finding specific restrictions that must be imposed on the likelihood ratio. In many cases these restrictions appear to be common to a variety of schemes of observations. Following [**40**], this fact was substantially used in this book. When discussing any statistical problem, we first deduce general results for observations of arbitrary nature and then modify them to the case of observations of martingales. Thereafter these results are applied to diffusion-type and counting processes, which are, in this book, the main models for demonstrating the efficiency and checking the correctness of our theory. The choice of these two particular schemes of observations is justified by their importance in solving various statistical problems [**12, 47, 50, 101, 120, 190**].

Chapter 1 contains general basic notions and results for stochastic processes, which are used throughout the rest of the book. The facts concerning the notion of a martingale and its generalizations are given. Certain classes of stochastic processes are introduced. Random measures; stochastic integrals with respect to local martingales, random measures, and semimartingales; and other notions are defined. The Itô formula for semimartingales and the Lenglart inequality are presented. Statistical experiments generated by observations of semimartingales are introduced, and formulas for the likelihood ratio are given. Limit theorems for semimartingales and, in particular, central limit theorems for local martingales are formulated.

Chapters 2 and 3 are devoted to the problem of distinguishing between two simple statistical hypotheses. In Chapter 2 a general scheme of statistical experiments is considered. Certain types of asymptotic distinguishability between families of hypotheses are introduced. These types are characterized in terms of the likelihood ratio, Hellinger integral of order ε, Kakutani–Hellinger distance, and the distance in variation between hypothetical measures, etc. The problem of complete asymptotic distinguishability is discussed. In the case of complete asymptotic distinguishability the behavior of error probabilities in the Neyman–Pearson test is investigated for various kinds of behavior of the likelihood ratio; namely under the following conditions: the law of large numbers is fulfilled, the theorem on large deviations of the logarithm of the likelihood ratio holds, properly centered and normalized logarithm of the likelihood ratio weakly converges. In the case of continuous families of hypotheses the behavior of error probabilities of the Neyman–Pearson test is investigated, provided the likelihood ratio weakly converges to some law under the null hypothesis. Asymptotic expansions of the likelihood ratio are considered for contiguous and noncontiguous families of hypothesis. At the end of Chapter 2 two reductions of the problem of hypothesis testing are considered. These reductions make it possible to take into account the behavior of the sets of singularity of hypothetical measures.

In Chapter 3 the results of Chapter 2 are used in statistical experiments generated by observations of semimartingales. All restrictions are formulated in predictable terms: either in terms of triples of predictable characteristics of semimartingales or in terms of Hellinger processes of order ε. Both the nonparametric case where there are two triples of characteristics and the parametric case where there is a family of predictable characteristics of a semimartingale that depend on an unknown parameter are considered. The results obtained are applied to studying various particular cases of diffusion-type and counting processes.

In Chapters 4 and 5 some problems of asymptotic estimation of unknown parameters are considered. In Chapter 4 the general limit theorems on asymptotic properties of maximum likelihood and Bayes estimates obtained by Ibragimov and Has'minskii [**40**] for observations of an arbitrary nature are applied to observations of semimartingales. The results obtained are used for studying diffusion-type and counting processes. In Chapter 5 an unknown parameter is assumed to be random, and under this condition certain information-theoretic problems of estimation of parameters are considered. The asymptotic behavior of the Shannon information contained in an observation of the unknown parameter is studied, and various methods of obtaining lower bounds for Shannon information contained in a statistical estimate for the unknown parameter are described. Based on these results, many informational inequalities for risk functions of statistical estimates analogous to the well-known Cramér–Rao and Hájek inequalities are derived. The results obtained are applied to general schemes of observations of semimartingales.

Clearly, it is impossible to consider all the problems of statistics of stochastic processes in one book. For example, we do not mention the problem of complex hypothesis testing, which is close to the results of Chapter 4. Some interesting results obtained in this area can be found in the works by Roussas [**110**], Ingster [**42, 44**], Burnashev [**11**], and others. In this connection the somewhat obsolete review [**6**] should also be mentioned.

The problems of statistics for the semimartingales that are not left-quasicontinuous have not been considered at all. The effect of the sets of singularity of hypothetical measures has not been discussed in detail: there is only one section devoted to the reduction of the problem of hypothesis testing. The important problem of applying the results of Chapter 5 to the study of asymptotic sufficiency has been omitted altogether.

The bibliography at the end of the book does not pretend to be complete. However the author tried to mention all the works that have played a significant part in the development of asymptotic methods in statistics of stochastic processes. For the English edition, the list of references was enlarged and the Bibliographical Notes revised.

Basic Notation

$A = (A^{ij})$ .... matrix with elements A^{ij}

$A' = (A^{ji})$ .... the transpose of a matrix A

$|A|$ .... the norm of a matrix A, $|A| = (\mathrm{Tr}\, AA')^{1/2}$

$a \wedge b$ .... the minimum of two numbers $a, b \in \mathbf{R}$, $a \wedge 0 = -a^-$

$a \vee b$ .... the maximum of two numbers $a, b \in \mathbf{R}$, $a \vee 0 = a^+$

$\mathcal{B}(A)$ .... the Borel σ-algebra of subsets of A, $\mathcal{B}(\mathbf{R}^k) = \mathcal{B}^k$, $\mathcal{B}(\mathbf{R}_0^k) = \mathcal{B}_0^k$, $\mathcal{B}_0^1 = \mathcal{B}_0$, $\mathcal{B}^1 = \mathcal{B}$, $\mathcal{B}(\mathbf{R}_+) = \mathcal{B}_+$

$\mathbf{D}$, $\widetilde{\mathbf{D}}$, $\mathbf{D}_y$ .... variances with respect to measures P, $\widetilde{\mathsf{P}}$, P_y

$\det A$ .... determinant of a matrix A

E, $\widetilde{\mathsf{E}}$, E_y .... expectation with respect to measures P, $\widetilde{\mathsf{P}}$, P_y

$f = (f_x)_{x \in X}$ .... a function defined on X

I_k .... the unit matrix of order k

$\mathcal{L}(Y|\mathsf{P})$ .... the distribution law of a vector Y with respect to a measure P

$\mathbf{N}$ .... the set of positive integers

$\mathcal{N}(a, B)$ .... the normal (Gaussian) law with vector of means a and covariance matrix B

$\mathsf{P} \ll \mathbf{Q}$ .... absolute continuity of a measure P with respect to a measure $\mathbf{Q}$

$\mathsf{P} \sim \mathbf{Q}$ .... equivalence (mutual absolute continuity) of measures P and $\mathbf{Q}$

$\mathsf{P} \perp \mathbf{Q}$ .... singularity of measures P and $\mathbf{Q}$

$P^t\text{-}\lim_{t\to\infty} Y_t = c$ .... means that $\lim_{t\to\infty} P^t\{|Y_t - c| > \varepsilon\} = 0$ for all $\varepsilon > 0$

$\mathbf{R}^k$ .... k-dimensional Euclidean space with a fixed orthonormal basis with points $x = (x^1, x^2, \ldots, x^k)'$, $\mathbf{R}^1 = \mathbf{R}$, $\mathbf{R}^k \setminus \{0\} = \mathbf{R}_0^k$, $\mathbf{R}_0^1 = \mathbf{R}_0$, $\mathbf{R}_+ = [0, \infty)$, $\overline{\mathbf{R}}_+ = [0, \infty]$

$\mathrm{Tr}\, A$ .... the trace of a matrix A

$|x|_p$ .... the norm of a vector $x \in \mathbf{R}^k$ for $p \in (0, \infty)$, $|x|_p = \left(\sum_{i=1}^k |x^i|^p\right)^{1/p}$, $|x|_2 = |x|$

(x, y) .... scalar product of vectors $x, y \in \mathbf{R}^k$, $x'y = (x, y)$

δ^{ij} .... the Kronecker delta

Δf_x .... a jump of a function at a point $x \in \mathbf{R}$, $\Delta f_x = f_x - f_{x-}$

$(\Omega, \mathcal{F}, \mathsf{P})$ .... a probability space, where Ω is a set of points ω; $\mathcal{F}$, a σ-algebra of subsets of Ω; and P, a probability measure on $\mathcal{F}$

χ_A .... the indicator function, $\chi_A = \chi(A) = (\chi_x(A))_{x \in X}$

$\chi_x(A)$ .... the indicator of a set A, $\chi_x(A) = \chi(A; x)$

$\square$ .... the symbol of the end of a proof (Q.E.D.)

Classes of stochastic processes

$\mathcal{M}$ .... class of all uniformly integrable martingales
$\overline{\mathcal{M}}$ .... class of all martingales
$\mathcal{M}^c$ .... class of all uniformly integrable continuous martingales
$\overline{\mathcal{M}}^c$ .... class of all continuous martingales
$\mathcal{M}^d$ .... class of all uniformly integrable purely discontinuous martingales
$\overline{\mathcal{M}}^d$ .... class of all purely discontinuous martingales
$\mathcal{M}^2 = \{\xi \in \mathcal{M}: \sup_{t\in\mathbf{R}_+} \mathsf{E}\,|\xi_t|^2 < \infty\}$ .. class of all uniformly square-integrable martingales
$\overline{\mathcal{M}}^2 = \{\xi \in \overline{\mathcal{M}}: \mathsf{E}\,|\xi_t|^2 < \infty\ \forall t \in \mathbf{R}_+\}$ .. class of all square-integrable martingales
$\mathcal{V}^+$ .... class of all adapted right-continuous nondecreasing processes A with $A_0 = 0$ and $A_t < \infty$ (P-a.s.) for all $t \in \mathbf{R}_+$
$\mathcal{V} = \{\xi - \eta: \xi \in \mathcal{V}^+, \eta \in \mathcal{V}^+\}$ ...class of all processes with a locally bounded variation
$\mathcal{V}^c$ .... class of all processes of $\mathcal{V}$ with (P-a.s.) continuous trajectories
$\mathcal{A}^+ = \{\xi \in \mathcal{V}^+: \mathsf{E}\,\xi_\infty < \infty\}$ ... class of all integrable nondecreasing processes
$\mathcal{A} = \{\xi - \eta: \xi \in \mathcal{A}^+, \eta \in \mathcal{A}^+\}$ ...class of all processes with integrable variation
$\mathcal{K}(\mathbf{F}, \mathsf{P})$ .... class of all processes with respect to a filtration $\mathbf{F}$ and measure P
$\mathcal{K}_{\mathrm{loc}}$ .... the local class
$\mathcal{M}_{\mathrm{loc}}$ .... class of all local martingales
$\mathcal{M}^2_{\mathrm{loc}}$ .... class of all square-integrable local martingales
$\mathcal{O}$ .... class of all optional processes
$\mathcal{P}$ .... class of all predictable processes
$\widetilde{\mathcal{O}}$ .... class of all $\mathcal{O} \times \mathcal{B}$-measurable functions
$\widetilde{\mathcal{P}}$ .... class of all $\mathcal{P} \times \mathcal{B}$-measurable functions
$\mathcal{S}$ .... class of all semimartingales
$\mathcal{S}_p$ .... class of all special semimartingales

Classes of integrable functions

$\mathcal{L}^2_{\mathrm{loc}}(\xi) = \{f: f \in \mathcal{P}, f^2 \circ \langle\xi\rangle \in \mathcal{A}_{\mathrm{loc}}\}$
$\mathcal{L}^1_{\mathrm{loc}}(\xi) = \{f: f \in \mathcal{P}, (f^2 \circ [\xi,\xi])^{1/2} \in \mathcal{A}^+_{\mathrm{loc}}\}$
$\mathcal{L}_{\mathrm{loc}}(\xi) = \{f: f \in \mathcal{L}^1_{\mathrm{loc}}(M), f \circ A \in \mathcal{V}\}$
$\mathcal{G}^i_{\mathrm{loc}}(\nu) = \{f: f \in \widetilde{\mathcal{P}}, |f|^i * \nu \in \mathcal{A}^+_{\mathrm{loc}}\}$, $i = 1, 2$
$\mathcal{G}_{\mathrm{loc}}(\nu) = \{f: f \in \widetilde{\mathcal{P}}, |f|^2(1+|f|)^{-1} * \nu \in \mathcal{A}^+_{\mathrm{loc}}\}$

Classes of loss functions

$\mathbf{W}$ .... class of all loss functions
$\mathbf{W}'$ .... a subclass of loss functions in $\mathbf{W}$

$\mathbf{W}_p$ .... class of all loss functions in $\mathbf{W}$ that have a polynomial majorant
$\mathbf{W}'_p$ .... class of all functions in $\mathbf{W}'$ that have a polynomial majorant
$\widetilde{\mathbf{W}}'_p$ .... class of all functions $l(u)$ in $\mathbf{W}'_p$ for which $\inf_{|u|\geq A} l(u) \geq \sup_{|u|\leq A^\gamma} l(u)$, $\gamma > 0$, for all A

Some other symbols

$\mathfrak{M}(\mathbf{F})$ .... class of all finite stopping times with respect to $\mathbf{F}$
$\overline{\mathfrak{M}(\mathbf{F})}$ .... class of all stopping time with respect to $\mathbf{F}$
$\Phi(K)$ .... class of all normalizing matrices
$H \circ A$ .... the Lebesgue–Stieltjes integral
$f\cdot\xi$ .... the stochastic integral over a process ξ
$f * \mu$ .... the integral with respect to a random measure μ
$f * (\mu - \nu)$ .... the stochastic integral with respect to a local martingale measure $\mu - \nu$
$\delta_t^{+,\alpha}$ .... the Neumann–Pearson test of level α
$\hat{\theta}_t$ .... the maximum likelihood estimate of a parameter θ
$\tilde{\theta}_t$ .... the Bayes estimate of a parameter θ
$I(\xi^t, \theta)$ .... the amount of Shannon information contained in an observation ξ^t of an unknown parameter θ
$I(\bar{\theta}^t, \theta)$ .... the amount of Shannon information contained in an estimate $\bar{\theta}^t$ of parameter θ
$H_t(\varepsilon) = H(\varepsilon; \widetilde{P}^t, P^t)$ ..... the Hellinger integral of order ε for measures $\widetilde{P}^t$ and P^t
$I(P^t|\widetilde{P}^t)$ .... the entropy of a measure P^t with respect to a measure $\widetilde{P}^t$, the Kullback–Leibler information between measures P^t and $\widetilde{P}^t$
$(H^t) \vartriangle (\widetilde{H}^t)$ .... completely asymptotically distinguishable families of hypotheses
$(H^t) \overline{\vartriangle} (\widetilde{H}^t)$ .... families of hypotheses that are not completely asymptotically distinguishable
$(H^t) \cong (\widetilde{H}^t)$ .... completely asymptotically indistinguishable families of hypotheses
$(\widetilde{H}^t) \lhd (H^t)$ .... a family of hypotheses $(\widetilde{H}^t)$ is contiguous to a family of hypotheses (H^t)
$(\widetilde{H}^t) \overline{\lhd} (H^t)$ .... a family of hypotheses $(\widetilde{H}^t)$ is noncontiguous to a family of hypotheses (H^t)
$(H^t) \lhd\rhd (\widetilde{H}^t)$ .... mutually contiguous families of hypotheses
$(H^t) \overline{\lhd}\overline{\rhd} (\widetilde{H}^t)$ .... mutually noncontiguous families of hypotheses
$(H^t) \overline{\lhd\rhd} (\widetilde{H}^t)$ .... families of hypotheses are not mutually contiguous
$(P^t) \vartriangle (\widetilde{P}^t)$ .... completely asymptotically separable families of measures
$(P^t) \overline{\vartriangle} (\widetilde{P}^t)$ .... families of measures are not completely asymptotically separable
$(P^t) \cong (\widetilde{P}^t)$ .... completely asymptotically inseparable families of measures

$(\widetilde{P}^t) \lhd (P^t)$ .... a family of measures $(\widetilde{P}^t)$ is contiguous to a family of measures (P^t)

$(\widetilde{P}^t)\, \overline{\lhd}\, (P^t)$ .... a family of measures $(\widetilde{P}^t)$ is noncontiguous to a family of measures (P^t)

$(P^t) \lhd\rhd (\widetilde{P}^t)$ .... mutually contiguous families of measures

$(P^t)\, \overline{\lhd}\,\overline{\rhd}\, (\widetilde{P}^t)$ .... mutually noncontiguous families of measures

$(P^t)\, \overline{\lhd\rhd}\, (\widetilde{P}^t)$ .... families of measures are not mutually contiguous

CHAPTER 1

Local Densities of Measures and Limit Theorems for Stochastic Processes

1.1. Basic notions of the theory of stochastic processes

Classes of stochastic processes. Let $(\Omega, \mathcal{F}, \mathsf{P})$ be a probability space and let a nondecreasing right-continuous family $\mathbf{F} = (\mathcal{F})_{t\in\mathbf{R}_+}$ of σ-algebras $\mathcal{F}_t \subset \mathcal{F}$, called a *filtration*, be defined on the space $(\Omega, \mathcal{F})$ so that $\mathcal{F} = \bigvee_{t\in\mathbf{R}_+} \mathcal{F}_t$. The triple $(\Omega, \mathcal{F}, \mathsf{P})$ is called a *space with filtration*, and the quadruple $(\Omega, \mathcal{F}, \mathbf{F}, \mathsf{P})$ is called a *stochastic basis*.

The stochastic basis $(\Omega, \mathcal{F}, \mathbf{F}, \mathsf{P})$ is said to be *complete* or *satisfying ordinary conditions with respect to the measure* P if the σ-algebra $\mathcal{F}$ is P-complete and every σ-algebra $\mathcal{F}_t$ contains all P-null sets of $\mathcal{F}$. In this case the filtration $\mathbf{F}$ is also said to *satisfy ordinary conditions with respect to the measure* P [**30**].

The stochastic basis $(\Omega, \mathcal{F}, \mathbf{F}, \mathsf{P})$ can always be completed in the following way. Let $\mathcal{F}^{\mathsf{P}}$ be the P-completion of the σ-algebra $\mathcal{F}$, $\mathcal{N}^{\mathsf{P}}$ the set of all P-null subsets of $\mathcal{F}^{\mathsf{P}}$, and $\mathcal{F}_t^{\mathsf{P}}$ the minimal σ-algebra containing $\mathcal{F}_t$ and $\mathcal{N}^{\mathsf{P}}$. It is clear that $(\Omega, \mathcal{F}^{\mathsf{P}}, \mathbf{F}^{\mathsf{P}} = (\mathcal{F}_t^{\mathsf{P}})_{t\in\mathbf{R}_+}, \mathsf{P})$ is a new stochastic basis, which is called the *completion* of $(\Omega, \mathcal{F}, \mathbf{F}, \mathsf{P})$.

A family $\xi = (\xi_t)_{t\in\mathbf{R}_+}$ of maps from Ω to a set E is called a *stochastic process* (or simply a *process*). The set E is called the *state space of the process*. In most cases $E = \mathbf{R}^m$ for some $m \in \mathbf{N}$, unless otherwise stated. A stochastic process can also be regarded as a map from $\Omega \times \mathbf{R}_+$ to E, so that $(\omega, t) \to \xi_t(\omega)$. For a fixed $\omega \in \Omega$ the map $t \to \xi_t(\omega)$ is called a *trajectory of the process* ξ. Subsets of $\Omega \times \mathbf{R}_+$ are called *random sets*. Therefore the indicator χ_A of a random set A is a process.

For the sake of simplicity, the expressions (ξ_t), (A_t), $(\mathcal{F}_t)$, etc., will stand for the families $(\xi_t)_{t\in\mathbf{R}_+}$, $(A_t)_{t\in\mathbf{R}_+}$, $(\mathcal{F}_t)_{t\in\mathbf{R}_+}$, etc., unless otherwise indicated. Similarly, instead of $(\tau_n)_{n\in\mathbf{N}}$, $(x_n)_{n\in\mathbf{N}}$, etc., we shall write (τ_n), (x_n), etc.

A random set A is called *nonessential* (P-*negligible* or simply *negligible*) if

$$\mathsf{P}\big\{\omega\colon \exists t\colon (\omega, t) \in A\big\} = 0.$$

Two processes ξ and η are said to be *indistinguishable* (or P-*indistinguishable*) if the random set $\{(\omega, t)\colon \xi_t(\omega) \neq \eta_t(\omega)\}$ is nonessential, that is, if almost all trajectories of ξ and η coincide. If $\mathsf{P}\{\omega\colon \xi_t(\omega) \neq \eta_t(\omega)\} = 0$ for all $t \in \mathbf{R}_+$ the processes ξ and η are called P-*equivalent* (in this case the process η is called a P-*modification* of the process ξ). Just as in the case of random variables, the relation $\xi = \eta$ (or $\xi \leq \eta$, etc.) between stochastic processes usually means "up to a nonessential process".

A process ξ is said to be *adapted to a filtration* $\mathbf{F}$ (briefly, $\mathbf{F}$-*adapted* or simply *adapted*) if ξ_t is $\mathcal{F}_t$-measurable for every $t \in \mathbf{R}_+$.

The σ-algebra $\mathcal{P}$ of subsets of $\Omega \times \mathbf{R}_+$ generated by all continuous adapted processes regarded as maps on $\Omega \times \mathbf{R}_+$ is called *predictable*. The σ-algebra $\mathcal{O}$ of subsets of $\Omega \times \mathbf{R}_+$ generated by all right-continuous adapted processes having left-hand limits and regarded as maps on $\Omega \times \mathbf{R}_+$ is called *optional*. It is well known that $\mathcal{P} \subset \mathcal{O}$ [**30**]. A $\mathcal{P}$-measurable ($\mathcal{O}$-measurable) process or random set is called *predictable* (respectively, *optional*). If it is necessary to emphasize that the above notions are related to the filtration $\mathbf{F}$, we shall write $\mathcal{P}(\mathbf{F})$, $\mathcal{O}(\mathbf{F})$, $\mathbf{F}$-*predictable*, or $\mathbf{F}$-*optional*. The set of all predictable (optional) processes will also be denoted by $\mathcal{P}$ (respectively, $\mathcal{O}$).

If τ and τ' are two maps from Ω into $\overline{\mathbf{R}}_+$, then the random set

$$[\![\tau, \tau']\!] = \{(\omega, t) : t \in \mathbf{R}_+, \tau(\omega) \le t \le \tau'(\omega)\}$$

is called a *stochastic interval*. Stochastic intervals $]\!]\tau, \tau']\!]$, $[\![\tau, \tau'[\![$, $]\!]\tau, \tau'[\![$, and $[\![\tau, \tau]\!] = [\![\tau]\!]$ are defined similarly. A map $\tau : \Omega \to \overline{\mathbf{R}}_+$ is said to be a *stopping time* (s.t.) with respect to the filtration $\mathbf{F}$ (briefly, $\mathbf{F}$-*s.t.* or simply *s.t.*) if $[\![0, \tau]\!] \in \mathcal{O}$. A s.t. τ is called *predictable* if $[\![0, \tau[\![\in \mathcal{P}$. Denote by $\overline{\mathfrak{M}}(\mathbf{F})$ (respectively, $\mathfrak{M}(\mathbf{F})$) or briefly $\overline{\mathfrak{M}}$ (respectively, $\mathfrak{M}$) the set of all (respectively, finite) $\mathbf{F}$-s.t.'s.

In this book we use the following notation for classes of processes defined on the stochastic basis $(\Omega, \mathcal{F}, \mathbf{F}, \mathsf{P})$:

$\overline{\mathcal{M}}$ (respectively, $\mathcal{M}$) is the class of all (uniformly integrable) martingales (by a martingale we always mean a martingale with regular trajectories, that is, with P-almost surely (a.s.) right-continuous trajectories that have left limits);

$\overline{\mathcal{M}}^c$ (respectively, $\mathcal{M}^c$) is the class of all (respectively, uniformly integrable) P-a.s. continuous martingales of $\overline{\mathcal{M}}$ (respectively, $\mathcal{M}$).

$\overline{\mathcal{M}}^d$ (respectively, $\mathcal{M}^d$) is the class of all (respectively, uniformly integrable) purely discontinuous martingales of $\overline{\mathcal{M}}$ (respectively, $\mathcal{M}$);

$\overline{\mathcal{M}}^2 = \{\xi \in \overline{\mathcal{M}} : \mathsf{E}\,|\xi_t|^2 < \infty \ \forall t \in \mathbf{R}_+\}$;

$\mathcal{M}^2 = \{\xi \in \mathcal{M} : \sup_{t \in \mathbf{R}_+} \mathsf{E}\,|\xi_t|^2 < \infty\}$;

$\mathcal{V}^+$ is the class of all adapted right-continuous nondecreasing processes A with $A_0 = 0$ and $A_t < \infty$ (P-a.s.) for all $t \in \mathbf{R}_+$;

$\mathcal{V} = \{\xi - \eta : \xi \in \mathcal{V}^+, \eta \in \mathcal{V}^+\}$;

$\mathcal{V}^c$ is the class of processes of $\mathcal{V}$ with (P-a.s.) continuous trajectories;

$\mathcal{A}^+ = \{\xi \in \mathcal{V}^+ : \mathsf{E}\,\xi_\infty < \infty\}$;

$\mathcal{A} = \{\xi - \eta : \xi \in \mathcal{A}^+, \eta \in \mathcal{A}^+\}$.

If $\mathcal{K}$ is a class of processes on $(\Omega, \mathcal{F}, \mathbf{F}, \mathsf{P})$, then $\mathcal{K}_{\mathrm{loc}}$ denotes the *local class* defined as follows: $\xi \in \mathcal{K}_{\mathrm{loc}}$ if and only if there exists an increasing sequence (τ_n) of s.t.'s (dependent on ξ) such that $\lim_{n \to \infty} \tau_n = \infty$ (P-a.s.) and the stopped process $\xi^{\tau_n} = (\xi_t^{\tau_n}) = (\xi_{t \wedge \tau_n})$ belongs to $\mathcal{K}$ for every n. The sequence (τ_n) is called a *localizing sequence for* ξ. It is obvious that $\mathcal{V}^+_{\mathrm{loc}} = \mathcal{V}^+$, $\mathcal{V}_{\mathrm{loc}} = \mathcal{V}$, and $\mathcal{V}^c_{\mathrm{loc}} = \mathcal{V}^c$. In what follows stochastic processes of the classes $\mathcal{M}_{\mathrm{loc}}$ and $\mathcal{M}^2_{\mathrm{loc}}$ are called *local martingales* and *square-integrable local martingales*, respectively.

For any class of processes $\mathcal{K}$ we introduce the subclass $\mathcal{K}_0 = \{\xi \in \mathcal{K} : \xi_0 = 0\}$ and put $\mathcal{K}_{0,\mathrm{loc}} = (\mathcal{K}_0)_{\mathrm{loc}}$. It should be observed that $(\mathcal{K}_0)_{\mathrm{loc}} = (\mathcal{K}_{\mathrm{loc}})_0 = \mathcal{K}_{\mathrm{loc},0}$, so that $\mathcal{K}_{0,\mathrm{loc}} = \mathcal{K}_{\mathrm{loc},0}$.

We introduce the following classes of processes:

$\mathcal{S}$ is the class of all *semimartingales*, that is, the class of the processes of the form $\xi = \xi_0 + M + A$, where $M \in \mathcal{M}_{\mathrm{loc},0}$ and $A \in \mathcal{V}$;

$\mathcal{S}_p$ is the class of all *special semimartingales*, that is, the class of the processes $\xi \in \mathcal{S}$ that admit a decomposition $\xi = \xi_0 + M + A$ with $A \in \mathcal{V} \cap \mathcal{P}$.

Hereafter $\mathcal{K}(\mathbf{F}, \mathsf{P})$ denotes a class of processes with respect to the filtration $\mathbf{F}$ and measure P. For instance, $\mathcal{M}(\mathbf{F}, \mathsf{P})$ is the class of uniformly integrable martingales with respect to $\mathbf{F}$ and P, etc. On the stochastic basis $(\Omega, \mathcal{F}, \mathbf{F}, \mathsf{P})$ let there be defined another stochastic measure $\mathbf{Q}$ such that $\mathsf{P} \ll \mathbf{Q}$. Next, let ξ be an $\mathbf{F}^{\mathsf{Q}}$-adapted process such that $\mathsf{E}\,|\xi_t| < \infty$ for all $t \in \mathbf{R}_+$ and $\mathsf{E}(\xi_t/\mathcal{F}_s^{\mathsf{Q}}) = \xi_s$ (P-a.s.) for $s \leq t$. It is well known [**46**] that in this case there exists a regular P-modification of the process ξ. Therefore we can introduce the class $\overline{\mathcal{M}}(\mathbf{F}^{\mathsf{Q}}, \mathsf{P})$ of martingales with respect to the filtration $\mathbf{F}^{\mathsf{Q}}$ and measure P (with P-a.s. regular trajectories). Now the designations $\mathcal{M}^2(\mathbf{F}^{\mathsf{Q}}, \mathsf{P})$, $\mathcal{V}^+(\mathbf{F}^{\mathsf{Q}}, \mathsf{P})$, etc., have a clear meaning.

The above results of the theory of stochastic processes and stochastic integration retain their validity with $(\mathbf{F}, \mathsf{P})$ in place of $(\mathbf{F}^{\mathsf{Q}}, \mathsf{P})$ for $\mathsf{P} \ll \mathbf{Q}$. It is common knowledge that in most cases the results obtained for the stochastic basis $(\Omega, \mathcal{F}^{\mathsf{P}}, \mathbf{F}^{\mathsf{P}}, \mathsf{P})$ hold also for the stochastic basis $(\Omega, \mathcal{F}, \mathbf{F}, \mathsf{P})$ ([**138**], Theorem IV-59; [**153**], Assertion 1.1).

THEOREM 1.1.1. 1. *$\tau \in \mathfrak{M}(\mathbf{F}^{\mathsf{P}})$ if and only if it is equal P-a.s. to some $\tau' \in \mathfrak{M}(\mathbf{F})$ and then $\mathcal{F}_\tau^{\mathsf{P}} = \mathcal{F}_{\tau'} \vee \mathcal{F}_0^{\mathsf{P}}$;*

2. *Every $\mathbf{F}^{\mathsf{P}}$-optional (respectively, $\mathbf{F}^{\mathsf{P}}$-predictable) process is P-indistinguishable from some $\mathbf{F}$-optional (respectively, $\mathbf{F}$-predictable) process; conversely, $\mathcal{O}(\mathbf{F}) \subset \mathcal{O}(\mathbf{F}^{\mathsf{P}})$, $\mathcal{P}(\mathbf{F}) \subset \mathcal{P}(\mathbf{F}^{\mathsf{P}})$.*

Now we present two more significant results. The first of them is the theorem on the existence of the compensator for increasing processes [**30, 98, 128, 156**].

THEOREM 1.1.2. *Let $A \in \mathcal{A}_{\text{loc}}$. Then there exists a unique up to indistinguishability, predictable process $\widetilde{A} \in \mathcal{A}_{\text{loc}}$, called the dual predictable projection, or the compensator, of the process A, such that $A - \widetilde{A} \in \mathcal{M}_{\text{loc}}$. In particular, if $A \in \mathcal{A}_{\text{loc}}^+$ then also $\widetilde{A} \in \mathcal{A}_{\text{loc}}^+$.*

The following assertion presents the well-known Lenglart inequality for dominated processes [**97, 98, 172**].

THEOREM 1.1.3. *Let ξ and η be two nonnegative adapted processes with regular trajectories such that $\xi_0 = \eta_0 = 0$ and $\eta \in \mathcal{A}_{\text{loc}}^+ \cap \mathcal{P}$, and let the process ξ be dominated by the process η in the sense that $\mathsf{E}\,\xi_\tau \leq \mathsf{E}\,\eta_\tau$ for any $\tau \in \mathfrak{M}$. Then for any $\tau \in \mathfrak{M}$ and $a > 0$ and $b > 0$ we have*

$$\mathsf{P}\left\{\sup_{t \leq \tau} \xi_t \geq a\right\} \leq \frac{1}{a}\,\mathsf{E}(\eta_\tau \wedge b) + \mathsf{P}\{\eta_\tau \geq b\}.$$

In what follows $\xi = (\xi^1, \dots, \xi^n) \in \mathcal{K}$ means that $\xi^i \in \mathcal{K}$ for all $i = 1, 2, \dots, n$.

Stochastic integrals. If $A \in \mathcal{V}$ and $H \in \mathcal{O}$, then the process

$$H \circ A = (H \circ A_t)_{t \in \mathbf{R}_+}$$

is well defined, where

$$H \circ A_t(\omega) = \int_0^t H_s(\omega)\, dA_s(\omega)$$

is the *Lebesgue–Stieltjes integral* for every $\omega \in \Omega$ [**30, 98, 128, 156**]. If $A \in \mathcal{A}_{\text{loc}}^+$ and H is a nonnegative predictable process, then $\mathsf{E}\, H \circ A_\tau = \mathsf{E}\, H \circ \widetilde{A}_\tau$ for any $\tau \in \mathfrak{M}$, where $\widetilde{A}$ is the compensator of the process A [**30**].

If $\xi \in \mathcal{M}^2_{\mathrm{loc}}$, then the Doob–Meyer decomposition implies [**98, 128, 153, 156**] that there exists a unique up to indistinguishability, predictable process $\langle \xi, \xi \rangle = \langle \xi \rangle \in \mathcal{A}^+_{\mathrm{loc}}$, called the *quadratic characteristic of* ξ, such that $\xi^2 - \langle \xi \rangle \in \mathcal{M}_{\mathrm{loc}}$. If $\xi, \eta \in \mathcal{M}^2_{\mathrm{loc}}$, then the process

$$\langle \xi, \eta \rangle = \frac{1}{4}\big(\langle \xi + \eta, \xi + \eta \rangle - \langle \xi - \eta, \xi - \eta \rangle\big),$$

called the *mutual quadratic characteristic of the processes* ξ *and* η, is unique up to indistinguishability of processes of $\mathcal{A}_{\mathrm{loc}} \cap \mathcal{P}$ and $\xi\eta - \langle \xi, \eta \rangle \in \mathcal{M}_{\mathrm{loc}}$.

For vector processes $\xi = (\xi^1, \dots, \xi^n) \in \mathcal{M}^2_{\mathrm{loc}}$ and $\eta = (\eta^1, \dots, \eta^m) \in \mathcal{M}^2_{\mathrm{loc}}$ we denote by $\langle \xi, \eta \rangle$ the matrix with elements $\langle \xi^i, \eta^j \rangle$, $i = 1, 2, \dots, n$, $j = 1, 2, \dots, m$.

In order to define a stochastic integral over $\xi \in \mathcal{M}^2_{\mathrm{loc}}$, we introduce the class of integrable functions

$$\mathcal{L}^2_{\mathrm{loc}}(\xi) = \{f \colon f \in \mathcal{P}, f^2 \circ \langle \xi \rangle \in \mathcal{A}_{\mathrm{loc}}\}.$$

THEOREM 1.1.4 [**98, 128, 153, 156**]. *If* $\xi \in \mathcal{M}^2_{\mathrm{loc}}$ *and* $f \in \mathcal{L}^2_{\mathrm{loc}}(\xi)$, *then the stochastic process*

$$f \cdot \xi = (f \cdot \xi_t)_{t \in \mathbf{R}_+},$$

where

$$f \cdot \xi_t = \int_0^t f_s \, d\xi_s, \qquad t \in \mathbf{R}_+,$$

is well defined. This integral, called the stochastic integral of f *over* ξ, *is a unique up to indistinguishability element of* $\mathcal{M}^2_{\mathrm{loc}}$ *such that* $\langle f \cdot \xi, \eta \rangle = f \circ \langle \xi, \eta \rangle$ *for any* $\eta \in \mathcal{M}^2_{\mathrm{loc}}$.

If $\xi \in \mathcal{M}^2_{\mathrm{loc}}$ and $f = (f^1, \dots, f^n) \in \mathcal{L}^2_{\mathrm{loc}}(\xi)$, then the integral of f over ξ is understood as the vector process $f \cdot \xi = (f^1 \cdot \xi, \dots, f^n \cdot \xi)$, where $f^i \cdot \xi$, $i = 1, 2, \dots, k$, are one-dimensional stochastic integrals defined in Theorem 1.1.4 and such that $\langle f \cdot \xi, f \cdot \xi \rangle$ is the matrix with elements $f^i f^j \circ \langle \xi \rangle$, $i, j = 1, 2, \dots, k$.

Now we turn to the definition of the stochastic integral over a local martingale. Any local martingale $\xi \in \mathcal{M}_{\mathrm{loc}}$ admits a unique decomposition $\xi = \xi_0 + \xi^c + \xi^d$, where $\xi^c \in \mathcal{M}^c_{\mathrm{loc},0}$ is a continuous part of ξ and $\xi^d \in \mathcal{M}^d_{\mathrm{loc},0}$ is a purely discontinuous part of ξ. With a local martingale $\xi \in \mathcal{M}_{\mathrm{loc}}$ we can associate the process $[\xi, \xi] \in \mathcal{V}$ called the *quadratic variation of* ξ and defined as

$$[\xi, \xi]_t = \langle \xi^c, \xi^c \rangle_t + \sum_{s \le t} (\Delta \xi_s)^2, \qquad t \in \mathbf{R}_+,$$

where $\Delta \xi_s = \xi_s - \xi_{s-}$, $\xi_{0-} = \xi_0$.

If $\xi, \eta \in \mathcal{M}_{\mathrm{loc}}$, then we put

$$[\xi, \eta] = \frac{1}{4}([\xi + \eta, \xi + \eta] - [\xi - \eta, \xi - \eta]).$$

For every $\xi \in \mathcal{M}_{\mathrm{loc}}$ we define the class of processes

$$\mathcal{L}^1_{\mathrm{loc}}(\xi) = \left\{f \colon f \in \mathcal{P}, (f^2 \circ [\xi, \xi])^{1/2} \in \mathcal{A}^+_{\mathrm{loc}}\right\}.$$

THEOREM 1.1.5 [**98, 128, 153, 156**]. *If $\xi \in \mathcal{M}_{\text{loc}}$ and $f \in \mathcal{L}^1_{\text{loc}}(\xi)$, then the process*

$$f \cdot \xi = (f \cdot \xi_t)_{t \in \mathbf{R}_+} \in \mathcal{M}_{\text{loc}},$$

called the stochastic integral of f over ξ, is well defined and has the following properties:

$$(f \cdot \xi)^{\tau_n} = f \cdot \xi^{\tau_n}, \qquad (f \cdot \xi)^c = f \cdot \xi^c,$$
$$\Delta(f \cdot \xi)_t = f_t \Delta \xi_t, \qquad [f \cdot \xi, \eta] = f \circ [\xi, \eta]$$

for any $\eta \in \mathcal{M}_{\text{loc}}$, where (τ_n) is a localizing sequence for ξ.

If $\xi \in \mathcal{M}_{\text{loc}}$ and $f = (f^1, f^2, \dots, f^k) \in \mathcal{L}^1_{\text{loc}}(\xi)$, then the stochastic integral of f over ξ is understood to be the vector process $f \cdot \xi = (f^1 \cdot \xi, f^2 \cdot \xi, \dots, f^k \cdot \xi)$, where $f^i \cdot \xi$ is the stochastic integral defined in Theorem 1.1.5.

Let $\xi \in \mathcal{S}$ have a decomposition $\xi = \xi_0 + M + A$, where $M \in \mathcal{M}_{\text{loc},0}$ and $A \in \mathcal{V}$. Now we introduce the class of processes

$$\mathcal{L}_{\text{loc}}(\xi) = \{f \colon f \in \mathcal{L}^1_{\text{loc}}(M), f \circ A \in \mathcal{V}\}.$$

THEOREM 1.1.6 [**98, 128, 153, 156**]. *If $\xi \in \mathcal{S}$ has a decomposition $\xi = \xi_0 + M + A$, where $M \in \mathcal{M}_{\text{loc},0}$ and $A \in \mathcal{V}$, and $f \in \mathcal{L}_{\text{loc}}(\xi)$, then the process $f \cdot \xi$, called the stochastic integral of f over the semimartingale ξ, is well defined and*

$$f \cdot \xi = f_0 \xi_0 + f \cdot M + f \circ A.$$

The stochastic integral thus defined does not depend on the decomposition $\xi = \xi_0 + M + A$.

Random measures and stochastic integrals over random measures. Let $(\Omega, \mathcal{F}, \mathbf{F}, \mathsf{P})$ be a given stochastic basis and let $(X, \mathcal{B})$ be a Lusin space (that is, a Borel subset of a compact metric space with Borel σ-algebra $\mathcal{B}$). The results presented here will be used only in the case $X = \mathbf{R}^m$ for some $m \in \mathbf{N}$. Put

$$\widetilde{\Omega} = \Omega \times \mathbf{R}_+ \times X, \qquad \widetilde{X} = \mathbf{R}_+ \times X, \qquad \widetilde{\mathcal{F}} = \mathcal{F} \times \mathcal{B}_+ \times \mathcal{B},$$
$$\widetilde{\mathcal{B}} = \mathcal{B}_+ \times \mathcal{B}, \qquad \widetilde{\mathcal{O}} = \mathcal{O} \times \mathcal{B}, \qquad \widetilde{\mathcal{P}} = \mathcal{P} \times \mathcal{B}.$$

A nonnegative function $\mu = (\mu(\omega; A); \omega \in \Omega, A \subset \widetilde{X})$ is called a *random measure* on X if the variable $\mu(\,\cdot\,; A)$ is $\mathcal{F}$-measurable for every $A \in \widetilde{\mathcal{B}}$ and, for every $\omega \in \Omega$, the random variable $\mu(\omega; \,\cdot\,)$ is a σ-finite measure on $(\widetilde{X}, \widetilde{\mathcal{B}})$ such that $\mu(\omega; \{0\} \times X) = 0$. If, in addition, $\mu(\omega; A)$ takes values $0, 1, 2, \dots, \infty$ for every $A \in \widetilde{\mathcal{B}}$ and

$$\mu(\omega; \{t\} \times X) \le 1$$

for all $\omega \in \Omega$ and $t \in \mathbf{R}_+$, then the random measure μ is called an *integral-valued measure*.

In what follows we write $\mu(\omega; A, B)$ instead of $\mu(\omega; A \times B)$ (we also use similar notation for other measures).

Let $U = U(\omega, t, x)$ be a nonnegative $\widetilde{\mathcal{F}}$-measurable function on $\widetilde{\Omega}$. We define a nondecreasing stochastic processes

$$U * \mu = (U * \mu_t)_{t \in \mathbf{R}_+},$$

where for every $\omega \in \Omega$

$$U * \mu_t(\omega) = \int_0^t \int_X U(\omega, s, x)\mu(\omega; ds, dx), \quad t \in \mathbf{R}_+.$$

A random measure μ is called an *optional measure* (respectively, a *predictable measure*) if the stochastic process $U * \mu$ is an *optional process* (respectively, a *predictable process*) for any nonnegative $\widetilde{\mathcal{O}}$-measurable (respectively, $\widetilde{\mathcal{P}}$-measurable) function U.

With every random measure μ and every probability measure P we can associate the *Doléans measure* $\mathbf{M}_\mu^{\mathsf{P}}$ on $(\widetilde{\Omega}, \widetilde{\mathcal{F}})$ by putting

$$\mathbf{M}_\mu^{\mathsf{P}}(d\omega, dt, dx) = \mathsf{P}(d\omega)\mu(\omega; dt, dx).$$

For every nonnegative $\widetilde{\mathcal{F}}$-measurable function f we put

$$\mathbf{M}_\mu^{\mathsf{P}}(f) = \int_{\widetilde{\Omega}} f(\omega, t, x)\mathbf{M}_\mu^{\mathsf{P}}(d\omega, dt, dx) = \mathsf{E} f * \mu_\infty.$$

We say that the measure $\mathbf{M}_\mu^{\mathsf{P}}$ is $\widetilde{\mathcal{O}}$-σ-finite (respectively, $\widetilde{\mathcal{P}}$-σ-finite) if there is a sequence of sets $(\widetilde{\Omega}_n)$ such that $\widetilde{\Omega}_n \uparrow \widetilde{\Omega}$ as $n \to \infty$, where $\widetilde{\Omega}_n \in \widetilde{\mathcal{O}}$ (respectively, $\widetilde{\mathcal{P}}$) and $\mathbf{M}_\mu^{\mathsf{P}}(\chi_{\widetilde{\Omega}_n}) < \infty$ for all $n \in \mathbf{N}$.

Let ξ be an adapted stochastic process with values in $\mathbf{R}^m$ and regular trajectories. We define an integral-valued random measure μ called the *jump measure* of the process ξ as

$$\mu(\omega; dt, dx) = \sum_{s \in \mathbf{R}_+} \chi(|\Delta\xi_s| \neq 0)\varepsilon_{(s, \Delta\xi_s)}(dt, dx),$$

where ε_a is the *Dirac measure* concentrated at the point a.

It is known ([**20**], pp. 186–187; [**153**], Example 3.21) that in this case the measure μ is optional and the measure $\mathbf{M}_\mu^{\mathsf{P}}$ is $\widetilde{\mathcal{P}}$-σ-finite.

The following result was proved in [**98, 153, 156**].

THEOREM 1.1.7. *For any random measure μ for which the measure $\mathbf{M}_\mu^{\mathsf{P}}$ is $\widetilde{\mathcal{P}}$-σ-finite there exists a unique up to indistinguishability predictable random measure ν on $\widetilde{X}$ such that*

$$\mathbf{M}_\mu^{\mathsf{P}}(f) = \mathbf{M}_\nu^{\mathsf{P}}(f) \tag{1.1.1}$$

for any nonnegative $\widetilde{\mathcal{P}}$-measurable function f. A random measure ν is called the dual predictable projection of the measure μ or the compensator of the measure μ. If μ is an optional measure, then for every nonnegative $\widetilde{\mathcal{P}}$-measurable function f such that $f * \mu \in \mathcal{A}_{\text{loc}}$, we have

$$f * \mu - f * \nu \in \mathcal{M}_{\text{loc}}. \tag{1.1.2}$$

The random measure $\mu - \nu$ is called a local martingale measure.

In the case of an integral-valued random measure μ with $\widetilde{\mathcal{P}}$-σ-finite measure $\mathbf{M}_\mu^{\mathsf{P}}$, there exists a version of the compensator ν such that $\nu(\omega;\{t\},X)\le 1$ for all $\omega\in\Omega$, $t\in\mathbf{R}_+$ (below we consider exactly this version).

Observe that in what follows we consider only integral-valued random measures μ with compensator ν that is continuous in t, namely, such that

$$\sum_{t\in\mathbf{R}_+}\nu(\omega;\{t\},X)=0 \tag{1.1.3}$$

for all $\omega\in\Omega$. In what follows the variable ω is always omitted.

Denote by $\widetilde{\mathcal{P}}$ (respectively, $\widetilde{\mathcal{O}}$) the set of all real $\widetilde{\mathcal{P}}$-measurable (respectively, $\widetilde{\mathcal{O}}$-measurable) functions on $\widetilde{\Omega}$.

Let μ be an optional integral-valued random measure with $\widetilde{\mathcal{P}}$-σ-finite measure $\mathbf{M}_\mu^{\mathsf{P}}$ and compensator ν, continuous in t. To define a stochastic integral with respect to the measure $\mu-\nu$ we introduce the following classes of integrable functions:

$$\mathcal{G}^i_{\mathrm{loc}}(\nu)=\left\{f\colon f\in\widetilde{\mathcal{P}},|f|^i*\nu\in\mathcal{A}^+_{\mathrm{loc}}\right\},\qquad i=1,2;$$
$$\mathcal{G}_{\mathrm{loc}}(\nu)=\left\{f\colon f\in\widetilde{\mathcal{P}},|f|^2(1+|f|)^{-1}*\nu\in\mathcal{A}^+_{\mathrm{loc}}\right\}.$$

Note that $\mathcal{G}^i_{\mathrm{loc}}(\nu)\subset\mathcal{G}_{\mathrm{loc}}(\nu)$ for $i=1,2$.

THEOREM 1.1.8 [**46, 98, 156**]. 1) *If $f\in\mathcal{G}_{\mathrm{loc}}(\nu)$, then there exists a unique up to indistinguishability stochastic process*

$$f*(\mu-\nu)=(f*(\mu-\nu)_t)\in\mathcal{M}^d_{\mathrm{loc}}$$

for which

$$\Delta(f*(\mu-\nu))_t=\int_X f(t,x)\mu(\{t\},dx).$$

2) $f\in\mathcal{G}^1_{\mathrm{loc}}(\nu)\ \Leftrightarrow\ \mathrm{Var}(f*(\mu-\nu))\in\mathcal{A}^+_{\mathrm{loc}}$ *(here $\mathrm{Var}\,\xi$ is the variation of the process ξ); if $f\in\mathcal{G}^1_{\mathrm{loc}}(\nu)$, then $f*(\mu-\nu)=f*\mu-f*\nu$.*

3) $f\in\mathcal{G}^2_{\mathrm{loc}}(\nu)\Leftrightarrow f*(\mu-\nu)\in\mathcal{M}^2_{\mathrm{loc}}$, $\langle f*(\mu-\nu)\rangle=f^2*\nu$.

4) *If $f\in\mathcal{G}_{\mathrm{loc}}(\nu)$, then there are $f_i\in\mathcal{G}^i_{\mathrm{loc}}(\nu)$, $i=1,2$, such that $f=f_1+f_2$ (in particular, we can put $f_1=f\chi(|f|>a)$, and $f_2=f\chi(|f|\le a)$ for every $a\in(0,\infty)$).*

For $f\in\mathcal{G}_{\mathrm{loc}}(\nu)$ the process $f*(\mu-\nu)$ is called the *stochastic integral of f with respect to the random measure $\mu-\nu$*. For $f*(\mu-\nu)_t$ we also use the notation

$$f*(\mu-\nu)_t=\int_0^t\!\!\int_X f(s,x)\,d(\mu-\nu)=\int_0^t\!\!\int_X f_{s,x}(\mu-\nu)(ds,dx).$$

If $f=(f^1,f^2,\dots,f^k)\in\mathcal{G}_{\mathrm{loc}}(\nu)$, then by the stochastic integral of f with respect to $\mu-\nu$ we understand the vector process

$$f*(\mu-\nu)=\big(f^1*(\mu-\nu),f^2*(\mu-\nu),\dots,f^k*(\mu-\nu)\big);$$

if $f \in \mathcal{G}^2_{\text{loc}}(\nu)$, then $\langle f * (\mu - \nu), f * (\mu - \nu)\rangle$ is the matrix with elements $f^i f^j * \nu$, $i, j = 1, 2, \dots, k$.

Since $|x|^2/(1 + |x|)$ is a convex even function monotone for $x \geq 0$, for any $a = (a^1, a^2, \dots, a^k)' \in \mathbf{R}^k$ the following implication holds:

$$f = (f^1, f^2, \dots, f^k)' \in \mathcal{G}_{\text{loc}}(\nu) \Rightarrow a'f \in \mathcal{G}_{\text{loc}}(\nu). \tag{1.1.4}$$

From this it follows that for any function $f = (f^1, f^2, \dots, f^k)' \in \mathcal{G}_{\text{loc}}(\nu)$ and any vector $a = (a^1, a^2, \dots, a^k)' \in \mathbf{R}^k$ the equality

$$a'(f * (\mu - \nu)) = (a'f) * (\mu - \nu) \tag{1.1.4$'$}$$

holds. The relation (1.1.4) also implies the following useful assertion:

$$f = (f^1, f^2, \dots, f^k)' \in \mathcal{G}_{\text{loc}}(\nu) \Leftrightarrow |f|_p \in \mathcal{G}_{\text{loc}}(\nu) \tag{1.1.5}$$

for any $p \in (0, \infty)$. Indeed, it suffices to note that

$$f \in \mathcal{G}_{\text{loc}}(\nu) \Leftrightarrow |f|_1 \in \mathcal{G}_{\text{loc}}(\nu) \Leftrightarrow |f|_p \in \mathcal{G}_{\text{loc}}(\nu)$$

for any $p \in (0, \infty)$, where the latter relation follows from the estimates

$$\begin{aligned} |f|_p \leq |f|_1 \leq k^{(p-1)/p}|f|_p, \qquad & p \in [1, \infty); \\ k^{(p-1)/p}|f|_p \leq |f|_1 \leq |f|_p, \qquad & p \in (0, 1]. \end{aligned}$$

Semimartingales. Let $\xi \in \mathcal{S}$ be a semimartingale defined on a stochastic basis $(\Omega, \mathcal{F}, \mathbf{F}, \mathsf{P})$. It is known [**46, 98, 156**] that the semimartingale ξ can be represented uniquely up to indistinguishability in the form

$$\xi = \xi_0 + B + \xi^c + x\chi(|x| \leq 1) * (\mu - \nu) + x\chi(|x| > 1) * \mu, \tag{1.1.6}$$

where $\mu = \mu(\omega; dt, dx)$ is the jump measure of the process ξ, ν is the compensator of the measure μ, and $\xi^c = (\xi^c_t) \in \mathcal{M}^c_{\text{loc},0}$ is the continuous martingale component of the semimartingale ξ with quadratic characteristic $\langle \xi^c \rangle = C$, $B = (B_t) \in \mathcal{A}_{\text{loc}} \cap \mathcal{P}$. For the vector semimartingale $\xi = (\xi^1, \xi^2, \dots, \xi^k)$ the representation (1.1.6) is also true, in which $B = (B^1, B^2, \dots, B^k)$, $\xi^c = (\xi^{1c}, \xi^{2c}, \dots, \xi^{kc})$, and $C = \langle \xi^c, \xi^c \rangle$ is the matrix with elements $\langle \xi^{ic}, \xi^{jc} \rangle$, $i, j = 1, 2, \dots, k$.

The representation (1.1.6) is called the *canonical representation*, whereas the set $T = (B, C, \nu)$ is called the *triple of predictable characteristics of the semimartingale* ξ. In what follows the compensator ν is assumed continuous in t, that is, satisfying the relation (1.1.3). Then the process B has continuous trajectories and the semimartingale ξ is *left-quasicontinuous* ([**98**], Theorem 4.1.1). If the triple $T = (B, C, \nu)$ is determinate, then the semimartingale ξ is a process with independent increments.

The following theorem provides a formula for the change of variables for a semimartingale, known as the *Itô formula*. The proof of this theorem can be found in almost all textbooks on stochastic integration [**98, 128, 153, 156**].

THEOREM 1.1.9. *Let $\xi = (\xi^1, \xi^2, \dots, \xi^k) \in \mathcal{S}$, and let F be a twice continuously differentiable function on $\mathbf{R}^k$. Then $F(\xi) = (F(\xi_t)) \in \mathcal{S}$ and the following Itô formula holds*:

$$
\begin{aligned}
F(\xi_t) = F(\xi_0) &+ \sum_{i=1}^{k} \frac{\partial F}{\partial x_i}(\xi_-) \cdot \xi_t^i + \frac{1}{2} \sum_{i,j=1}^{k} \frac{\partial^2 F}{\partial x_i \, \partial x_j}(\xi_-) \circ \langle \xi^{ic}, \xi^{jc} \rangle_t \\
&+ \sum_{0<s\le t} \left[F(\xi_s) - F(\xi_{s-}) - \sum_{i=1}^{k} \frac{\partial F}{\partial x_i}(\xi_{s-}) \Delta \xi_s^i \right],
\end{aligned} \tag{1.1.7}
$$

where the sum $\sum_{0<s\le t}$ converges (P*-almost surely*) *for all $t \in \mathbf{R}_+$*.

Note that the stochastic integral with respect to the semimartingale ξ^i is always well defined since $\frac{\partial F}{\partial x_i}(\xi_-)$ is a predictable locally bounded process (see Theorem 1.1.6 and also Problem 2.3.1 in [**98**]).

1.2. Statistic experiments generated by stochastic processes

Basic definitions and assumptions. Let $(\Omega, \mathcal{F}, \mathsf{P})$ be a complete probability space, $(X^t, \mathfrak{B}^t)$, $t \in \mathbf{R}_+$, a family of measurable spaces, ξ^t a measurable map of the space $(\Omega, \mathcal{F})$ to the space $(X^t, \mathfrak{B}^t)$ for all $t \in \mathbf{R}_+$, and P_{ξ^t} the distribution of the random element ξ^t, that is, P_{ξ^t} is the probability measure on $\mathfrak{B}^t$ defined by the equality $P_{\xi^t}(A) = \mathsf{P}\{\omega : \xi^t(\omega) \in A\}$, $A \in \mathfrak{B}^t$. The random element ξ^t is accessible to observation and is called an *observation*. As for the distribution P_{ξ^t}, it is only known that it belongs to some class of distributions $\mathcal{P}^t$. The triple $(X^t, \mathfrak{B}^t, \mathcal{P}^t)$ is called a *statistical experiment generated by the observation* ξ^t. In what follows we consider families $\mathcal{P}^t$ of the following two types: $\mathcal{P}^t = \{P^t, \widetilde{P}^t\}$ and $\mathcal{P}^t = \{P^t_\theta, \theta \in \Theta\}$, where θ is a parameter with values in Θ, and Θ is an open set in $\mathbf{R}^k$, $k \ge 1$. The triple $(X^t, \mathfrak{B}^t, (P^t, \widetilde{P}^t))$ is also called a *binary statistical experiment*.

In what follows we consider only those statistical experiments $(X^t, \mathfrak{B}^t, \mathcal{P}^t)$ for which the family $\mathcal{P}^t$ is dominated by some σ-finite measure Q^t, that is, every measure in $\mathcal{P}^t$ is absolutely continuous with respect to the dominating measure Q^t. In what follows we often use a measure from the family $\mathcal{P}^t$ as the dominating measure Q^t.

The main objects a statistician deals with are spaces $(X^t, \mathfrak{B}^t)$ and families of distributions $\mathcal{P}^t$, whereas the probability space $(\Omega, \mathcal{F}, \mathsf{P})$ plays a secondary role and can be chosen in many ways. A natural way to construct a suitable probability space $(\Omega, \mathcal{F}, \mathsf{P})$ is as follows. Take the space X^t to be the space of elementary events Ω and the σ-algebra $\mathfrak{B}^t$ to be the σ-algebra of events $\mathcal{F}$. Put $\mathsf{P}(A) = P_{\xi^t}(A)$ for $A \in \mathfrak{B}^t$ and take $\xi^t(x) = x$ for all $x \in X^t$, so that ξ^t is a random element with values in X^t defined on $(X^t, \mathfrak{B}^t, P_{\xi^t})$ and distributed according to P_{ξ^t}. Applying this rule for all $P^t \in \mathcal{P}^t$, we obtain the family of probability spaces $(X^t, \mathfrak{B}^t, P^t)$, $P^t \in \mathcal{P}^t$, on which the observations ξ^t are defined by the rule $\xi^t(x) = x$, $x \in X^t$, so that the distribution of the observation ξ^t is given by the measure P^t in $\mathcal{P}^t$. Other ways of constructing auxiliary probability spaces $(\Omega, \mathcal{F}, \mathsf{P})$ will be described while constructing stochastic experiments generated by stochastic processes.

The observation ξ^t reduces to a random choice of a point x in X^t in accordance with some distribution P^t in $\mathcal{P}^t$. The measurable space $(X^t, \mathfrak{B}^t)$ is often called a *sample space*. In statistics problems considered below it is necessary to make certain conclusions concerning the distribution $P^t \in \mathcal{P}^t$ from the observation ξ^t. In the case where $\mathcal{P}^t = \{P^t, \widetilde{P}^t\}$ this is the *problem of testing two simple hypotheses*, and in the case where $\mathcal{P}^t = \{P^t_\theta, \theta \in \Theta\}$ this is the *problem of estimating parameter* θ, or the *problem of testing some hypotheses for parameter* θ.

In what follows observations ξ^t are considered to be trajectories of stochastic processes on the time interval $[0,t]$, so that the stochastic experiment $(X^t, \mathfrak{B}^t, \mathcal{P}^t)$ is generated by the stochastic process $\xi^t = (\xi_s)_{0\le s\le t}$, X^t is the space of functions $x = (x_s)_{0\le s\le t}$ containing the trajectories of the process ξ^t, and $\mathcal{P}^t$ is the family of probability measures on the σ-algebra $\mathfrak{B}^t$ of subsets of the space X^t generated by the stochastic process ξ^t.

Stochastic experiments generated by semimartingales. Let $\Omega = \mathbf{D}$ be the set of real functions $\omega = (x_t)_{t\in\mathbf{R}_+}$ that are right-continuous and have left limits, let $\mathcal{G}_t$ be the least σ-algebra in the space Ω containing the cylinder sets

$$\{\omega = (x_s): x_{t_1} \in A_1, \dots, x_{t_n} \in A_n\},$$

where $0 \le t_1 < \dots < t_n \le t$, $A_i \in \mathcal{B}^1$, $i = 1, 2, \dots, n$, $n = 1, 2, \dots$, let $\mathcal{G} = \bigvee_{t\in\mathbf{R}_+} \mathcal{G}_t$ be the least σ-algebra containing all σ-algebras $\mathcal{G}_t$, $t \in \mathbf{R}_+$, and let P and $\widetilde{\mathsf{P}}$ be two probability measures on $(\Omega, \mathcal{G})$. We introduce one more probability measure $\mathbf{Q} = (\mathsf{P}+\widetilde{\mathsf{P}})/2$ on $(\Omega, \mathcal{G})$. Evidently, each of the measures P and $\widetilde{\mathsf{P}}$ can be extended to the σ-algebra $\mathcal{F} = \mathcal{G}^{\mathbf{Q}}$ which is the completion of the σ-algebra $\mathcal{G}$ with respect to the measure $\mathbf{Q}$; the corresponding extensions are denoted by the same letters, P and $\widetilde{\mathsf{P}}$. Put $\mathcal{F}_t = \mathcal{G}^{\mathbf{Q}}_{t+}$ for all $t \in \mathbf{R}_+$, where $\mathcal{G}^{\mathbf{Q}}_{t+} = \bigcap_{s>t} \mathcal{G}^{\mathbf{Q}}_s$ (evidently, $\mathcal{F} = \bigvee_{t\in\mathbf{R}_+} \mathcal{F}_t$). By P^t and $\widetilde{P}^t$ we denote the restrictions of (extended) measures P and $\widetilde{\mathsf{P}}$ from $\mathcal{F}$ to $\mathcal{F}_t$, $t \in \mathbf{R}_+$.

Thus we have a complete probability space $(\Omega, \mathcal{F}, \mathbf{Q})$ with the family of σ-algebras $\mathbf{F} = (\mathcal{F}_t)_{t\in\mathbf{R}_+}$ satisfying the ordinary Dellacherie conditions, that is, we have a completed stochastic basis $(\Omega, \mathcal{F}, \mathbf{F}, \mathbf{Q})$. We also consider the stochastic basis $(\Omega, \mathcal{F}, \mathbf{F}, \mathsf{P})$, which is not completed. Since $\mathsf{P} \ll \mathbf{Q}$, according to the remark in Section 1.1, on the stochastic basis $(\Omega, \mathcal{F}, \mathbf{F}, \mathsf{P})$ we can consider a semimartingale $\xi = (\xi_t)_{t\in\mathbf{R}_+} \in \mathcal{S}(\mathbf{F}, \mathsf{P})$. Then from [**46, 98, 156**] it follows that any semimartingale $\xi \in \mathcal{S}(\mathbf{F}, \mathsf{P})$ admits a unique up to P-indistinguishability canonical representation

$$\xi = \xi_0 + \alpha + m + x\chi(|x| \le 1) * (\mu - \nu) + x\chi(|x| > 1) * \mu, \tag{1.2.1}$$

where $\alpha = (\alpha_t) \in \mathcal{A}_{\mathrm{loc}}(\mathbf{F}, \mathsf{P}) \cap \mathcal{P}(\mathbf{F})$, $m = (m_t) \in \mathcal{M}^c_{\mathrm{loc},0}(\mathbf{F}, \mathsf{P})$ with quadratic characteristic $\langle m\rangle = (\langle m_t\rangle)$, and μ is the jump measure of the process ξ with $(\mathbf{F}, \mathsf{P})$-compensator ν. The set $(\alpha, \langle m\rangle, \nu)$ is the triple of predictable characteristics of the semimartingale $\xi \in \mathcal{S}(\mathbf{F}, \mathsf{P})$. The semimartingale $\xi \in \mathcal{S}(\mathbf{F}, \widetilde{\mathsf{P}})$ with the triple of predictable characteristics $(\widetilde{\alpha}, \langle\widetilde{m}\rangle, \widetilde{\nu})$ admits a similar canonical representation. In what follows we consider left-quasicontinuous semimartingales ξ that satisfy the condition

$$\sum_{s\in\mathbf{R}_+} \nu(\{s\}, \mathbf{R}_0) = 0 \quad (\mathsf{P}\text{-a.s.}), \qquad \sum_{s\in\mathbf{R}_+} \widetilde{\nu}(\{s\}, \mathbf{R}_0) = 0 \quad (\widetilde{\mathsf{P}}\text{-a.s.}). \tag{1.2.2}$$

We say that a measure $\widetilde{\mathsf{P}}$ is *locally absolutely continuous with respect to the measure* P (denoted $\widetilde{\mathsf{P}} \overset{\mathrm{loc}}{\ll} \mathsf{P}$) if $\widetilde{P}^t \overset{\mathrm{loc}}{\ll} P^t$ for all $t \in \mathbf{R}_+$. If $\widetilde{\mathsf{P}} \overset{\mathrm{loc}}{\ll} \mathsf{P}$ and $\mathsf{P} \overset{\mathrm{loc}}{\ll} \widetilde{\mathsf{P}}$, we say that the measures $\widetilde{\mathsf{P}}$ and P are *locally equivalent* (in this case we write $\widetilde{\mathsf{P}} \overset{\text{loc}}{\sim} \mathsf{P}$). If $\widetilde{\mathsf{P}} \overset{\text{loc}}{\ll} \mathsf{P}$, then there exists [**46, 153, 156**] a unique up to **Q**-indistinguishability **F**-adopted process $z = (z_t)_{t \in \mathbf{R}_+}$ with regular trajectories such that $z \in \mathcal{M}_{\text{loc}}(\mathbf{F}, \mathsf{P})$ and for any stopping time $\tau \in \overline{\mathfrak{M}}(\mathbf{F})$ for which $\widetilde{P}^\tau \ll P^\tau$, we have $z_\tau = d\widetilde{P}^\tau / dP^\tau$ (P-a.s.). The process z is called the *local density of the measure* $\widetilde{\mathsf{P}}$ *with respect to the measure* P.

It is known [**46, 98, 153, 156**] that if $\xi \in \mathcal{S}(\mathbf{F}, \mathsf{P})$ and $\widetilde{\mathsf{P}} \overset{\text{loc}}{\ll} \mathsf{P}$, then $\xi \in \mathcal{S}(\mathbf{F}, \widetilde{\mathsf{P}})$. The following lemma, known as the Girsanov theorem for semimartingales ([**46**], Section 5, Subsection 5; [**156**], Theorem III.3.24), allows one to calculate characteristics of ξ with respect to the measure $\widetilde{\mathsf{P}}$ in terms of characteristics of ξ with respect to the measure P.

LEMMA 1.2.1. *Let* $\widetilde{\mathsf{P}} \overset{\text{loc}}{\ll} \mathsf{P}$ *and* $\xi \in \mathcal{S}(\mathbf{F}, \widetilde{\mathsf{P}})$ *with a triple of predictable characteristics* $(\alpha, \langle m \rangle, \nu)$. *Then there exist a* $\mathcal{B} \times \mathcal{B}_0$*-measurable function* λ *and a predictable process* γ *satisfying* ($\widetilde{\mathsf{P}}$*-a.s.*) *the conditions*

$$|x\chi(|x| \le 1)(\lambda - 1)| * \nu_t < \infty, \qquad |\gamma| \circ \langle m \rangle_t < \infty, \qquad |\gamma|^2 \circ \langle m \rangle_t < \infty$$

for all $t \in \mathbf{R}_+$ *and such that the predictable characteristics of* ξ *with respect to* $\widetilde{\mathsf{P}}$ *are* ($\widetilde{\mathsf{P}}$*-a.s.*) *of the form*

$$\begin{gathered} \widetilde{\alpha} = \alpha + \gamma \circ \langle m \rangle + x\chi(|x| \le 1)(\lambda - 1) * \nu, \\ \langle \widetilde{m} \rangle = \langle m \rangle, \quad \widetilde{\nu} = \lambda * \nu. \end{gathered}$$

Let $\xi \in \mathcal{S}(\mathbf{F}, \widetilde{\mathsf{P}})$, and let (τ_n) be a sequence of stopping times $\tau_n \in \overline{\mathfrak{M}}(\mathbf{F})$ such that $\tau_n \uparrow \infty$ ($\widetilde{\mathsf{P}}$-a.s.). Then, evidently, $\xi^{\tau_n} = (\xi_{t \wedge \tau_n}) \in \mathcal{S}(\mathbf{F}, \widetilde{\mathsf{P}})$ for any n with triples of **F**-predictable characteristics $(\widetilde{\alpha}^{\tau_n}, \langle \widetilde{m} \rangle^{\tau_n}, \widetilde{\nu}^{\tau_n})$, where

$$\widetilde{\alpha}^{\tau_n} = (\widetilde{\alpha}_{t \wedge \tau_n}), \qquad \langle \widetilde{m} \rangle^{\tau_n} = (\langle \widetilde{m} \rangle_{t \wedge \tau_n}), \qquad \widetilde{\nu}^{\tau_n} = \chi([\![0, \tau_n]\!] \times \mathbf{R}_0)\widetilde{\nu}.$$

We say that the measure $\widetilde{\mathsf{P}}$ possesses the (τ_n)*-uniqueness property* if the triples $(\widetilde{\alpha}^{\tau_n}, \langle \widetilde{m} \rangle^{\tau_n}, \widetilde{\nu}^{\tau_n})$ uniquely define the restriction $\widetilde{P}^{\tau_n}$ of the measure $\widetilde{\mathsf{P}}$ to the σ-algebra $\mathcal{F}_{\tau_n}$.

The following theorem provides conditions for local absolute continuity of the measure $\widetilde{\mathsf{P}}$ with respect to the measure P and describes the form of the local density z in terms of **F**-predictable characteristics of the semimartingale ξ.

THEOREM 1.2.1. *Let the following conditions be fulfilled:*

I) $\widetilde{P}^0 = P^0$;

II) $\nu(\{t\}, \mathbf{R}_0) = 0$ ($\widetilde{\mathsf{P}}$*-a.s.*) *for all* $t \in \mathbf{R}_+$;

III) *there exists a nonnegative* $\mathcal{P}(\mathbf{F}) \times \mathcal{B}_0$*-measurable function*

$$\lambda = (\lambda_{t,x})_{t \in \mathbf{R}_+, x \in \mathbf{R}_0}$$

such that $d\widetilde{\nu}/d\nu = \lambda$ ($\widetilde{\mathsf{P}}$*-a.s.*);

IV) $\langle m \rangle = \langle \widetilde{m} \rangle$ ($\widetilde{\mathsf{P}}$*-a.s.*);

V) *there exists an* $\mathbf{F}$*-predictable process* $\gamma = (\gamma_t)_{t \in \mathbf{R}_+}$ *such that*

$$\widetilde{\alpha}_t - \alpha_t - x\chi(|x| \le 1)(\lambda - 1) * \nu_t = \gamma \circ \langle m \rangle_t \quad (\widetilde{\mathsf{P}}\text{-}a.s.)$$

for all $t \in \mathbf{R}_+$;

VI) $C_t = \gamma^2 \circ \langle m \rangle_t + (\gamma^{1/2} - 1)^2 * \nu_t < \infty$ $(\widetilde{\mathsf{P}}\text{-}a.s.)$ *for all* $t \in \mathbf{R}_+$;

VII) *the measure* $\widetilde{\mathsf{P}}$ *possesses the property of* (τ_n)*-uniqueness, where*

$$\tau_n = \inf\{t \colon C_t > n\}.$$

Then $\widetilde{\mathsf{P}} \overset{\text{loc}}{\ll} \mathsf{P}$. *If, in addition, conditions* I)–VII) *with* $(\alpha, \langle m \rangle, \nu)$ *and* P *replaced by* $(\widetilde{\alpha}, \langle \widetilde{m} \rangle, \widetilde{\nu})$ *and* $\widetilde{\mathsf{P}}$ *and* $(\widetilde{\alpha}, \langle \widetilde{m} \rangle, \widetilde{\nu})$ *and* $\widetilde{\mathsf{P}}$ *replaced by* $(\alpha, \langle m \rangle, \nu)$ *and* P *hold and condition*

VIII) $(\lambda - 1 - \ln \lambda) * \nu_t < \infty$ $(\mathsf{P}\text{-}a.s.)$ *for all* $t \in \mathbf{R}_+$

also holds, then the local density z *is* $(\mathsf{P}\text{-}a.s.)$ *of the form*

$$z = \exp\{M - V\}, \tag{1.2.3}$$

where

$$M = \gamma \cdot m + \ln \lambda * (\mu - \nu) \in \mathcal{M}_{\text{loc}}(\mathbf{F}, \mathsf{P}), \tag{1.2.4}$$

$$V = \frac{1}{2}\gamma^2 \circ \langle m \rangle + (\lambda - 1 - \ln \lambda) * \nu \in \mathcal{V}_{\text{loc}}(\mathbf{F}, \mathsf{P}). \tag{1.2.5}$$

PROOF. Let conditions I)–VII) be fulfilled. Put

$$\lambda^{(n)}_{t,x} = \chi_t([0, \tau_n])\lambda_{t,x} + \chi_t((\tau_n, \infty)), \qquad \gamma^{(n)} = \chi_t([0, \tau_n])\gamma_t, \tag{1.2.6}$$

$$C^{(n)}_t = \left(\gamma^{(n)}\right)^2 \circ \langle m \rangle_t + \left(\left(\lambda^{(n)}\right)^{1/2} - 1\right)^2 * \nu_t. \tag{1.2.7}$$

Observe that by virtue of condition VI) we have $\tau_n \uparrow \infty$ $(\widetilde{\mathsf{P}}\text{-a.s.})$ as $n \to \infty$. Moreover, it is clear that $C^{(n)}_t = C_{t \wedge \tau_n}$ and $C^{(n)}_\infty = C_{\tau_n} \le n$. From this, in view of Lemma 7 of [**46**], it follows that the equation

$$z^{(n)} = 1 + z^{(n)}_- \gamma^{(n)} \cdot m + z^{(n)}_- \left(\lambda^{(n)} - 1\right) * (\mu - \nu) \tag{1.2.8}$$

has a unique up to P-indistinguishability solution $z^{(n)} \in \mathcal{M}(\mathbf{F}, \mathsf{P})$ with $\mathsf{E}\, z^{(n)}_\infty = 1$ in the class of nonnegative local martingales $\mathcal{M}^+_{\text{loc}}(\mathbf{F}, \mathsf{P})$.

We introduce probability measures $\overline{P}^{(n)}$ with $d\overline{P}^{(n)} = z^{(n)}_\infty d\mathsf{P}$. In view of Lemma 1.2.1, from (1.2.6) and (1.2.7) it follows that the measures $\overline{P}^{(n)}$ are such that $\overline{P}^{(n),0} = P^0$ and $\xi^{\tau_n} \in \mathcal{S}(\mathbf{F}, \overline{P}^{(n)})$ with the triple of predictable characteristics $(\widetilde{\alpha}^{\tau_n}, \langle \widetilde{m} \rangle^{\tau_n}, \widetilde{\nu}^{\tau_n})$. Then, by condition VII), $\overline{P}^{(n),\tau_n} = \widetilde{P}^{\tau_n}$ for all n and hence $\widetilde{P}^{\tau_n} \ll P^{\tau_n}$. In view of condition VI), we have $\lim_{n\to\infty} \widetilde{\mathsf{P}}(\tau_n < \infty) = 0$. Therefore $\widetilde{\mathsf{P}} \overset{\text{loc}}{\ll} \mathsf{P}$.

Now let conditions I)–VII) be fulfilled, with $(\alpha, \langle m\rangle, \nu)$ and P replaced by $(\widetilde{\alpha}, \langle\widetilde{m}\rangle, \widetilde{\nu})$ and $\widetilde{\mathsf{P}}$, and $(\widetilde{\alpha}, \langle\widetilde{m}\rangle, \widetilde{\nu})$ and $\widetilde{\mathsf{P}}$ replaced by $(\alpha, \langle m\rangle, \nu)$ and P. Then $\mathsf{P} \overset{\text{loc}}{\ll} \widetilde{\mathsf{P}}$ and hence $\widetilde{\mathsf{P}} \overset{\text{loc}}{\sim} \mathsf{P}$. Using this local equivalence of the measures and condition VIII), we obtain the representation (1.2.3)–(1.2.5) for the local density z.

Consider the stochastic process

$$N = \gamma \cdot m + (\lambda - 1) * (\mu - \nu).$$

Since $\widetilde{\mathsf{P}} \overset{\text{loc}}{\sim} \mathsf{P}$, condition VI) implies that $C_t < \infty$ (P-a.s.) for all $t \in \mathbf{R}_+$. But $C \in \mathcal{P}(\mathbf{F})$ and hence $C \in \mathcal{V}_{\text{loc}}(\mathbf{F}, \mathsf{P}) \cap \mathcal{P}(\mathbf{F})$. Taking into account the inclusion

$$\mathcal{V}_{\text{loc}}(\mathbf{F}, \mathsf{P}) \cap \mathcal{P}(\mathbf{F}) \subset \mathcal{A}_{\text{loc}}(\mathbf{F}, \mathsf{P}),$$

we get $C \in \mathcal{A}_{\text{loc}}(\mathbf{F}, \mathsf{P})$. Now, in view of Assertion 1 of Lemma 7 in [**46**], we have $N \in \mathcal{M}_{\text{loc}}(\mathbf{F}, \mathsf{P})$. Considering that $\Delta N \geq -1$, from [**98, 153, 156**] we conclude that in the class $\mathcal{M}^+_{\text{loc}}(\mathbf{F}, \mathsf{P})$ the equation

$$z = 1 + z_- \cdot N \tag{1.2.9}$$

has a unique up to P-indistinguishability solution

$$z_t = \exp\Big\{N_t - \frac{1}{2}\gamma^2 \circ \langle m\rangle_t\Big\} \prod_{0<s\leq t} (1 + \Delta N_s) \exp(-\Delta N_s), \tag{1.2.10}$$

where the product converges (P-a.s.) for all $t \in \mathbf{R}_+$. Taking into consideration that μ is integer-valued and putting $\ln 0 = -\infty$ and $e^{-\infty} = 0$, we get

(1.2.11)

$$\begin{aligned}
\prod_{0<s\leq t} (1 + \Delta N_s) \exp(-\Delta N_s) &= \exp\Big\{\sum_{0<s\leq t} \Big[\ln(1 + \Delta N_s) - \Delta N_s\Big]\Big\} \\
&= \exp\Big\{\sum_{0<s\leq t} \Big[\ln\Big(1 + \int (\lambda_{s,x} - 1)\mu(\{s\}, dx)\Big) - \int (\lambda_{s,x} - 1)\mu(\{s\}, dx)\Big]\Big\} \\
&= \exp\{-(\lambda - 1 - \ln\lambda) * \mu_t\}.
\end{aligned}$$

In view of condition VII), we have $\lambda - 1 - \ln\lambda \in \mathcal{G}^1_{\text{loc}}(\nu, \mathbf{F}, \mathsf{P})$, and, according to Theorem 1.1.8, the stochastic integral $(\lambda - 1 - \ln\lambda) * (\mu - \nu) \in \mathcal{M}^d_{\text{loc}}(\mathbf{F}, \mathsf{P})$ is well defined and (P-a.s.)

$$(\lambda - 1 - \ln\lambda) * (\mu - \nu) = (\lambda - 1 - \ln\lambda) * \mu - (\lambda - 1 - \ln\lambda) * \nu. \tag{1.2.12}$$

As shown above, $C \in \mathcal{A}_{\text{loc}}(\mathbf{F}, \mathsf{P})$. Since $x^2/(1+|x|) \asymp (1-\sqrt{1+x})^2$ for $x \geq -1$, we have $\lambda - 1 \in \mathcal{G}_{\text{loc}}(\nu, \mathbf{F}, \mathsf{P})$. Taking into account that $\lambda - 1 - \ln\lambda \in \mathcal{G}^1_{\text{loc}}(\nu, \mathbf{F}, \mathsf{P})$, we get $\ln\lambda \in \mathcal{G}_{\text{loc}}(\nu, \mathbf{F}, \mathsf{P})$ and hence (P-a.s.)

$$(\lambda - 1) * (\mu - \nu) - (\lambda - 1 - \ln\lambda) * (\mu - \nu) = \ln\lambda * (\mu - \nu) \in \mathcal{M}^d_{\text{loc}}(\mathbf{F}, \mathsf{P}). \tag{1.2.13}$$

Thus (1.2.10)–(1.2.13) imply that the process z, which is a solution of the equation (1.2.9), admits (P-a.s.) the representation

$$z = \exp\Big\{\gamma \cdot m - \frac{1}{2}\gamma^2 \circ \langle m\rangle + \ln\lambda * (\mu - \nu) - (\lambda - 1 + \ln\lambda) * \nu\Big\}. \tag{1.2.14}$$

Similarly, we get that the process $z^{(n)}$, which is a solution of the equation (1.2.8), admits (P-a.s.) for every n the representation

$$z^{(n)} = \exp\Big\{\gamma^{(n)} \cdot m - \frac{1}{2}(\gamma^{(n)})^2 \circ \langle m \rangle + \ln \lambda^{(n)} * (\mu - \nu) - (\lambda^{(n)} - 1 - \ln \lambda^{(n)}) * \nu\Big\}. \tag{1.2.15}$$

From (1.2.14) and (1.2.15) it follows that $z_{t\wedge\tau_n} = z_t^{(n)}$ (P-a.s.) for all $t \in \mathbf{R}_+$ and every n. Since $z_{\tau_n}^{(n)} = d\widetilde{P}^{\tau_n}/dP^{\tau_n}$ (P-a.s.), we have $z_{\tau_n} = d\widetilde{P}^{\tau_n}/dP^{\tau_n}$ (P-a.s.) for all n. Therefore the process z is a local density of the measure $\widetilde{\mathsf{P}}$ with respect to the measure P. Since the local density is unique, it has the form (1.2.14) up to P-indistinguishability. From this the representation (1.2.3)–(1.2.5) follows. □

COROLLARY 1.2.1. *If all conditions of Theorem* 1.2.1, *except condition* VIII), *are fulfilled and instead of condition* VIII) *the following condition holds*:
VIII)′ $\ln^2 \lambda * \nu_t < \infty$ (P-*a.s.*) *for all* $t \in \mathbf{R}_+$,
then the representation (1.2.3)–(1.2.5) *is true, where* $M \in \mathcal{M}^2_{\mathrm{loc}}(\mathbf{F}, \mathsf{P})$ *with quadratic characteristic*

$$\langle M \rangle = \gamma^2 \circ \langle m \rangle + \ln^2 \lambda * \nu. \tag{1.2.16}$$

PROOF. It is evident that if condition VI) is fulfilled, then, in view of the local equivalence of measures $\widetilde{\mathsf{P}}$ and P, established in the proof of Theorem 1.2.1, conditions VIII) and VIII′) are equivalent to the conditions

$$\chi\left(\lambda \le \frac{1}{2}\right) |\ln \lambda| * \nu \in \mathcal{V}_{\mathrm{loc}}(\mathbf{F}, \mathsf{P}),$$
$$\chi\left(\lambda \le \frac{1}{2}\right) \ln^2 \lambda * \nu \in \mathcal{V}_{\mathrm{loc}}(\mathbf{F}, \mathsf{P}),$$

respectively. Hence all the conditions of Theorem 1.2.1 are fulfilled, whence the representation (1.2.3)–(1.2.5) follows. Then, by condition VIII′), it is evident that $M \in \mathcal{M}^2_{\mathrm{loc}}(\mathbf{F}, \mathsf{P})$ with characteristic (1.2.16). □

REMARK 1.2.1. If all conditions of Theorem 1.2.1 except condition VIII) are fulfilled, then, similarly to the proof of Theorem 1.2.1, we get (P-a.s.)

$$z = \exp\{\widetilde{M} - \widetilde{V}\}, \tag{1.2.17}$$

where

$$\widetilde{M} = \gamma \cdot m + (\lambda - 1) * (\mu - \nu) \in \mathcal{M}_{\mathrm{loc}}(\mathbf{F}, \mathsf{P}), \tag{1.2.18}$$
$$\widetilde{V} = 2^{-1}\gamma^2 \circ \langle m \rangle + (\lambda - 1 - \ln \lambda) * \mu \in \mathcal{V}_{\mathrm{loc}}(\mathbf{F}, \mathsf{P}). \tag{1.2.19}$$

To prove (1.2.19) we note that, according to assumption (7.14) in [**153**], the local equivalence $\widetilde{\mathsf{P}} \overset{\mathrm{loc}}{\sim} \mathsf{P}$, proved in the theorem, implies $\mathsf{P}\{R < \infty\} = 0$, where $R =$

$\inf\{t: z_t = 0\}$. Now from (1.2.10), (1.2.11), and the proof of the first part of Theorem 6.2 in [**153**] we conclude that $(\lambda - 1 - \ln \lambda) * \mu \in \mathcal{V}_{\text{loc}}(\mathbf{F}, \mathsf{P})$, whence (1.2.19) follows. The difference between the representations (1.2.3)–(1.2.5) and (1.2.17)–(1.2.19) consists in that $V \in \mathcal{P}(\mathbf{F})$, while the process $\widetilde{V}$ does not possess this property.

Now we proceed to the construction of statistical experiments generated by observations of a semimartingale. Let $(\Omega, \mathcal{G}, \mathsf{P})$ and $(\Omega, \mathcal{G}, \widetilde{\mathsf{P}})$ be two probability spaces with a fixed family $(\mathcal{G}_t)$ of σ-algebras $\mathcal{G}_t \subset \mathcal{G}$, described at the beginning of this subsection, and let $\xi = (\xi_t)$ be a $(\mathcal{G}_t)$-adapted process defined on the space $(\Omega, \mathcal{G})$ and such that $\xi_t(\omega) = x_t$ for $\omega = (x_s)_{s\in\mathbf{R}_+} \in \Omega$ and all $t \in \mathbf{R}_+$. In this case the observation is of the form $\xi^t = (\xi_s)_{0\le s\le t}$, whereas the sample space is of the form $(\mathbf{D}^t, \mathfrak{B}^t)$, where $\mathbf{D}^t$ is the space of functions $(x_s)_{0\le s\le t}$ that are right-continuous for $s \in [0, t)$ and have left limits for $s \in (0, t]$, and $\mathfrak{B}^t$ is the Borel σ-algebra of subsets of $\mathbf{D}^t$ generated by cylinder sets. Denote by P^t and $\widetilde{P}^t$ the images of measures P and $\widetilde{\mathsf{P}}$ under the map $\xi^t: \Omega \to \mathbf{D}^t$ such that $P^t(A) = \mathsf{P}\{\omega: \xi^t(\omega) \in A\}$ and $\widetilde{P}^t(A) = \widetilde{\mathsf{P}}\{\omega: \xi^t(\omega) \in A\}$ for any $A \in \mathfrak{B}^t$. Thus we have the *family of binary statistical experiments* $(\mathbf{D}^t, \mathfrak{B}^t, P^t, \widetilde{P}^t)$, $t \in \mathbf{R}_+$, *generated by observations* ξ^t *of the* $(\mathcal{G}_t)$*-adapted process* ξ defined on $(\Omega, \mathcal{G})$.

Consider the problem of equivalence of measures P^t and $\widetilde{P}^t$ for $t \in \mathbf{R}_+$. To this end we consider stochastic bases $(\Omega, \mathcal{F}, \mathbf{F}, \mathsf{P})$ and $(\Omega, \mathcal{F}, \mathbf{F}, \widetilde{\mathsf{P}})$, introduced above, and assume that on each of these bases the process ξ is a semimartingale with triples of predictable characteristics $(\alpha, \langle m\rangle, \nu)$ and $(\widetilde{\alpha}, \langle \widetilde{m}\rangle, \widetilde{\nu})$ with respect to $(\mathbf{F}, \mathsf{P})$ and $(\mathbf{F}, \widetilde{\mathsf{P}})$, respectively. The following theorem provides conditions for the equivalence of measures P^t and $\widetilde{P}^t$ for all $t \in \mathbf{R}_+$ in terms of these predictable characteristics of the semimartingale and describes the form of the corresponding density.

THEOREM 1.2.2. *If the conditions of Theorem* 1.2.1 *are fulfilled, then* $\widetilde{P}^t \sim P^t$ *for each* $t \in \mathbf{R}_+$ *and* (P*-a.s.*) *we have*

$$\rho_t\left(\xi^t\right) = \mathsf{E}\left\{z_t/\mathcal{F}_t^\xi\right\}, \tag{1.2.20}$$

where $\rho_t(x) = d\widetilde{P}^t/dP^t(x)$, $x \in \mathbf{D}^t$, $z = (z_t)$ *is the local density of the measure* $\widetilde{\mathsf{P}}$ *with respect to the measure* P, *and* $\mathcal{F}_t^\xi = \sigma\{\xi_s, 0 \le s \le t\}$.

If, in addition,

IX) *for all* $t \in \mathbf{R}_+$ *and any set* $A \in \mathcal{G}_{t+}$ *there exists a set* $B \in \mathcal{G}_t$ *such that* $\mathsf{P}(B) = \mathsf{P}(A)$,

then for each $t \in \mathbf{R}_+$ *we have*

$$\rho_t\left(\xi^t\right) = z_t \ (\mathsf{P}\text{-}a.s.). \tag{1.2.21}$$

PROOF. For any $A \in \mathfrak{B}^t$ we have

$$\begin{aligned}\widetilde{P}^t(A) &= \widetilde{P}^t\left\{\omega: \xi^t(\omega) \in A\right\} = \int_{(\xi^t)^{-1}(A)} z_t(\omega) P^t(d\omega) = \mathsf{E}\,\chi_A\left(\xi^t\right) z_t \\ &= \mathsf{E}\,\chi_A\left(\xi^t\right) \mathsf{E}\left\{z_t/\mathcal{F}_t^\xi\right\} = \int_A \mathsf{E}\left\{z_t/\mathcal{F}_t^\xi\right\}\Big|_{\xi^t=x} P^t(dx)\end{aligned}$$

(here we used that, in view of Theorem 1.2.1, $\widetilde{P}^t \ll P^t$ with density z_t and $\mathsf{E}\{z_t/\mathcal{F}_t^\xi\}$ is a measurable function of ξ^t). From this it follows that $\widetilde{P}^t \ll P^t$ for all $t \in \mathbf{R}_+$ and the equality (1.2.20) holds (P-a.s.). Reasoning similarly, we see that $P^t \ll \widetilde{P}^t$ for all $t \in \mathbf{R}_+$.

In view of condition IX), from the definition of conditional mathematical expectation it follows that (P-a.s.)

$$\mathsf{E}\left\{z_t/\mathcal{F}_t^\xi\right\} = \mathsf{E}\left\{z_t/\mathcal{F}_{t+}^\xi\right\}. \tag{1.2.22}$$

Evidently, $\mathcal{F}_t^\xi = \mathcal{G}_t$ and $\mathcal{F}_{t+}^\xi = \mathcal{F}_{t+}$. And similarly, since $\mathcal{F}_t = \mathcal{G}_{t+}^{\mathbf{Q}}$, we get ($\mathsf{P}$-a.s.)

$$\mathsf{E}\left\{z_t/\mathcal{F}_{t+}^\xi\right\} = \mathsf{E}\{z_t/\mathcal{F}_t\}. \tag{1.2.23}$$

Considering that the random variable z_t is $\mathcal{F}_t$-measurable, from (1.2.20) and (1.2.23) we get (1.2.21). □

COROLLARY 1.2.2. *If the conditions of Theorems* 1.2.1 *and* 1.2.2 *are fulfilled, then* $\widetilde{P}^t \sim P^t$ *for all* $t \in \mathbf{R}_+$ *and* (P-*a.s.*)

$$\Lambda = \ln \rho = M - V, \tag{1.2.24}$$

where $\Lambda = (\Lambda_t)$, $\rho = (\rho_t)$, *and the processes* M *and* V *are defined by formulas* (1.2.4) *and* (1.2.5) (*here we write* Λ_t *and* ρ_t *instead of* $\Lambda_t(\xi^t)$ *and* $\rho_t(\xi^t)$, *respectively*).

The representation (1.2.24) is obtained by uniting the relations (1.2.3)–(1.2.5) and (1.2.21).

REMARK 1.2.2. It is most difficult to check conditions VII) and IX). The (τ_n)-uniqueness (condition VII)) has been established so far for semimartingales with determined triples of predictable characteristics, diffusion-type processes, counting processes [**155, 156**], and Markov processes [**113, 159**]. As for condition IX), it should be observed that it is related to the problem of representation of martingales ([**13**], Proposition 3), considered in [**26, 98, 152, 153, 155, 156**].

Now let us assume that on a measurable space $(\Omega, \mathcal{G})$ with filtration $(\mathcal{G}_t)$ we are given a parametric family of probability measures $(\mathsf{P}_\theta, \theta \in \Theta)$, where $\Theta \subset \mathbf{R}^k$ is an open set. Reasoning as above, we obtain a *parametric family of statistical experiments* $(\mathbf{D}_t, \mathfrak{B}^t, (P_\theta^t, \theta \in \Theta))$, $t \in \mathbf{R}_+$, generated by observations ξ^t of the process ξ defined on the space with filtration $(\Omega, \mathcal{G}, (\mathcal{G}_t))$, where $P_\theta^t(A) = \mathsf{P}_\theta\{\omega: \xi^t(\omega) \in A\}$, $A \in \mathfrak{B}^t$. We assume that for every $\theta \in \Theta$ the process ξ considered on the stochastic basis $(\Omega, \mathcal{F}, \mathbf{F}, \mathsf{P}_\theta)$ is a semimartingale with the triple of predictable characteristic $(\alpha(\theta), \langle m(\theta)\rangle, \nu(\theta))$.

Statistical experiments generated by diffusion-type processes. Let $(\Omega, \mathcal{F}, \mathbf{F}, \mathsf{P})$ be a stochastic basis satisfying ordinary conditions with respect to the measure P, let $w = (w_t)$ be a *standard Wiener process* on $(\Omega, \mathcal{F}, \mathbf{F}, \mathsf{P})$, that is, $w \in \mathcal{M}_{\mathrm{loc},0}^c(\mathbf{F}, \mathsf{P})$ with $\langle w\rangle_t = t$, $t \in \mathbf{R}_+$, and let $(\mathbf{C}, \mathcal{G})$ be a measurable space of continuous functions $x = (x_t)$ with Borel σ-algebra $\mathcal{G} = \bigvee_{t \in \mathbf{R}_+} \mathcal{G}_t$, where $\mathcal{G}_t$ is the smallest σ-algebra of subsets of $\mathbf{C}$ generated by the cylinder sets with bases in $[0, t]$.

Let $(a(t,x))_{t\in\mathbf{R}_+,\,x\in\mathbf{C}}$, $(\widetilde{a}(t,x))_{t\in\mathbf{R}_+,\,x\in\mathbf{C}}$ and $(b(t,x))_{t\in\mathbf{R}_+,\,x\in\mathbf{C}}$ be functions defining measurable maps from $(\mathbf{R}_+\times\mathbf{C},\ \mathfrak{B}_+\times\mathcal{G})$ to $(\mathbf{R}^1,\mathcal{B}^1)$ and adapted to the family $(\mathcal{G}_t)$ (that is, $a(t,x)$, $\widetilde{a}(t,x)$ and $b(t,x)$ are $\mathcal{G}_t$-measurable for every $t\in\mathbf{R}_+$). Note that $(\mathcal{G}_t)$-adapted functions on $\mathbf{R}_+\times\mathbf{C}$ are called *nonanticipating functionals* on $\mathbf{C}$.

Consider *diffusion-type stochastic processes* $\xi=(\xi_t)$ and $\widetilde{\xi}=(\widetilde{\xi}_t)$ that are strong solutions of the stochastic differential equations

$$d\xi_t = a(t,\xi)\,dt + b(t,\xi)\,dw_t, \qquad \xi_0=\eta, \tag{1.2.25}$$

$$d\widetilde{\xi}_t = \widetilde{a}(t,\widetilde{\xi})\,dt + b(t,\widetilde{\xi})\,dw_t, \qquad \widetilde{\xi}_0=\eta, \tag{1.2.26}$$

respectively, where η is an $\mathcal{F}_0$-measurable random variable, independent of w. We assume that the coefficients of the equations (1.2.25) and (1.2.26) satisfy the conditions of existence and uniqueness of solutions on every interval $[0,t]$, $t\in\mathbf{R}_+$.

Stochastic processes ξ and $\widetilde{\xi}$ that are solutions of the equations (1.2.25) and (1.2.26) generate probability measures P and $\widetilde{\mathsf{P}}$, respectively, on the space $(\mathbf{C},\mathcal{G})$. Denote by P^t and $\widetilde{P}^t$ the restrictions of the measures P and $\widetilde{\mathsf{P}}$ to $\mathcal{G}_t$, $t\in\mathbf{R}_+$. Theorem 7.19 in [**96**] provides conditions for the equivalence of measures P^t and $\widetilde{P}^t$ for all $t\in\mathbf{R}_+$, as well as the form of the corresponding density $\rho_t(x)=d\widetilde{P}^t/dP^t(x)$, $x\in\mathbf{C}$.

THEOREM 1.2.3. *Let the following conditions hold for all $t\in\mathbf{R}_+$.*

I) *The coefficients $a(s,x)$, $\widetilde{a}(s,x)$, and $b(s,x)$ satisfy the conditions of existence and uniqueness of strong solutions of the equations* (1.2.25) *and* (1.2.26) *on* $[0,t]$.

II) *The equation* (*with respect to* $\gamma(t,\xi)$)

$$b(t,\xi)\gamma(t,\xi)=\widetilde{a}(t,\xi)-a(t,\xi) \tag{1.2.27}$$

has (P-*a.s.*) *a finite solution.*

III) *We have*

$$\mathsf{P}\Big\{\int_0^t \left(b^{\oplus}(s,\xi)\right)^2\left[\widetilde{a}^2(s,\xi)+a^2(s,\xi)\right]ds<\infty\Big\}$$
$$=\mathsf{P}\Big\{\int_0^t \left(b^{\oplus}(s,\widetilde{\xi})\right)^2\left[\widetilde{a}^2(s,\widetilde{\xi})+a^2(s,\widetilde{\xi})\right]ds<\infty\Big\}=1,$$

where $b^{\oplus}=b^{-1}$ for $b\neq 0$ and $b^{\oplus}=0$ for $b=0$.

Then $\widetilde{P}^t\sim P^t$ for all $t\in\mathbf{R}_+$ and (P-a.s.)

$$\rho_t(\xi)=\exp\left\{b^{\oplus}\gamma\cdot\xi_t-2^{-1}\left(b^{\oplus}\right)^2\left(\widetilde{a}^2-a^2\right)\circ I_t\right\}, \tag{1.2.28}$$

where $a=(a(t,\xi))$, $\widetilde{a}=(\widetilde{a}(t,\xi))$, $b=(b(t,\xi))$, $b^{\oplus}=(b^{\oplus}(t,\xi))$, $\gamma=(\gamma(t,\xi))$, $I=(I_t)$, $I_t=t$.

REMARK 1.2.3. It is not hard to see that condition II) is equivalent to the condition

$$\mathsf{P}\left[\left(\{b=0\}\setminus\{\widetilde{a}-a=0\}\right)\cup\{|b|=\infty\}\cup\{|\widetilde{a}-a|=\infty\}\right]=0.$$

For this reason the solution of the equation (1.2.27) can be represented in the form

$$\gamma(t,\xi)=b^{\oplus}(t,\xi)[\widetilde{a}(t,\xi)-a(t,\xi)], \tag{1.2.29}$$

which is demonstrated in the proof of equality (1.2.28).

REMARK 1.2.4. The stochastic integral $b^{\oplus}\gamma \cdot \xi$ in (1.2.28) is a stochastic integral with respect to the semimartingale ξ. According to (1.2.22), condition III), and Theorem 1.1.6, this integral is well defined and (P-a.s.)

$$b^{\oplus}\gamma \cdot \xi = b^{\oplus}\gamma a \circ I + b^{\oplus} b\gamma \cdot w. \tag{1.2.30}$$

The following theorem provides a helpful form of the local density $\rho(\xi) = (\rho_t(\xi))$.

THEOREM 1.2.4. *If conditions* I)–III) *of Theorem* 1.2.3 *are fulfilled for all* $t \in \mathbf{R}_+$, *then* $\widetilde{P}^t \sim P^t$ *for all* $t \in \mathbf{R}_+$ *and* (P-*a.s.*)

$$\rho_t(\xi) = \exp\left\{\gamma \cdot w_t - 2^{-1}\gamma^2 \circ I_t\right\}. \tag{1.2.31}$$

PROOF. Using (1.2.29) and (1.2.30), from (1.2.28) we get that for all $t \in \mathbf{R}_+$ (P-a.s.)

$$\ln \rho_t(\xi) = b^{\oplus} b\gamma \cdot w_t - 2^{-1}\left(b^{\oplus} b\gamma\right)^2 \circ I_t + a\gamma b^{\oplus}\left(1 - b^{\oplus} b\right) \circ I_t. \tag{1.2.32}$$

According to Remark 1.2.3, condition II) implies that (P-a.s.)

$$a(s,\xi)\gamma(s,\xi)b^{\oplus}(s,\xi)\big(1 - b^{\oplus}(s,\xi)b(s,\xi)\big) = 0$$

for all $s \in \mathbf{R}_+$. From this it follows that

$$a\gamma b^{\oplus}\left(1 - b^{\oplus} b\right) \circ I_t = 0 \quad \text{(P-a.s.)} \tag{1.2.33}$$

for all $t \in \mathbf{R}_+$. Then, evidently,

$$b^{\oplus} b\gamma \cdot w_t = \gamma \cdot w_t + \gamma\left(b^{\oplus} b - 1\right) \cdot w_t, \tag{1.2.34}$$

$$\left(b^{\oplus} b\gamma\right)^2 \circ I_t = \gamma^2 \circ I_t + \gamma^2\left(\left(b^{\oplus} b\right)^2 - 1\right) \circ I_t. \tag{1.2.35}$$

Again, in view of Remark 1.2.3, we get (P-a.s.)

$$\gamma(b^{\oplus} b - 1) \cdot w_t = 0, \qquad \gamma^2((b^{\oplus} b)^2 - 1) \circ I_t = 0 \tag{1.2.36}$$

for all $t \in \mathbf{R}_+$. Now (1.2.32)–(1.2.36) imply (1.2.31). □

Observations ξ^t and $\widetilde{\xi}^t$ of the diffusion-type processes, described above, generate on the interval $[0,t]$ the family of binary statistical experiments $(\mathbf{C}^t, \mathfrak{B}^t, P^t, \widetilde{P}^t)$, $t \in \mathbf{R}_+$, where $\mathbf{C}^t$ is the space of continuous functions $x = (x_s)_{0 \le s \le t}$, $\mathfrak{B}^t$ is the Borel σ-algebra of subsets of $\mathbf{C}^t$, and P^t and $\widetilde{P}^t$ are the distributions of observations ξ and $\widetilde{\xi}^t$ of the diffusion-type processes, that is, $P^t(A) = \mathsf{P}\{\omega : \xi^t(\omega) \in A\}$ and $\widetilde{P}^t(A) = \mathsf{P}\{\omega : \widetilde{\xi}^t(\omega) \in A\}$, $A \in \mathfrak{B}^t$. If the conditions of Theorem 1.2.3 are fulfilled, then $P^t \sim \widetilde{P}^t$ for all $t \in \mathbf{R}_+$ and (P-a.s.)

$$\frac{d\widetilde{P}^t}{dP^t}(\xi^t) = \rho_t(\xi). \tag{1.2.37}$$

Now we proceed to the parametric case. Let $\xi^\theta = (\xi_t^\theta)$ be diffusion-type processes that depend on a parameter $\theta \in \Theta \subset \mathbf{R}^k$, $k \geq 1$, and are, for fixed θ, strong solutions of the stochastic differential equations

$$d\xi_t^\theta = a\left(t, \xi^\theta, \theta\right) dt + b\left(t, \xi^\theta\right) dw_t, \qquad \xi_0^\theta = \eta, \tag{1.2.38}$$

where η is an $\mathcal{F}_0$-measurable random variable independent of θ and w, and $a(t, x, \theta)$ and $b(t, x)$ are nonanticipating functionals on $\mathbf{C}$ for every $\theta \in \Theta$ satisfying the conditions for existence and uniqueness of a strong solution of the equation (1.2.38) on every interval $[0, t]$, $t \in \mathbf{R}_+$.

The process ξ^θ generates a probability measure P_θ on $(\mathbf{C}, \mathcal{G})$. By P_θ^t we denote the restriction of the measure P_θ to $\mathcal{G}_t$, $t \in \mathbf{R}_+$. In what follows the superscript θ of the processes ξ^θ is omitted; in order to emphasize that a certain expression for ξ^θ depends on θ, the symbol of the probability measure P will be provided with the subscript θ. For instance, instead of $\mathsf{P}\{\xi^\theta \in A\}$, $\mathsf{E}\, f(\xi^\theta)$ we shall write $\mathsf{P}_\theta\{\xi \in A\}$, $\mathsf{E}_\theta\, f(\xi)$, etc.

Observations $\xi^t = (\xi_s)_{0 \leq s \leq t}$ of the process ξ, which is the solution of the equation (1.2.38) for some $\theta \in \Theta$, generate the parametric family of statistical experiments $(\mathbf{C}^t, \mathfrak{B}^t, (P_\theta^t, \theta \in \Theta))$, $t \in \mathbf{R}_+$, where P_y^t is the distribution of the observation ξ^t of the process ξ, which is the solution of the equation (1.2.38) for $\theta = y$, that is, $P_y^t(A) = \mathsf{P}_y\{\omega : \xi^t(\omega) \in A\}$, $A \in \mathfrak{B}^t$. Putting $\widetilde{P}^t = P_y^t$ and $P^t = P_\theta^t$, from Theorem 1.2.3 we get the condition of equivalence of the measures P_y^t and P_θ^t and from formula (1.2.37) we get the form of the corresponding density.

Statistical experiments generated by counting processes. Here we proceed as in the case of semimartingales. Let $\Omega = X$ be the space of nondecreasing right-continuous step functions $x = (x_t)$ such that $x_0 = 0$, $x_t < \infty$ for all $t \in \mathbf{R}_+$, and $\Delta x_t = 0$ or $\Delta x_t = 1$ for all $t \in \mathbf{R}_+$, and let $\mathcal{G}_t$ be the smallest σ-algebra of subsets of X generated by the cylinder sets with bases in $[0, t]$, $\mathcal{G} = \bigvee_{t \in \mathbf{R}_+} \mathcal{G}_t$. The family of σ-algebras $\mathcal{G}_t$ is known [**98, 152, 156**] to be left continuous.

Given two probability measures P and $\widetilde{\mathsf{P}}$ on $(\Omega, \mathcal{G})$, let $\mathbf{Q} = 2^{-1}(\mathsf{P} + \widetilde{\mathsf{P}})$ be another probability measure on $(\Omega, \mathcal{G})$. Put $\mathcal{F}_t = \mathcal{G}_t^{\mathbf{Q}}$, $t \in \mathbf{R}_+$, $\mathcal{F} = \bigvee_{t \in \mathbf{R}_+} \mathcal{F}_t$ and denote by P^t and $\widetilde{P}^t$ the restrictions of the measures P and $\widetilde{\mathsf{P}}$ to the σ-algebra $\mathcal{F}_t$, $\mathbf{F} = (\mathcal{F}_t)$.

On the stochastic basis $(\Omega, \mathcal{F}, \mathbf{F}, \mathsf{P})$ we consider the counting process $\xi = (\xi_t)$ that has the coordinate representation $\xi_t(\omega) = x_t$ for all $t \in \mathbf{R}_+$, where $\omega = (x_t) \in \Omega$. It is not hard to see that $\xi \in \mathcal{A}_{\text{loc}}^+(\mathbf{F}, \mathsf{P})$, whence, in view of Theorem 1.1.2, it follows that there exists a unique up to P-indistinguishability process $\nu = (\nu_t) \in \mathcal{A}_{\text{loc}}^+(\mathbf{F}, \mathsf{P}) \cap \mathcal{P}(\mathbf{F})$, called the *compensator of the process* ξ, such that $\xi - \nu \in \mathcal{M}_{\text{loc}}(\mathbf{F}, \mathsf{P})$. Similarly, on the stochastic basis $(\Omega, \mathcal{F}, \mathbf{F}, \widetilde{\mathsf{P}})$ the counting process ξ has the compensator $\widetilde{\nu} = (\widetilde{\nu}_t) \in \mathcal{A}_{\text{loc}}^+(\mathbf{F}, \widetilde{\mathsf{P}}) \cap \mathcal{P}(\mathbf{F})$. In what follows we assume the compensators ν and $\widetilde{\nu}$ to be continuous processes with respect to P and $\widetilde{\mathsf{P}}$, respectively.

THEOREM 1.2.5. *Let the following conditions be fulfilled:*

I) $\Delta \nu_t = 0$ ($\widetilde{\mathsf{P}}$-*a.s.*) *for all* $t \in \mathbf{R}_+$;

II) *there exists a nonnegative predictable process* $\lambda = (\lambda_t)$ *such that* $\widetilde{\nu}_t = \lambda \circ \nu_t$ ($\widetilde{\mathsf{P}}$-*a.s.*) *for all* $t \in \mathbf{R}_+$;

III) $(1 - \lambda^{1/2})^2 \circ \nu_t < \infty$ ($\widetilde{\mathsf{P}}$-*a.s.*) *for all* $t \in \mathbf{R}_+$.

Then $\widetilde{\mathsf{P}} \overset{\text{loc}}{\ll} \mathsf{P}$. *If, in addition, conditions* I)–III) *are fulfilled with* ν *and* P *replaced by* $\widetilde{\nu}$ *and* $\widetilde{\mathsf{P}}$, *and* $\widetilde{\nu}$ *and* $\widetilde{\mathsf{P}}$ *replaced by* ν *and* P, *then* $\widetilde{\mathsf{P}} \overset{\text{loc}}{\sim} \mathsf{P}$ *and the local density* z *of the measure* $\widetilde{\mathsf{P}}$ *with respect to the measure* P *is* (P-*a.s.*) *of the form*

$$z = \exp\{\ln \lambda * \xi - (\lambda - 1) \circ \nu\}, \tag{1.2.39}$$

where

$$|\ln \lambda| * \xi \in \mathcal{V}_{\text{loc}}(\mathbf{F}, \mathsf{P}), \tag{1.2.40}$$

$$|\lambda - 1| \circ \nu \in \mathcal{V}_{\text{loc}}(\mathbf{F}, \mathsf{P}). \tag{1.2.41}$$

If, in addition,

IV) $(\lambda - 1 - \ln \lambda) \circ \nu \in \mathcal{V}_{\text{loc}}(\mathbf{F}, \mathsf{P})$,

then (P-*a.s.*) *we have*

$$z = \exp\{M - V\}, \tag{1.2.42}$$

where

$$M = \ln \lambda * (\xi - \nu) \in \mathcal{M}^d_{\text{loc}}(\mathbf{F}, \mathsf{P}), \tag{1.2.43}$$

$$V = (\lambda - 1 - \ln \lambda) \circ \nu \in \mathcal{V}_{\text{loc}}(\mathbf{F}, \mathsf{P}). \tag{1.2.44}$$

PROOF. The representation (1.2.42)–(1.2.44) follows immediately from Theorem 1.2.1. By condition III) and local equivalence $\widetilde{\mathsf{P}} \overset{\text{loc}}{\sim} \mathsf{P}$ we obtain (1.2.41), and hence $|\lambda - 1| * \xi \in \mathcal{V}_{\text{loc}}(\mathbf{F}, \mathsf{P})$ and (P-a.s.)

$$(\lambda - 1) * (\xi - \nu) = (\lambda - 1) * \xi - (\lambda - 1) \circ \nu,$$
$$(\lambda - 1 - \ln \lambda) * \xi = (\lambda - 1) * \xi - \ln \lambda * \xi.$$

From this and Remark 1.2.1 we get (1.2.39)–(1.2.41). □

REMARK 1.2.5. In view of the obvious inequalities

$$\left(\widetilde{\nu}_t^{1/2} - \nu_t^{1/2}\right)^2 \leq \left(1 - \lambda^{1/2}\right)^2 \circ \nu_t \leq 2(\widetilde{\nu}_t + \nu_t),$$

condition III) is equivalent to the condition

III′) $\nu_t < \infty$ ($\widetilde{\mathsf{P}}$-a.s.) for all $t \in \mathbf{R}_+$.

From this it follows that $|\lambda - 1| \circ \nu \in \mathcal{V}_{\text{loc}}(\mathbf{F}, \mathsf{P})$. This fact was used in the proof of Theorem 1.2.5.

REMARK 1.2.6. In view of (1.2.41), condition IV) is equivalent to the condition

IV′) $|\ln \lambda| \circ \nu \in \mathcal{V}_{\text{loc}}(\mathbf{F}, \mathsf{P})$.

From this it follows that $\ln \lambda \in \mathcal{G}^1_{\text{loc}}(\nu, \mathbf{F}, \mathsf{P})$ and hence, by virtue of Theorem 1.1.8,

$$\ln \lambda * (\xi - \nu) = \ln \lambda * \xi - \ln \lambda \circ \nu.$$

Note that formula (1.2.39) was proved in [**32**] (Theorem 14; see also [**152**], Proposition 4.3).

The observation $\xi^t = (\xi_s)_{0\le s\le t}$ of the counting process ξ described above, generates the family of binary statistical experiments $(X^t, \mathfrak{B}^t, (P^t, \widetilde{P}^t))$, $t \in \mathbf{R}_+$, where X^t is the space of nondecreasing right-continuous step functions

$$x = (x_s)_{0\le s\le t}$$

such that $x_0 = 0$, $x_s < \infty$ for all $s \le t$, $\Delta x_s = 0$ or $\Delta x_s = 1$ for all $s \in (0,t]$, $\mathfrak{B}^t$ is the least σ-algebra of subsets of X^t generated by cylinder sets, and P^t and $\widetilde{P}^t$ are the distributions of the observation ξ^t, that is, $P^t(A) = \mathsf{P}\{\xi^t \in A\}$ and $\widetilde{P}^t(A) = \widetilde{\mathsf{P}}\{\xi^t \in A\}$, $A \in \mathfrak{B}^t$. If the conditions of Theorem 1.2.5 are fulfilled, then, evidently, $\widetilde{P}^t \sim P^t$ for all $t \in \mathbf{R}_+$ and (P-a.s.)

$$\rho_t(\xi^t) = \frac{d\widetilde{P}^t}{dP^t}(\xi^t) = z_t.$$

Now let a parametric family of probability measures $(\mathsf{P}_\theta, \theta \in \Theta)$, where $\Theta \subset \mathbf{R}^k$, $k \ge 1$, be given on a measurable space $(\Omega, \mathcal{G})$ with filtration $(\mathcal{G}_t)$. Reasoning as above, we get the parametric family of statistical experiments

$$\left(X^t, \mathfrak{B}^t, (P^t_\theta, \theta \in \Theta)\right), \qquad t \in \mathbf{R}_+,$$

generated by the observations ξ^t of the counting process ξ with

$$P^t_\theta(A) = \mathsf{P}_\theta\left\{\xi^t \in A\right\}, \qquad A \in \mathfrak{B}^t.$$

We assume that for every θ the counting process ξ, considered on the stochastic basis $(\Omega, \mathcal{F}, \mathbf{F}, \mathsf{P}_\theta)$, has the compensator $\nu(\theta)$, which is assumed to be (P_θ-a.s.) continuous in what follows.

1.3. Limit theorems for semimartingales

Central limit theorem for local martingales. For every $t \in \mathbf{R}_+$ we consider a complete stochastic basis $(\Omega, \mathcal{F}^t, \mathbf{F}^t, P^t)$, where $\mathbf{F}^t = (\mathcal{F}^t_s)_{s\in\mathbf{R}_+}$ and $\mathcal{F}^t_s \subset \mathcal{F}^t$, and on this basis we consider a local martingale $Y^t \in \mathcal{M}_{\mathrm{loc},0}(\mathbf{F}^t, P^t)$ of the form

$$Y^t = M^t + U^t * \left(\mu^t - \nu^t\right), \tag{1.3.1}$$

where $M^t \in \mathcal{M}^c_{\mathrm{loc},0}(\mathbf{F}^t, P^t)$, μ^t is an $\mathbf{F}^t$-optional integral-valued random measure on $\mathbf{R}_+ \times \mathbf{R}_0$ for which the measure $\mathbf{M}^{P^t}_{\mu^t}$ satisfies the property of $\widetilde{\mathcal{P}}(\mathbf{F}^t)$-$\sigma$-finiteness, ν^t is the $\mathbf{F}^t$-compensator of the measure μ^t such that

$$\sum_{s\in\mathbf{R}_+} \nu^t(\{s\}, \mathbf{R}_0) = 0,$$

and $U^t = (U^t_{s,x})_{s\in\mathbf{R}_+,\, x\in\mathbf{R}_0} \in \mathcal{G}_{\mathrm{loc}}(\nu^t, \mathbf{F}^t, P^t)$.

The following central limit theorem is valid for the distributions of Y^t at the moment st for fixed $s \in (0,\infty)$ as $t \to \infty$.

THEOREM 1.3.1. *For fixed $s \in (0,\infty)$ and any $\varepsilon \in (0,1]$ let the following conditions be fulfilled:*

1) $P^t\text{-}\lim_{t\to\infty} \chi(|U^t| > \varepsilon)|U^t| * \nu^t_{st} = 0$;
2) $P^t\text{-}\lim_{t\to\infty} \left[\langle M^t\rangle_{st} + \chi(|U^t| \le \varepsilon)|U^t|^2 * \nu^t_{st}\right] = c(s) \in [0,\infty)$, *where $c(s)$ is some nonrandom number dependent on s.*

Then

$$\mathcal{L}\left(Y^t_{st}|P^t\right) \xrightarrow{w} \mathcal{N}(0,c(s)), \qquad t\to\infty. \tag{1.3.2}$$

In the case $c(s) = 0$ by $\mathcal{N}(0,c(s))$ we mean the *degenerate distribution* concentrated at the point 0.

PROOF. Denote by $\overline{\mu}^t$ the jump measure of the process Y^t and by $\overline{\nu}^t$ its $\mathbf{F}^t$-compensator:

$$\overline{\mu}^t((0,u],A) = \mu^t\big((s,x)\in(0,u]\times\mathbf{R}_0 : U^t_{s,x}\in A\big) = \chi(U^t\in A) * \mu^t_u, \tag{1.3.3}$$

$$\overline{\nu}^t((0,u],A) = \nu^t\big((s,x)\in(0,u]\times\mathbf{R}_0 : U^t_{s,x}\in A\big) = \chi(U^t\in A) * \nu^t_u \tag{1.3.4}$$

for all $u \in \mathbf{R}_+$ and $A \in \mathcal{B}_0$, where $\chi(U^t\in A) = (\chi(U^t_{s,x}\in A))_{s\in\mathbf{R}_+,\,x\in\mathbf{R}_0}$.

Introduce the stochastic process $\widetilde{Y}^t = (Y^t_{st})_{s\in\mathbf{R}_+}$. It is evident that $\widetilde{Y}^t \in \mathcal{M}_{\text{loc},0}(\mathbf{F}^t,P^t)$ and

$$\widetilde{Y}^t = \widetilde{M}^t + x * (\widetilde{\mu}^t - \widetilde{\nu}^t),$$

where $\widetilde{M}^t = (\widetilde{M}^t_s) \in \mathcal{M}^c_{\text{loc},0}(\mathbf{F}^t,P^t)$, $\widetilde{\mu}^t$ is the jump measure of the process $\widetilde{Y}^t$, and $\widetilde{\nu}^t$ is its $\mathbf{F}^t$-compensator. Besides,

$$\widetilde{M}^t_s = M^t_{st}, \qquad \left\langle\widetilde{M}^t\right\rangle_s = \left\langle M^t\right\rangle_{st}, \qquad s\in\mathbf{R}_+, \tag{1.3.5}$$

$$\begin{aligned}&\widetilde{\mu}^t((0,u],A) = \overline{\mu}^t((0,ut],A),\\ &\widetilde{\nu}^t((0,u],A) = \overline{\nu}^t((0,ut],A), \qquad u\in\mathbf{R}_+,\ A\in\mathcal{B}_0.\end{aligned} \tag{1.3.6}$$

The relations (1.3.3)–(1.3.6) imply

$$\chi\left(\left|U^t\right| > \varepsilon\right)\left|U^t\right| * \nu^t_{st} = \chi(x>\varepsilon)|x| * \widetilde{\nu}^t_s, \tag{1.3.7}$$

$$\left\langle M^t\right\rangle_{st} + \chi\left(\left|U^t\right|\le\varepsilon\right)\left|U^t\right|^2 * \nu^t_{st} = \left\langle\widetilde{M}^t\right\rangle_s + \chi(x\le\varepsilon)|x|^2 * \widetilde{\nu}^t_s. \tag{1.3.8}$$

Using Corollary 2 and Remarks 1 and 5 from [**97**], we get

$$\mathcal{L}\left(\widetilde{Y}^t_s|P^t\right) \xrightarrow{w} \mathcal{N}(0,c(s)), \qquad t\to\infty.$$

But this relation is equivalent to (1.3.2). □

REMARK 1.3.1. The results of [**97**] were proved for the case where the measure P^t does not depend on t. However, repeating the argument from [**97**], it is not hard to see that the assertions of [**97**] used in the proof of Theorem 1.3.1 are also valid in the case where the measures P^t may depend on t. Moreover, it is easy to see that in the case $c(s) = 0$ the assertion of Theorem 1.3.1 follows easily from the Lenglart inequality.

Now we prove a multi-dimensional variant of Theorem 1.3.1.

Consider a k-dimensional local martingale

$$Y^t = \left(Y^{t,1}, \dots, Y^{t,k}\right)' \in \mathcal{M}_{\mathrm{loc},0}\left(\mathbf{F}^t, P^t\right)$$

of the form

$$Y^t = M^t + U^t * \left(\mu^t - \nu^t\right), \tag{1.3.9}$$

where

$$M^t = \left(M^{t,1}, \dots, M^{t,k}\right)' \in \mathcal{M}^c_{\mathrm{loc},0}\left(\mathbf{F}^t, P^t\right),$$

μ^t and ν^t are the random measures as in (1.3.1),

$$U^t = \left(U^{t,1}, \dots, U^{t,k}\right)' \in \mathcal{G}_{\mathrm{loc}}\left(\nu^t, \mathbf{F}^t, P^t\right),$$

and

$$U^t * \left(\mu^t - \nu^t\right) = \left(U^{t,1} * \left(\mu^t - \nu^t\right), \dots, U^{t,k} * \left(\mu^t - \nu^t\right)\right)'.$$

THEOREM 1.3.2. *Suppose the following conditions be fulfilled for a fixed* $s \in (0, \infty)$ *and any* $\varepsilon \in (0, 1]$:

1) $P^t\text{-}\lim_{t\to\infty} \chi(|U^t| > \varepsilon)|U^t| * \nu^t_{st} = 0$;
2) $P^t\text{-}\lim_{t\to\infty}[\langle M^t, M^t\rangle_{st} + \chi(|U^t| \le \varepsilon)U^t(U^t)' * \nu^t_{st}] = I(s)$, *where* $I(s)$ *is some symmetric nonrandom* $k \times k$*-matrix dependent on* s *with finite elements.*

Then

$$\mathcal{L}\left(Y^t_{st}|P^t\right) \xrightarrow{w} \mathcal{N}(0, I(s)), \qquad t \to \infty. \tag{1.3.10}$$

PROOF. In accordance with the Cramér–Wold [**7**] method, to prove (1.3.10) it suffices to show that

$$\mathcal{L}\left(\lambda' Y^t_{st}|P^t\right) \xrightarrow{w} \mathcal{N}(0, \lambda' I(s)\lambda), \qquad t \to \infty, \tag{1.3.11}$$

for any $\lambda \in \mathbf{R}^k$ with $|\lambda| = 1$. Evidently,

$$\lambda' Y^t = \lambda' M^t + \lambda'\left(U^t * \left(\mu^t - \nu^t\right)\right),$$
$$\lambda' M^t \in \mathcal{M}^c_{\mathrm{loc},0}\left(\mathbf{F}^t, P^t\right)$$

and (P^t-a.s.)

$$\left\langle \lambda' M^t \right\rangle = \lambda' \left\langle M^t, M^t \right\rangle \lambda. \tag{1.3.12}$$

From (1.1.4) and (1.1.4′) we get $\lambda' U^t \in \mathcal{G}_{\mathrm{loc}}(\nu^t, \mathbf{F}^t, P^t)$ and

$$\lambda'\left(U^t * \left(\mu^t - \nu^t\right)\right) = \left(\lambda' U^t\right) * \left(\mu^t - \nu^t\right).$$

Hence

$$\lambda' Y^t = \lambda' M^t + (\lambda' U^t) * (\mu^t - \nu^t). \tag{1.3.13}$$

From condition 1), for any $\varepsilon \in (0,1]$ and $\lambda \in \mathbf{R}^k$ with $|\lambda| = 1$ we have

$$P^t\text{-}\lim_{t\to\infty} \chi\left(\left|\lambda' U^t\right| > \varepsilon\right) \left|\lambda' U^t\right| * \nu^t_{st} = 0. \tag{1.3.14}$$

For any $\varepsilon \in (0,1]$ and $\lambda \in \mathbf{R}^k$ with $|\lambda| = 1$ we have

(1.3.15)

$$\begin{aligned}&\chi\left(\left|\lambda' U^t\right| \le \varepsilon\right) \left|\lambda' U^t\right|^2 * \nu^t \\ &\quad = \lambda' \chi\left(\left|U^t\right| \le \varepsilon\right) U^t (U^t)' * \nu^t \lambda + \chi\left(\left|U^t\right| > \varepsilon,\, \left|\lambda' U^t\right| \le \varepsilon\right) \left|\lambda' U^t\right|^2 * \nu^t,\end{aligned}$$

where

$$\chi\left(\left|U^t\right| > \varepsilon,\, \left|\lambda' U^t\right| \le \varepsilon\right) \left|\lambda' U^t\right|^2 * \nu^t \le \varepsilon \chi\left(\left|U^t\right| > \varepsilon\right) \left|U^t\right| * \nu^t. \tag{1.3.16}$$

Taking into account conditions 1) and 2) and estimate (1.3.16), from the equalities (1.3.12) and (1.3.15) we get

$$P^t\text{-}\lim_{t\to\infty} \left[\left\langle \lambda' M^t \right\rangle_{st} + \chi\left(\left|\lambda' U^t\right| \le \varepsilon\right) \left|\lambda' U^t\right|^2 * \nu^t_{st}\right] = \lambda' I(s) \lambda \tag{1.3.17}$$

for any $\varepsilon \in (0,1]$ and $\lambda \in \mathbf{R}^k$ with $|\lambda| = 1$.

From (1.3.13), (1.3.14) and (1.3.17) it follows that the conditions of Theorem 1.3.1 are fulfilled for the process $\lambda' Y^t$ for any $\lambda \in \mathbf{R}^k$ with $|\lambda| = 1$. Now the relation (1.3.10) follows from Theorem 1.3.1. □

REMARK 1.3.2. The matrix $I(s)$ from Theorem 1.3.2 can be degenerate. If the rank of the matrix $I(s)$ is zero, then all elements of this matrix are zeros and by $\mathcal{N}(0, I(s))$ is meant the distribution in $\mathbf{R}^k$ concentrated at the point $0 \in \mathbf{R}^k$.

Now we formulate the central limit theorem for local square-integrable martingales Y^t in the multi-dimensional case as a corollary of Theorem 1.3.2.

Let the process $Y^t \in \mathcal{M}^2_{\text{loc}}(\mathbf{F}^t, P^t)$ be of the form (1.3.9), where

$$U^t = \left(U^{t,1}, \ldots, U^{t,x}\right)' \in \mathcal{G}^2_{\text{loc}}\left(\nu^t, \mathbf{F}^t, P^t\right).$$

THEOREM 1.3.3. *Let the following conditions be fulfilled for a fixed* $s \in (0,\infty)$ *and any* $\varepsilon \in (0,1]$:

1) $P^t\text{-}\lim_{t\to\infty} \chi(|U^t| > \varepsilon)|U^t|^2 * \nu^t_{st} = 0$;
2) $P^t\text{-}\lim_{t\to\infty} \left[\langle M^t, M^t\rangle_{st} + U^t (U^t)' * \nu^t_{st}\right] = I(s)$, *where* $I(s)$ *is some symmetric nonrandom matrix depending on* s *with finite elements.*

Then

$$\mathcal{L}\left(Y^t_{st} | P^t\right) \xrightarrow{w} \mathcal{N}(0, I(s)), \qquad t \to \infty.$$

REMARK 1.3.3. Theorem 1.3.3 can be proved by repeating the proof of Theorem 1.3.2 with corresponding changes and using Corollary 2 from [**97**].

REMARK 1.3.4. Condition 1) of Theorem 1.3.3 is the Lindeberg condition. In the work [**80**] Theorem 1.3.3 was proved in the case where, instead of condition 1), the condition

1′) $\lim_{t\to\infty} \mathsf{E}^t\, |U^t|^2 \wedge |U^t|^3 * \nu^t_{st} = 0$

is fulfilled. This condition can be called the Lyapunov condition since for its fulfillment it is sufficient that for some $\delta \in (0,1]$ condition

1″) $\lim_{t\to\infty} \mathsf{E}^t\, |U^t|^{2+\delta} * \nu^t_{st} = 0$

be fulfilled.

Uniform variants of the central limit theorem for local martingales. Let P^t_θ, P_θ, $t \in \mathbf{R}_+$, $\theta \in V$, be probability measures on the space $(\mathbf{R}^k, \mathcal{B}^k)$, where V is a set. We say that the measures P^t_θ weakly converge as $t \to \infty$ to the measure P_θ uniformly with respect to $\theta \in V$ (denoted $P^t_\theta \xrightarrow{w} \mathsf{P}_\theta$ uniformly with respect to $\theta \in V$) if for any continuous bounded function $g\colon \mathbf{R}^k \to \mathbf{R}^1$

$$\lim_{t\to\infty} \int_{\mathbf{R}^k} g(x) P^t_\theta(dx) = \int_{\mathbf{R}^k} g(x) \mathsf{P}_\theta(dx)$$

uniformly with respect to $\theta \in V$. The following lemma proved in [**40**] (Appendix 1, Theorem 7) provides sufficient conditions for this uniform convergence.

LEMMA 1.3.1. *Let the following conditions be fulfilled:*
1) $\lim_{N\to\infty}\big[\sup_{t\in\mathbf{R}_+,\theta\in V} P^t_\theta(S^{k,c}_N) + \sup_{\theta\in V} \mathsf{P}_\theta(S^{k,c}_N)\big] = 0$, *where*

$$S^{k,c}_N = \mathbf{R}^k \backslash S^k_N \quad \textit{and} \quad S^k_N = \big\{x \in \mathbf{R}^k \colon |x| \le N\big\};$$

2)

$$\lim_{t\to\infty} \int_{\mathbf{R}^k} \exp(i\lambda' x) P^t_\theta(dx) = \int_{\mathbf{R}^k} \exp(i\lambda' x) \mathsf{P}_\theta(dx)$$

uniformly with respect to $\theta \in V$ *for all* $\lambda \in \mathbf{R}^k$.
Then

$$P^t_\theta \xrightarrow{w} \mathsf{P}_\theta$$

uniformly with respect to $\theta \in V$ *as* $t \to \infty$.

Now, suppose for every $t \in \mathbf{R}_+$ on the stochastic basis $(\Omega, \mathcal{F}^t, \mathbf{F}^t, P^t)$ there is given a family $(Y^{t,\theta},\ \theta \in V)$ of local martingales $Y^{t,\theta} \in \mathcal{M}_{\mathrm{loc},0}(\mathbf{F}^t, P^t)$ of the form

$$Y^{t,\theta} = M^{t,\theta} + U^{t,\theta} * \big(\mu^{t,\theta} - \nu^{t,\theta}\big), \tag{1.3.18}$$

where $M^{t,\theta} \in \mathcal{M}^c_{\mathrm{loc},0}(\mathbf{F}^t, \mathsf{P}^t)$, $\mu^{t,\theta}$ is an $\mathbf{F}^t$-optional integer-valued random measure for which the measure $\mathbf{M}^{P^t}_{\mu^{t,\theta}}$ possesses the property of $\widetilde{\mathcal{P}}(\mathbf{F}^t)$-$\sigma$-finiteness, $\nu^{t,\theta}$ is an $\mathbf{F}^t$-compensator of the measure $\mu^{t,\theta}$ such that

$$\sum_{s\in\mathbf{R}_+} \nu^{t,\theta}(\{s\}, \mathbf{R}_0) = 0,$$

and $U^{t,\theta} \in \mathcal{G}_{\mathrm{loc}}(\nu^{t,\theta}, \mathbf{F}^t, P^t)$. The following uniform variant of Theorem 1.3.1 is valid.

THEOREM 1.3.4. *Let the following conditions be fulfilled for fixed* $s \in (0,\infty)$ *and all* $\varepsilon \in (0,1]$:

1) $P^t\text{-}\lim_{t\to\infty} \chi(|U^{t,\theta}|>\varepsilon)|U^{t,\theta}| * \nu^{t,\theta}_{st} = 0$ *uniformly with respect to* $\theta \in V$;
2) $P^t\text{-}\lim_{t\to\infty}\big[\langle M^{t,\theta}\rangle_{st} + \chi(|U^{t,\theta}|\le\varepsilon)|U^{t,\theta}|^2 * \nu^{t,\theta}_{st}\big] = c(s) \in [0,\infty)$ *uniformly with respect to* $\theta\in V$, *where* $c(s)$ *is a nonrandom quantity dependent on* s;
3)

$$\lim_{N\to\infty}\Big[\sup_{t\in\mathbf{R}_+,\theta\in V} P^t\left\{\chi\left(\left|U^{t,\theta}\right|>\varepsilon\right)\left|U^{t,\theta}\right| * \nu^{t,\theta}_{st} > N\right\}$$
$$+\sup_{t\in\mathbf{R}_+,\,\theta\in V} P^t\left\{\left\langle M^{t,\theta}\right\rangle_{st} + \chi\left(\left|U^{t,\theta}\right|\le\varepsilon\right)\left|U^{t,\theta}\right|^2 * \nu^{t,\theta}_{st} > N\right\}\Big] = 0.$$

Then

$$\mathcal{L}\left(Y^{t,\theta}_{st}|\mathsf{P}^t\right) \xrightarrow{w} \mathcal{N}(0,c(s)) \tag{1.3.19}$$

uniformly with respect to $\theta\in V$ *as* $t\to\infty$.

PROOF. Since $U^{t,\theta}\in\mathcal{G}_{\text{loc}}(\nu^{t,\theta},\mathbf{F}^t,P^t)$, in view of Theorem 1.1.8 it is evident that for any $\varepsilon\in(0,1]$ we have

$$\chi\left(\left|U^{t,\theta}\right|>\varepsilon\right)U^{t,\theta}\in\mathcal{G}^1_{\text{loc}}\left(\nu^{t,\theta},\mathbf{F}^t,P^t\right),$$
$$\chi\left(\left|U^{t,\theta}\right|\le\varepsilon\right)U^{t,\theta}\in\mathcal{G}^2_{\text{loc}}\left(\nu^{t,\theta},\mathbf{F}^t,P^t\right).$$

From this, using relations (1.1.4) and (1.1.4′) and Theorem 1.1.8 we get

$$Y^{t,\theta} = \overline{Y}^{t,\theta} + \widetilde{Y}^{t,\theta}, \tag{1.3.20}$$

where

$$\overline{Y}^{t,\theta} = M^{t,\theta} + \chi\left(\left|U^{t,\theta}\right|\le\varepsilon\right)U^{t,\theta} * \left(\mu^{t,\theta}-\nu^{t,\theta}\right)\in\mathcal{M}^2_{\text{loc}}\left(\mathbf{F}^t,P^t\right)$$

with quadratic characteristic

$$\left\langle\overline{Y}^{t,\theta}\right\rangle = \left\langle M^{t,\theta}\right\rangle + \chi\left(\left|U^{t,\theta}\right|\le\varepsilon\right)\left|U^{t,\theta}\right|^2 * \nu^{t,\theta}, \tag{1.3.21}$$

and

$$\widetilde{Y}^{t,\theta} = \chi\left(\left|U^{t,\theta}\right|>\varepsilon\right)U^{t,\theta} * \left(\mu^{t,\theta}-\nu^{t,\theta}\right)\in\mathcal{M}^d_{\text{loc}}\left(\mathbf{F}^t,P^t\right)$$

and (P^t-a.s.)

$$\widetilde{Y}^{t,\theta} = \chi\left(\left|U^{t,\theta}\right|>\varepsilon\right)U^{t,\theta} * \mu^{t,\theta} - \chi\left(\left|U^{t,\theta}\right|>\varepsilon\right)U^{t,\theta} * \nu^{t,\theta}. \tag{1.3.22}$$

Now we apply the Lenglart inequality (Theorem 1.1.3), putting $\xi = |\overline{Y}^{t,\theta}|^2$ and $\eta = \langle\overline{Y}^{t,\theta}\rangle$. Since $\mathsf{E}^t\,|\overline{Y}^{t,\theta}_\tau|^2 \le \mathsf{E}^t\langle\overline{Y}^{t,\theta}\rangle_\tau$ for any stopping time $\tau\in\mathfrak{M}(\mathbf{F}^t)$ **[98]** (here $\mathsf{E}^t\varphi = \int\varphi(\omega)P^t(d\omega)$), in view of Theorem 1.1.3 for any $H>0$, $N>0$ and $u\in\mathbf{R}_+$

$$P^t\left\{\left|\overline{Y}^{t,\theta}_u\right|>H\right\}\le NH^{-2} + P^t\left\{\left\langle\overline{Y}^{t,\theta}\right\rangle_u > N\right\}.$$

By condition 3) and (1.3.21), we get

$$\lim_{H\to\infty} \sup_{t\in\mathbf{R}_+,\,\theta\in V} P^t\left\{\left|\overline{Y}^{\,t,\theta}_{st}\right| > H\right\} = 0. \tag{1.3.23}$$

From (1.3.22) it follows that (P^t-a.s.)

$$\left|Y^{\,t,\theta}\right| \le \chi\left(\left|U^{t,\theta}\right| > \varepsilon\right)\left|U^{t,\theta}\right| * \mu^{t,\theta} + \chi\left(\left|U^{t,\theta}\right| > \varepsilon\right)\left|U^{t,\theta}\right| * \nu^{t,\theta},$$

whence, taking into account the inclusion $\chi(|U^{t,\theta}| > \varepsilon)U^{t,\theta} \in \mathcal{G}^1_{\mathrm{loc}}(\nu^{t,\theta}, \mathbf{F}^t, P^t)$ and Theorem 1.1.8, for any stopping time $\tau \in \mathfrak{M}(\mathbf{F}^t)$ we have

$$\mathsf{E}^t\left|\widetilde{Y}^{t,\theta}_\tau\right| \le 2\,\mathsf{E}^t\,\chi\left(\left|U^{t,\theta}\right| > \varepsilon\right)\left|U^{t,\theta}\right| * \nu^{t,\theta}_\tau. \tag{1.3.24}$$

Now we put $\xi = |\widetilde{Y}^{t,\theta}|$, $\eta = 2\chi(|U^{t,\theta}| > \varepsilon)|U^{t,\theta}| * \nu^{t,\theta}$ in Theorem 1.1.3 and apply the Lenglart inequality once again. In view of (1.3.24), for every $H > 0$, $N > 0$, and $u \in \mathbf{R}_+$ we have

$$P^t\left\{\left|\widetilde{Y}^{t,\theta}_u\right| > H\right\} \le NH^{-1} + P^t\left\{2\chi\left(\left|U^{t,\theta}\right| > \varepsilon\right)\left|U^{t,\theta}\right| * \nu^{t,\theta}_u > N\right\},$$

whence, by condition 3),

$$\lim_{H\to\infty} \sup_{t\in\mathbf{R}_+,\,\theta\in V} P^t\left\{\left|\widetilde{Y}^{t,\theta}_{st}\right| > H\right\} = 0. \tag{1.3.25}$$

From (1.3.20), (1.3.23), and (1.3.25) we get

$$\lim_{H\to\infty} \sup_{t\in\mathbf{R}_+,\,\theta\in V} P^t\left\{\left|Y^{t,\theta}_{st}\right| > H\right\} = 0. \tag{1.3.26}$$

Repeating the arguments of Theorem 1.3.1 and taking into account conditions 1) and 2), we obtain that for any $\lambda \in \mathbf{R}^1$

$$\lim_{t\to\infty} \mathsf{E}^t \exp\left\{i\lambda Y^{t,\theta}_{st}\right\} = \exp\left\{-2^{-1}c(s)\lambda^2\right\} \tag{1.3.27}$$

uniformly with respect to $\theta \in V$. Applying Lemma 1.3.1, from (1.3.26) and (1.3.27) we get (1.3.19). □

Now we shall obtain a multi-dimensional variant of Theorem 1.3.3.

Consider local martingales

$$Y^{t,\theta} = \left(Y^{t,\theta,1}, \dots, Y^{t,\theta,k}\right)' \in \mathcal{M}_{\mathrm{loc},0}\left(\mathbf{F}^t, P^t\right), \qquad t \in \mathbf{R}_+,\ \theta \in V,$$

of the form

$$Y^{t,\theta} = M^{t,\theta} + U^{t,\theta} * \left(\mu^{t,\theta} - \nu^{t,\theta}\right), \tag{1.3.28}$$

where

$$M^{t,\theta} = \left(M^{t,\theta,1}, \dots, M^{t,\theta,k}\right)' \in \mathcal{M}^c_{\mathrm{loc},0}\left(\mathbf{F}^t, P^t\right),$$

$\mu^{t,\theta}$ and $\nu^{t,\theta}$ are the random measures as in (1.3.18),

$$U^{t,\theta} = \left(U^{t,\theta,1}, \dots, U^{t,\theta,k}\right)' \in \mathcal{G}_{\mathrm{loc}}\left(\nu^{t,\theta}, \mathbf{F}^t, P^t\right),$$

and

$$U^{t,\theta} * \left(\mu^{t,\theta} - \nu^{t,\theta}\right) = \left(U^{t,\theta,1} * \left(\mu^{t,\theta} - \nu^{t,\theta}\right), \dots, U^{t,\theta,k} * \left(\mu^{t,\theta} - \nu^{t,\theta}\right)\right)'.$$

THEOREM 1.3.5. *Let the following conditions be fulfilled for fixed $s \in (0,\infty)$ and any $\varepsilon \in (0,1]$:*

1) $P^t\text{-}\lim_{t\to\infty} \chi(|U^{t,\theta}| > \varepsilon)|U^{t,\theta}| * \nu_{st}^{t,\theta} = 0$ *uniformly with respect to $\theta \in V$;*
2) $P^t\text{-}\lim_{t\to\infty} \left[\langle M^{t,\theta}, M^{t,\theta}\rangle_{st} + \chi(|U^{t,\theta}| \le \varepsilon) U^{t,\theta}(U^{t,\theta})' * \nu_{st}^{t,\theta}\right] = I(s)$ *uniformly with respect to $\theta \in V$, where $I(s)$ is some symmetric nonrandom matrix with finite elements that depends on s;*
3)

$$\lim_{N\to\infty}\Bigg[\sup_{t\in\mathbf{R}_+,\,\theta\in V} P^t\left\{\chi\left(\left|U^{t,\theta}\right| > \varepsilon\right)\left|U^{t,\theta}\right| * \nu_{st}^{t,\theta} > N\right\} + \sup_{t\in\mathbf{R}_+,\,\theta\in V} P^t\Big\{\mathrm{Tr}\Big(\langle M^{t,\theta}, M^{t,\theta}\rangle_{st} + \chi\left(\left|U^{t,\theta}\right| \le \varepsilon\right) U^{t,\theta}\left(U^{t,\theta}\right)' * \nu_{st}^{t,\theta}\Big) > N\Big\}\Bigg] = 0.$$

Then

$$\mathcal{L}\left(Y_{st}^{t,\theta}|\mathsf{P}^t\right) \xrightarrow{w} \mathcal{N}(0, I(s)) \tag{1.3.29}$$

uniformly with respect to $\theta \in V$ as $t \to \infty$.

PROOF. Repeating the proof of Theorem 1.3.2, from conditions 1) and 2) we obtain that for any $\lambda \in \mathbf{R}^k$

$$\lim_{t\to\infty} \mathsf{E}^t \exp\left(i\lambda' Y^{t,\theta}\right) = \exp\left(-2^{-1}\lambda' I(s)\lambda\right) \tag{1.3.30}$$

uniformly with respect to $\theta \in V$.

Since $U^{t,\theta} \in \mathcal{G}_{\mathrm{loc}}(\nu^{t,\theta}, \mathbf{F}^t, P^t)$, we can use (1.1.4), (1.1.5), and Theorem 1.1.8 to obtain

$$\chi\left(\left|U^{t,\theta}\right| \le \varepsilon\right) U^{t,\theta} \in \mathcal{G}^2_{\mathrm{loc}}\left(\nu^{t,\theta}, \mathbf{F}^t, P^t\right), \tag{1.3.31}$$

$$\chi\left(\left|U^{t,\theta}\right| > \varepsilon\right) U^{t,\theta} \in \mathcal{G}^1_{\mathrm{loc}}\left(\nu^{t,\theta}, \mathbf{F}^t, P^t\right) \tag{1.3.32}$$

for any $\varepsilon \in (0,1]$. Actually, the following implications take place:

$$\begin{aligned} U^{t,\theta} \in \mathcal{G}_{\mathrm{loc}}\left(\nu^{t,\theta}, \mathbf{F}^t, P^t\right) &\Rightarrow \left|U^{t,\theta}\right| \in \mathcal{G}_{\mathrm{loc}}\left(\nu^{t,\theta}, \mathbf{F}^t, P^t\right) \\ &\Rightarrow \chi\left(\left|U^{t,\theta}\right| \le \varepsilon\right)\left|U^{t,\theta}\right| \in \mathcal{G}^2_{\mathrm{loc}}\left(\nu^{t,\theta}, \mathbf{F}^t, P^t\right) \\ &\Rightarrow \chi\left(\left|U^{t,\theta}\right| \le \varepsilon\right) U^{t,\theta} \in \mathcal{G}^2_{\mathrm{loc}}\left(\nu^{t,\theta}, \mathbf{F}^t, P^t\right), \\ U^{t,\theta} \in \mathcal{G}_{\mathrm{loc}}\left(\nu^{t,\theta}, \mathbf{F}^t, P^t\right) &\Rightarrow \left|U^{t,\theta}\right| \in \mathcal{G}_{\mathrm{loc}}\left(\nu^{t,\theta}, \mathbf{F}^t, P^t\right) \\ &\Rightarrow \chi\left(\left|U^{t,\theta}\right| > \varepsilon\right)\left|U^{t,\theta}\right| \in \mathcal{G}^1_{\mathrm{loc}}\left(\nu^{t,\theta}, \mathbf{F}^t, P^t\right) \\ &\Rightarrow \chi\left(\left|U^{t,\theta}\right| > \varepsilon\right) U^{t,\theta} \in \mathcal{G}^1_{\mathrm{loc}}\left(\nu^{t,\theta}, \mathbf{F}^t, P^t\right), \end{aligned}$$

which imply (1.3.1) and (1.3.32).

Therefore for any $\varepsilon \in (0,1]$ we have

$$Y^{t,\theta} = \overline{Y}^{t,\theta} + \widetilde{Y}^{t,\theta}, \tag{1.3.33}$$

where

(1.3.34)
$$\overline{Y}^{t,\theta} = M^{t,\theta} + \chi\left(\left|U^{t,\theta}\right| \le \varepsilon\right) U^{t,\theta} * \left(\mu^{t,\theta} - \nu^{t,\theta}\right) \in \mathcal{M}^2_{\mathrm{loc}}\left(\mathbf{F}^t, P^t\right),$$

(1.3.35)
$$\begin{aligned}\widetilde{Y}^{t,\theta} &= \chi\left(\left|U^{t,\theta}\right| > \varepsilon\right) U^{t,\theta} * \left(\mu^{t,\theta} - \nu^{t,\theta}\right)\\ &= \chi\left(\left|U^{t,\theta}\right| > \varepsilon\right) U^{t,\theta} * \mu^{t,\theta} - \chi\left(\left|U^{t,\theta}\right| > \varepsilon\right) U^{t,\theta} * \nu^{t,\theta} \in \mathcal{M}^d_{\mathrm{loc}}(\mathbf{F}^t, P^t).\end{aligned}$$

In view of representations (1.3.33)–(1.3.35), the proof is similar to the proof of Theorem 1.3.4, with the corresponding changes caused by the application of the Lenglart inequality in the multi-dimensional case. These changes are discussed below.

Putting in Theorem 1.1.3

$$\xi = \left|\overline{Y}^{t,\theta}\right|^2, \qquad \eta = \operatorname{Tr}\left\langle \overline{Y}^{t,\theta}, \overline{Y}^{t,\theta}\right\rangle$$

and taking into account that for any stopping time $\tau \in \mathfrak{M}(\mathbf{F}^t)$ [**98**]

$$\mathsf{E}^t \left|\overline{Y}^{t,\theta}_\tau\right|^2 \le \mathsf{E}^t \operatorname{Tr}\left\langle \overline{Y}^{t,\theta}, \overline{Y}^{t,\theta}\right\rangle_\tau,$$

for any $H > 0$, $N > 0$ and $u \in \mathbf{R}_+$ we have

$$P^t\left\{\left|\overline{Y}^{t,\theta}_u\right| > H\right\} \le NH^{-2} + P^t\left\{\operatorname{Tr}\left\langle \overline{Y}^{t,\theta}, \overline{Y}^{t,\theta}\right\rangle_u > N\right\},$$

whence, by the equality

$$\left\langle \overline{Y}^{t,\theta}, \overline{Y}^{t,\theta}\right\rangle = \left\langle M^{t,\theta}, M^{t,\theta}\right\rangle + \chi\left(\left|U^{t,\theta}\right| \le \varepsilon\right) U^{t,\theta}\left(U^{t,\theta}\right)' * \nu^{t,\theta}$$

and condition 3), the relation (1.3.23) follows.

Putting

$$\xi = \left|\widetilde{Y}^{t,\theta}\right|, \qquad \eta = 2\sqrt{k}\,\chi\left(\left|U^{t,\theta}\right| > \varepsilon\right)\left|U^{t,\theta}\right| * \nu^{t,\theta}$$

in Theorem 1.1.3 and taking into account that due to (1.3.32), (1.3.35), and inequality $|x| \le |x|_1 \le \sqrt{k}|x|$, $x \in \mathbf{R}^k$, for any stopping time $\tau \in \mathfrak{M}(\mathbf{F}^t)$ we have

$$\begin{aligned}\mathsf{E}^t \left|\widetilde{Y}^{t,\theta}_\tau\right| &\le \mathsf{E}^t \left|\widetilde{Y}^{t,\theta}_\tau\right|_1 \le 2\,\mathsf{E}^t \chi\left(\left|U^{t,\theta}\right| > \varepsilon\right)\left|U^{t,\theta}\right|_1 * \nu^{t,\theta}_\tau\\ &\le 2\sqrt{k}\,\mathsf{E}^t \chi\left(\left|U^{t,\theta}\right| > \varepsilon\right)\left|U^{t,\theta}\right| * \nu^{t,\theta}_\tau\end{aligned}$$

for any $H > 0$, $N > 0$ and $u \in \mathbf{R}_+$, we get

$$P^t\left\{\left|\widetilde{Y}^{t,\theta}_u\right| > H\right\} \le NH^{-1} + P^t\left\{2\sqrt{k}\,\chi\left(\left|U^{t,\theta}\right| > \varepsilon\right)\left|U^{t,\theta}\right| * \nu^{t,\theta}_u > N\right\},$$

whence, by condition 3), the relation (1.3.25) follows.

Now (1.3.23), (1.3.25), and (1.3.33) imply the relation (1.3.26). From (1.3.26), (1.3.30), and Lemma 1.3.1 we get (1.3.29). □

Now we formulate a uniform variant of the central limit theorem for local square-integrable martingales $Y^{t,\theta}$ in the multi-dimensional case as a corollary to Theorem 1.3.5.

Let the process $Y^{t,\theta} \in \mathcal{M}^2_{\mathrm{loc}}(\mathbf{F}^t, P^t)$ be of the form (1.3.28), where $U^{t,\theta} \in \mathcal{G}^2_{\mathrm{loc}}(\nu^{t,\theta}, \mathbf{F}^t, P^t)$, $\theta \in V$.

THEOREM 1.3.6. *Let the following conditions be fulfilled for fixed $s \in (0,\infty)$ and any $\varepsilon \in (0,1]$:*

1) $P^t\text{-}\lim_{t\to\infty} \chi(|U^{t,\theta}| > \varepsilon)|U^{t,\theta}|^2 * \nu_{st}^{t,\theta} = 0$ *uniformly with respect to $\theta \in V$;*
2) $P^t\text{-}\lim_{t\to\infty} \big[\langle M^{t,\theta}, M^{t,\theta}\rangle_{st} + U^{t,\theta}(U^{t,\theta})' * \nu_{st}^{t,\theta}\big] = I(s)$ *uniformly with respect to $\theta \in V$, where $I(s)$ is some nonrandom symmetric matrix with finite elements that depends on s;*
3) $\lim_{N\to\infty} \sup_{t\in\mathbf{R}_+,\,\theta\in V} P^t\big\{\mathrm{Tr}\big[\langle M^{t,\theta}, M^{t,\theta}\rangle_{st} + U^{t,\theta}(U^{t,\theta})' * \nu_{st}^{t,\theta}\big] > N\big\} = 0.$

Then

$$\mathcal{L}\left(Y_{st}^{t,\theta} | \mathsf{P}^t\right) \xrightarrow{w} \mathcal{N}(0, I(s))$$

uniformly with respect to $\theta \in V$ as $t \to \infty$.

Weak convergence of semimartingales. Let $(\Omega, \mathcal{F}, \mathsf{P})$ be a complete probability space with the σ-algebras $\mathbf{F} = (\mathcal{F}_s)_{s\in\mathbf{R}_+}$ and $\mathbf{F}^t = (\mathcal{F}_s^t)_{s\in\mathbf{R}_+}$, $t \in \mathbf{R}_+$, satisfying ordinary conditions. On the stochastic basis $(\Omega, \mathcal{F}, \mathbf{F}^t, \mathsf{P})$ we consider the semimartingale $Y^t = \mathcal{S}(\mathbf{F}^t, \mathsf{P})$ with the canonical representation

$$Y^t = Y_0^t + B^t + M^t + x\chi(|x| \le 1) * (\mu^t - \nu^t) + x\chi(|x| > 1) * \mu^t,$$

where $B^t \in \mathcal{A}_{\mathrm{loc}}(\mathbf{F}^t, \mathsf{P}) \cap \mathcal{P}(\mathbf{F}^t)$, $M^t \in \mathcal{M}_{\mathrm{loc},0}^c(\mathbf{F}^t, \mathsf{P})$ with quadratic characteristic C^t, μ^t is the jump measure of the semimartingale Y^t with compensator ν^t satisfying the condition $\sum_{s\in\mathbf{R}_+} \nu^t(\{s\}, \mathbf{R}_0) = 0$, that is, the semimartingales Y^t are left-quasicontinuous.

Furthermore, on the stochastic basis $(\Omega, \mathcal{F}, \mathbf{F}, \mathsf{P})$ we consider a semimartingale $Y^t = \mathcal{S}(\mathbf{F}, \mathsf{P})$ with canonical representation

$$Y = Y_0 + B + M + x\chi(|x| \le 1) * (\mu - \nu) + x\chi(|x| > 1) * \mu,$$

where $B \in \mathcal{A}_{\mathrm{loc}}(\mathbf{F}, \mathsf{P}) \cap \mathcal{P}(\mathbf{F})$, $M \in \mathcal{M}_{\mathrm{loc},0}^c(\mathbf{F}, \mathsf{P})$ with quadratic characteristic C, and μ is the jump measure of the semimartingale Y with compensator ν satisfying the condition $\sum_{s\in\mathbf{R}_+} \nu(\{s\}, \mathbf{R}_0) = 0$, that is, the semimartingale Y is also left-quasicontinuous. The semimartingale Y is assumed to be a process with independent increments. Thus the triple (B, C, ν) of the semimartingale Y is determinate and continuous in time.

To formulate the following theorem, for $\delta \in (0,1]$ and $t \in \mathbf{R}_+$ we define a local square-integrable martingale $M^{t,\delta} \in \mathcal{M}_{\mathrm{loc}}^2(\mathbf{F}^t, \mathsf{P})$ of the form

$$M^{t,\delta} = M^t + x\chi(|x| \le \delta) * (\mu^t - \nu^t).$$

Evidently, the quadratic characteristic $\langle M^{t,\delta}\rangle$ is defined by the formula

$$\langle M^{t,\delta}\rangle = C^t + x^2\chi(|x| \le \delta) * \nu^t.$$

If S is a nonempty subset of $\mathbf{R}_+$, then $Y^t \xrightarrow{w_f(S)} Y$, $t \to \infty$, means the weak convergence of distributions of vectors $(Y_{s_1}^t, \dots, Y_{s_m}^t)$, $t \to \infty$, to the distribution of the vector $(Y_{s_1}, \dots, Y_{s_m})$ for any finite subset $\{s_1, s_2, \dots, s_m\} \subset S$. In the case $S = \mathbf{R}_+$ we write w_f instead of $w_f(S)$.

THEOREM 1.3.7. *Let S be a nonempty subset in $\mathbf{R}_+$ and let the following conditions be fulfilled:*

1) *for any bounded function $f = f(x)$ continuous on $\mathbf{R}_0 \backslash (\{-1\} \cup \{1\})$ and such that $f(x) = \chi(|x| > \varepsilon) f(x)$ for some $\varepsilon > 0$, we have*

$$\mathsf{P}\text{-}\lim_{t\to\infty} f * \nu_s^t = f * \nu_s, \qquad s \in S;$$

2) $\mathsf{P}\text{-}\lim_{t\to\infty} B_s^t = B_s$, $s \in S$;
3) $\lim_{\delta\to 0} \overline{\lim}_{t\to\infty} \mathsf{P}\{|\langle M^{t,\delta}\rangle_s - C_s| > a\} = 0$ *for any* $a > 0$ *and* $s \in S$;
4) $\mathcal{L}(Y_0^t|\mathsf{P}) \xrightarrow{w} \mathcal{L}(Y_0|\mathsf{P})$ *as* $t \to \infty$.

Then the relation $Y^t \xrightarrow{w_f(S)} Y$, $t \to \infty$, *is valid.*

PROOF. The proof of this theorem can be found in [**98, 99, 154**]. □

REMARK 1.3.5. Theorems 1.3.1 and 1.3.3 can be easily obtained from Theorem 1.3.7.

Consider a local square-integrable martingale $Y \in \mathcal{M}^2_{\mathrm{loc}}(\mathbf{F}, \mathsf{P})$ of the form $Y = M + U * (\mu - \nu)$, where $M \in \mathcal{M}_{\mathrm{loc},0}(\mathbf{F}, \mathsf{P})$, $U \in \mathcal{G}^2_{\mathrm{loc}}(\nu, \mathbf{F}, \mathsf{P})$, and μ is an optional integral-valued random measure with time-continuous compensator ν.

THEOREM 1.3.8. *Suppose that, under certain normalization of ψ_t ($\psi_t \to \infty$ as $t \to \infty$, ψ_t is a continuous function), there exists a nonrandom nondecreasing bounded function $K(y)$ on $\overline{\mathbf{R}}^1$, $K(-\infty) = 0$, for which the following relation holds at every continuity point (including $y = \infty$):*

$$\mathsf{P}\text{-}\lim_{t\to\infty} \psi_t^{-2} \left[U^2 \chi(U < y\psi_t) * \nu_{st} + \chi_{\overline{\mathbf{R}}_+}(y) \langle M\rangle_{st}\right] = sK(y) \tag{1.3.36}$$

for every $s \in \mathbf{R}_+$. Then $(\psi_t^{-1} Y_{st}) \xrightarrow{w_f} X$ as $t \to \infty$, where $X = (X_s)$ is a homogeneous process with independent increments, $X_0 = 0$, and characteristic function

$$\mathsf{E} \exp(i\lambda X_s) = \exp\left\{ s \int \left(e^{i\lambda y} - 1 - i\lambda y\right) y^{-2}\, dK(y)\right\}.$$

Proof of Theorem 1.3.8 can be found in [**118**].

REMARK 1.3.6. If in Theorem 1.3.8 the following conditions are fulfilled instead of condition (1.3.36):

1) $\mathsf{P}\text{-}\lim_{t\to\infty} \psi_t^{-2} \langle Y\rangle_{st} = s\sigma^2$, where $\sigma \in (0, \infty)$;
2) $\mathsf{P}\text{-}\lim_{t\to\infty} \psi_t^{-2} U^2 \chi(|U| > \varepsilon\psi_t) * \nu_{st} = 0$ for any $s \in \mathbf{R}_+$ and $\varepsilon > 0$,

then $(\psi_t^{-1} Y_{st}) \xrightarrow{w_f} \sigma w$, where w is the standard Wiener process, that is, $w \in \mathcal{M}^c_{\mathrm{loc},0}$ with quadratic characteristic $\langle w\rangle_t = t$. This assertion follows from Theorem 1.3.8 and Theorem 1.3.7.

CHAPTER 2

Asymptotic Distinguishing between Simple Hypotheses in the Scheme of General Statistical Experiments

2.1. Statistical hypotheses and tests

Let $(X^t, \mathfrak{B}^t, P^t, \widetilde{P}^t)$, $t \in \mathbf{R}_+$, be a family of statistical experiments generated by observations ξ^t of an arbitrary nature, and let H^t and $\widetilde{H}^t$ be the *statistical hypotheses* consisting in that the distribution of the observation ξ^t is defined by the measures P^t and $\widetilde{P}^t$, respectively. Denote by $\delta_t = (\delta_t(x))_{x \in X^t}$ a measurable map of the space $(X^t, \mathfrak{B}^t)$ to the space $([0,1], \mathcal{B}([0,1]))$. The map δ_t (as well as the random variable $\delta_t(\xi^t)$, denoted by the same letter δ_t) is called the *test for distinguishing between the hypotheses H^t and $\widetilde{H}^t$ from the observation ξ^t*, $\delta_t(x)$ being the conditional probability to reject the hypothesis H^t under the condition that $\xi^t = x$. Denote by Σ^t the totality of all the tests δ_t for distinguishing between the hypotheses H^t and $\widetilde{H}^t$. For any $\delta_t \in \Sigma^t$ we introduce the *probabilities of errors of the first* and the *second kinds*:

$$\alpha(\delta_t) = \mathsf{E}^t \, \delta_t, \qquad \beta(\delta_t) = \widetilde{\mathsf{E}}^t (1 - \delta_t),$$

where E^t and $\widetilde{\mathsf{E}}^t$ stand for the mathematical expectations with respect to the distributions P^t and $\widetilde{P}^t$, respectively. The probability of the error of the first kind $\alpha(\delta_t)$ is the probability to accept the hypothesis $\widetilde{H}^t$ with the help of the test δ_t under the condition that the hypothesis H^t is true. Every so often $\alpha(\delta_t)$ is called the *level of the test* δ_t. The probability of the error of the second kind $\beta(\delta_t)$ is the probability to accept the hypothesis H^t with the help of the test δ_t under the condition that the hypothesis $\widetilde{H}^t$ is true. The quantity $1 - \beta(\delta_t)$ is often called the *power of a test* δ_t. For any $\alpha \in [0,1]$ we denote by Σ^t_α the totality of all tests δ_t in Σ^t that satisfy the condition $\alpha(\delta_t) \le \alpha$.

Let $Q^t = (P^t + \widetilde{P}^t)/2$ be one more probability measure on the measurable space $(X^t, \mathfrak{B}^t)$, and let $\mathfrak{z}_t = dP^t/dQ^t$ and $\tilde{\mathfrak{z}}_t = d\widetilde{P}^t/dQ^t$ be the finite versions of the Radon–Nikodym derivatives of the measures P^t and $\widetilde{P}^t$, respectively, with respect to the measure Q^t. Now we introduce the *likelihood ratio*

$$z_t = \tilde{\mathfrak{z}}_t / \mathfrak{z}_t, \qquad \widetilde{z}_t = \mathfrak{z}_t / \tilde{\mathfrak{z}}_t, \tag{2.1.1}$$

putting $a/0 = \infty$ for $a > 0$ and $0/0 = 0$. Since $Q^t\{\mathfrak{z}_t = 0, \tilde{\mathfrak{z}}_t = 0\} = 0$, the likelihood ratios z_t and $\widetilde{z}_t$ are well defined by the relation (2.1.1) with respect to the measures Q^t, P^t, and $\widetilde{P}^t$. Note that $\mathfrak{z}_t + \tilde{\mathfrak{z}}_t = 2$.

In what follows we use a lemma that provides the Lebesgue decomposition of one of the measures P^t or $\widetilde{P}^t$ with respect to the other measure [**95, 127, 146**].

LEMMA 2.1.1. *For any $A \in \mathfrak{B}^t$ the following decompositions are valid:*

$$\widetilde{P}^t(A) = \int_A z_t \, dP^t + \widetilde{P}^t(A \cap \{z_t = \infty\}), \tag{2.1.2}$$

$$P^t(A) = \int_A \widetilde{z}_t \, d\widetilde{P}^t + P^t(A \cap \{\widetilde{z}_t = \infty\}). \tag{2.1.3}$$

Put

$$\overline{\alpha}_t = P^t(\widetilde{\mathfrak{z}}_t > 0), \qquad \overline{\beta}_t = \widetilde{P}^t(\mathfrak{z}_t > 0).$$

It is evident that the following equalities hold:

$$\overline{\alpha}_t = P^t(z_t > 0) = P^t(\widetilde{z}_t < \infty), \tag{2.1.4}$$

$$\overline{\beta}_t = \widetilde{P}^t(\widetilde{z}_t > 0) = \widetilde{P}^t(z_t < \infty). \tag{2.1.5}$$

Then the Lebesgue decompositions (2.1.2) and (2.1.3) imply

$$\widetilde{P}^t \ll P^t \Leftrightarrow \overline{\beta}_t = 1, \tag{2.1.6}$$

$$P^t \ll \widetilde{P}^t \Leftrightarrow \overline{\alpha}_t = 1. \tag{2.1.7}$$

Also, the Lebesgue decompositions imply the following lemma [**95, 146**].

LEMMA 2.1.2. *For any nonnegative function η defined on $(X^t, \mathfrak{B}^t)$ and such that $\widetilde{\mathsf{E}}^t\eta < \infty$ the following equality holds:*

$$\widetilde{\mathsf{E}}^t\eta = \mathsf{E}^t\,\eta z_t + \widetilde{\mathsf{E}}^t\eta\chi(z_t = \infty). \tag{2.1.8}$$

If $\mathsf{E}^t\,\eta < \infty$, then

$$\mathsf{E}^t\,\eta = \widetilde{\mathsf{E}}^t\eta\widetilde{z}_t + \mathsf{E}^t\,\eta\chi(\widetilde{z}_t = \infty). \tag{2.1.9}$$

To test the hypotheses H^t and $\widetilde{H}^t$, we introduce the test

$$\delta_t^{c,\varepsilon} = \chi(z_t > c) + \varepsilon\chi(z_t = c), \tag{2.1.10}$$

where $c \in [0, \infty]$ and $\varepsilon \in [0,1]$ are parameters of the test.

The following lemma is known as the *fundamental Neyman–Pearson lemma* [**9, 58**].

LEMMA 2.1.3. 1. *For any $\alpha \in [0,1]$ there exists a test $\delta_t^{c(\alpha),\varepsilon(\alpha)}$ of level α, where $(c(\alpha), \varepsilon(\alpha))$ is some solution of the equation $\alpha(\delta_t^{c,\varepsilon}) = \alpha$ with respect to (c, ε).*

2. *For any $\alpha \in [0,1]$ the test $\delta_t^{c(\alpha),\varepsilon(\alpha)}$ with $\varepsilon(0) = 1$ is the most powerful test in the class Σ_α^t.*

DEFINITION 2.1.1. A test $\delta_t^{c(\alpha),\varepsilon(\alpha)}$ with $\varepsilon(0) = 1$ is called the *Neyman–Pearson test* of level α for distinguishing between hypotheses H^t and $\widetilde{H}^t$ (in what follows this test is denoted by $\delta_t^{+,\alpha}$).

REMARK 2.1.1. It is not difficult to show that the functions $c(\alpha)$ and $\varepsilon(\alpha)$ can be chosen to be

$$c(\alpha) = \begin{cases} \infty, & \alpha = 0, \\ \overline{c}(\alpha), & 0 < \alpha < \overline{\alpha}_t, \\ 0, & \overline{\alpha}_t \le \alpha \le 1, \end{cases} \qquad \varepsilon(\alpha) = \begin{cases} 1, & \alpha = 0, \\ \overline{\varepsilon}(\alpha), & 0 < \alpha < \overline{\alpha}_t, \\ \widetilde{\varepsilon}(\alpha), & \overline{\alpha}_t \le \alpha \le 1, \end{cases} \tag{2.1.11}$$

where $\overline{c}(\alpha)$ is the least c satisfying the inequalities

$$P^t(z_t > c) \le \alpha \le P^t(z_t \ge c),$$
$$\overline{\varepsilon}(\alpha) = \frac{\alpha - P^t(z_t > \overline{c}(\alpha))}{P^t(z_t = \overline{c}(\alpha))}, \qquad \widetilde{\varepsilon}(\alpha) = \frac{\alpha - \overline{\alpha}_t}{1 - \overline{\alpha}_t}.$$

If $P^t(z_t = \overline{c}(\alpha)) = 0$, then, evidently, $P^t(z_t > \overline{c}(\alpha)) = \alpha$ and hence $\overline{\varepsilon}(\alpha) = 0/0$. In this case $\overline{\varepsilon}(\alpha)$ can be chosen from $[0, 1]$ in an arbitrary manner. If $\overline{\alpha}_t = 1$, then $[\overline{\alpha}_t, 1] = \{1\}$ and hence $\widetilde{\varepsilon}(\alpha) = 0/0$. In this case $\widetilde{\varepsilon}(\alpha)$ can also be chosen to be any number in $[0, 1]$. From (2.1.11) we immediately obtain

$$\beta(\delta_t^{+,\alpha}) = \begin{cases} \overline{\beta}_t, & \alpha = 0, \\ \widetilde{P}^t(z_t < \overline{c}(\alpha)) + (1 - \overline{\varepsilon}(\alpha))\widetilde{P}^t(z_t = \overline{c}(\alpha)), & 0 < \alpha < \overline{\alpha}_t, \\ 0, & \overline{\alpha}_t \le \alpha \le 1. \end{cases} \tag{2.1.12}$$

Now we introduce the set $\mathfrak{N}^t = \{(\alpha(\delta_t), \beta(\delta_t)) \colon \delta_t \in \Sigma^t\}$. It is well known [**58**] that the set $\mathfrak{N}^t$ is convex, centrally symmetric about the point $(\frac{1}{2}, \frac{1}{2})$, contains the points $(0, 1)$ and $(1, 0)$, and $\mathfrak{N}^t \subseteq [0, 1] \times [0, 1]$. Moreover, by (2.1.6) and (2.1.7),

$$P^t \sim \widetilde{P}^t \Leftrightarrow \bar{\alpha}_t = \bar{\beta}_t = 1.$$

Therefore, the set $\mathfrak{N}^t$ contains the only two points, $(0, 1)$ and $(1, 0)$, of the boundary of the set $[0, 1] \times [0, 1]$ only in the case $P^t \sim \widetilde{P}^t$. Note that, in view of the fundamental Neyman–Pearson lemma, the point $(\alpha, \beta(\delta_t^{+,\alpha}))$ lies on the boundary of the set $\mathfrak{N}^t$ for any $\alpha \in [0, 1]$. This fact suggests that the behavior of $\beta(\delta_t^{+,\alpha})$, as $t \to \infty$, depends on the relation between α and t and is similar to the behavior of the boundary of the set $\mathfrak{N}^t$ as $t \to \infty$.

In what follows we consider the Neyman–Pearson test δ_t^{+,α_t} of level α_t that may depend on t. In this case we use the notation: $c_t = c(\alpha_t)$ and $\varepsilon_t = \varepsilon(\alpha_t)$. Moreover, we put $\Lambda_t = \ln z_t$, $d_t = \ln c_t$, and $\ln 0 = -\infty$. In this notation the test δ_t^{+,α_t} is of the form

$$\delta_t^{+,\alpha_t} = \chi(\Lambda_t > d_t) + \varepsilon_t \chi(\Lambda_t = d_t). \tag{2.1.13}$$

Now we introduce the nonrandomized test

$$\delta_t^{*,c} = \delta_t^{c,1} = \chi(z_t \ge c), \qquad c \ge 0. \tag{2.1.14}$$

It is well known [**9**] that for every $c \in [0, \infty)$

$$\inf\{c\alpha(\delta_t) + \beta_t(\delta_t) \colon \delta_t \in \Sigma^t\} = c\alpha(\bar{\delta}_t) + \beta_t(\bar{\delta}_t) \tag{2.1.15}$$

for any test $\bar{\delta}_t \in \Sigma^t$ coinciding with the test $\delta_t^{*,c}$ on the set $\{z_t \ne c\}$ and arbitrarily defined on the set $\{z_t = c\}$. From this it follows that for any $c \in [0, \infty)$ and $\varepsilon \in [0, 1]$

$$\inf\{c\alpha(\delta_t) + \beta_t(\delta_t) \colon \delta_t \in \Sigma^t\} = c\alpha\left(\delta_t^{*,c}\right) + \beta_t\left(\delta_t^{*,c}\right) = c\alpha\left(\delta_t^{c,\varepsilon}\right) + \beta_t\left(\delta_t^{c,\varepsilon}\right). \tag{2.1.16}$$

Note that the test $\bar{\delta}_t$ satisfying the equality (2.1.15) is called the *Bayes test* with respect to an a priori distribution $(c/(c+1), 1/(c+1))$.

The equalities (2.1.10), (2.1.14)–(2.1.16), and the Neyman–Pearson lemma establish the relation between the Neyman–Pearson test and Bayes tests.

This allows us to study the behavior of Bayes tests with the help of Neyman–Pearson tests δ_t^{+,α_t}.

2.2. Types of asymptotic distinguishability between families of hypotheses and their characterization

Consider families of statistical hypotheses (H^t) and $(\widetilde{H}^t)$ (here and in what follows (H^t) and $(\widetilde{H}^t)$ mean $(H^t)_{t\in\mathbf{R}_+}$ and $(\widetilde{H}^t)_{t\in\mathbf{R}_+}$, respectively; this remark concerns also all the other families, random variables, etc.) and the corresponding families of probability measures (P^t) and $(\widetilde{P}^t)$. In this section we describe types of asymptotic distinguishability between families of hypotheses (H^t) and $(\widetilde{H}^t)$ as $t \to \infty$. The description is based on the asymptotic behavior of the sets $\mathfrak{N}^t$ as $t \to \infty$.

Complete asymptotic distinguishability.

DEFINITION 2.2.1. Families of hypotheses (H^t) and $(\widetilde{H}^t)$ are said to be *completely asymptotically distinguishable* (denoted $(H^t) \vartriangle (\widetilde{H}^t)$), if there exists a sequence $t_n \uparrow \infty$, $n \to \infty$, and tests $\delta_{t_n} \in \Sigma^{t_n}$ such that

$$\alpha(\delta_{t_n}) \to 0 \quad \text{and} \quad \beta(\delta_{t_n}) \to 0, \qquad n \to \infty. \tag{2.2.1}$$

Otherwise, that is, when there are no sequence $t_n \uparrow \infty$, $n \to \infty$, and tests $\delta_{t_n} \in \Sigma^{t_n}$ satisfying condition (2.2.1), we say that the families of hypotheses (H^t) and $(\widetilde{H}^t)$ are *not completely asymptotically distinguishable* (denoted $(H^t) \overline{\vartriangle} (\widetilde{H}^t)$).

DEFINITION 2.2.2. Families of measures (P^t) and $(\widetilde{P}^t)$ are called *completely asymptotically separable* (denoted $(P^t) \vartriangle (\widetilde{P}^t)$) if there exist a sequence $t_n \uparrow \infty, n \to \infty$, and sets $A_n \in \mathfrak{B}^{t_n}$ such that

$$P^{t_n}(A_n) \to 0 \quad \text{and} \quad \widetilde{P}^{t_n}(A_n) \to 1, \qquad n \to \infty. \tag{2.2.2}$$

Otherwise, that is, when there are no sequence $t_n \uparrow \infty$, $n \to \infty$, and sets $A_n \in \mathfrak{B}^{t_n}$ satisfying condition (2.2.2), we say that the families of measures (P^t) and $(\widetilde{P}^t)$ are *not completely asymptotically separable* (denoted $(P^t) \overline{\vartriangle} (\widetilde{P}^t)$).

In order to formulate the main theorem of this section, we need the following definitions.

DEFINITION 2.2.3. The *distance in variation between measures* P^t *and* $\widetilde{P}^t$ (denoted $\|\widetilde{P}^t - P^t\|$) is the total variation of the measure $\widetilde{P}^t - P^t$, namely:

$$\|\widetilde{P}^t - P^t\| = \mathsf{E}_Q^t \left|\widetilde{\mathfrak{z}}_t - \mathfrak{z}_t\right|, \tag{2.2.3}$$

where E_Q^t is the mathematical expectation with respect to the measure Q^t.

DEFINITION 2.2.4. *The Kakutani–Hellinger distance between measures P^t and $\widetilde{P}^t$* is a nonnegative quantity $\rho(P^t, \widetilde{P}^t)$ such that

$$\rho^2(P^t, \widetilde{P}^t) = \frac{1}{2}\,\mathsf{E}_Q^t\,|\tilde{\mathfrak{z}}_t^{1/2} - \mathfrak{z}_t^{1/2}|^2. \tag{2.2.4}$$

DEFINITION 2.2.5. *The Hellinger integral $H(\varepsilon; \widetilde{P}^t, P^t)$ of $\widetilde{P}^t$ and P^t of order* $\varepsilon \in \mathbf{R}^1$ is defined as

$$H(\varepsilon; \widetilde{P}^t, P^t) = \mathsf{E}_Q^t\,\tilde{\mathfrak{z}}_t^{\varepsilon}\mathfrak{z}_t^{1-\varepsilon}, \tag{2.2.5}$$

where for $\varepsilon = 0$ the expression $\tilde{\mathfrak{z}}_t^{\varepsilon}\mathfrak{z}_t^{1-\varepsilon}$ is equal to $\mathfrak{z}_t\chi(\tilde{\mathfrak{z}}_t > 0)$, whereas for $\varepsilon < 0$ and $\tilde{\mathfrak{z}}_t = 0$ it equals 0 or ∞ depending on whether $\mathfrak{z}_t = 0$ or $\mathfrak{z}_t > 0$, respectively; for $\varepsilon = 1$ the expression $\tilde{\mathfrak{z}}_t^{\varepsilon}\mathfrak{z}_t^{1-\varepsilon}$ is equal to $\tilde{\mathfrak{z}}_t\chi(\mathfrak{z}_t > 0)$; and for $\varepsilon > 1$ and $\mathfrak{z} = 0$ it equals 0 or ∞ for $\tilde{\mathfrak{z}}_t = 0$ or $\tilde{\mathfrak{z}}_t > 0$, respectively. The quantity $H(\frac{1}{2}; \widetilde{P}^t, P^t)$ is simply called *the Hellinger integral for measures $\widetilde{P}^t$ and P^t.*

Note that the quantities $\|\widetilde{P}^t - P^t\|$, $\rho(\widetilde{P}^t, P^t)$, and $H(\varepsilon; \widetilde{P}^t, P^t)$ defined by equalities (2.2.3), (2.2.4), and (2.2.5), respectively, do not depend on the choice of measure Q^t. More precisely, any dominant measure such that $P^t \ll Q^t$ and $\widetilde{P}^t \ll Q^t$ can be taken as measure Q^t.

The following lemma ([**127**], Theorem 3.9.1) describes some relations among the above quantities.

LEMMA 2.2.1. *The following inequalities hold:*

$$2\left[1 - H\left(\frac{1}{2}; \widetilde{P}^t, P^t\right)\right] \le \|\widetilde{P}^t - P^t\| \le \sqrt{8\left[1 - H\left(\frac{1}{2}; \widetilde{P}^t, P^t\right)\right]}, \tag{2.2.6}$$

$$\|\widetilde{P}^t - P^t\| \le 2\sqrt{1 - H^2\left(\frac{1}{2}; \widetilde{P}^t, P^t\right)}. \tag{2.2.7}$$

In particular,

$$2\rho^2\left(P^t, \widetilde{P}^t\right) \le \left\|\widetilde{P}^t - P^t\right\| \le \sqrt{8}\,\rho\left(P^t, \widetilde{P}^t\right). \tag{2.2.8}$$

Moreover, we would like to mention the following useful equality:

$$\rho^2(P^t, \widetilde{P}^t) = 1 - H\left(\frac{1}{2}; \widetilde{P}^t, P^t\right). \tag{2.2.9}$$

Now we introduce one more quantity, $\|P^t \wedge \widetilde{P}^t\|$, equal to the left-hand side of the equality (2.1.15) for $c = 1$, namely:

$$\|P^t \wedge \widetilde{P}^t\| = \inf\{\alpha(\delta_t) + \beta(\delta_t) \colon \delta_t \in \Sigma^t\}. \tag{2.2.10}$$

In view of (2.1.14)–(2.1.16), the quantity $\|P^t \wedge \widetilde{P}^t\|/2$ is the probability of error for the Bayes test in the case of equiprobable hypotheses H^t and $\widetilde{H}^t$.

LEMMA 2.2.2. *The following equality is valid:*

$$\|P^t \wedge \widetilde{P}^t\| = 1 - \frac{1}{2}\|\widetilde{P}^t - P^t\|. \tag{2.2.11}$$

PROOF. In view of (2.1.14) and (2.1.15), from (2.2.10) we have

$$\begin{aligned}\|P^t \wedge \widetilde{P}^t\| &= \mathsf{E}^t\,\chi(z_t \geq 1) + \widetilde{\mathsf{E}}^t\big(1 - \chi(z_t \geq 1)\big)\\ &= 1 + \mathsf{E}_Q^t\,\mathfrak{z}_t\chi(z_t \geq 1) - \mathsf{E}_Q^t\,\tilde{\mathfrak{z}}_t\chi(z_t \geq 1)\\ &= 1 - \mathsf{E}_Q^t(\tilde{\mathfrak{z}}_t - \mathfrak{z}_t)\chi(z_t \geq 1).\end{aligned} \tag{2.2.12}$$

Since $\mathsf{E}_Q^t(\tilde{\mathfrak{z}}_t - \mathfrak{z}_t) = 0$, from (2.2.3) we similarly obtain

$$\begin{aligned}\|\widetilde{P}^t - P^t\| &= \mathsf{E}_Q^t\,(\tilde{\mathfrak{z}}_t - \mathfrak{z}_t)\,\chi(z_t \geq 1) + \mathsf{E}_Q^t(\mathfrak{z}_t - \tilde{\mathfrak{z}}_t)\chi(z_t < 1)\\ &= 2\,\mathsf{E}_Q^t(\tilde{\mathfrak{z}}_t - \mathfrak{z}_t)\chi(z_t \geq 1).\end{aligned} \tag{2.2.13}$$

Now the required equality (2.2.11) follows from (2.2.12) and (2.2.13). □

REMARK 2.2.1. Similarly to (2.2.12), we have

$$\begin{aligned}\|P^t \wedge \widetilde{P}^t\| &= \mathsf{E}^t\,\chi(z_t \geq 1) + \widetilde{\mathsf{E}}^t\chi(z_t < 1)\\ &= \mathsf{E}_Q^t\,\mathfrak{z}_t\chi(z_t \geq 1) + \mathsf{E}_Q^t\,\tilde{\mathfrak{z}}\chi(z_t < 1) = \mathsf{E}_Q^t(\mathfrak{z}_t \wedge \tilde{\mathfrak{z}}_t).\end{aligned}$$

This equality explains the notation $\|P^t \wedge \widetilde{P}^t\|$.

Now we use the quantities introduced above and relations (2.2.6)–(2.2.9) and (2.2.11) among them to prove the following theorem, which yields the characterization of complete asymptotic distinguishability $(H^t) \vartriangle (\widetilde{H}^t)$ in various terms.

THEOREM 2.2.1. *The following conditions are equivalent*:

a) $(H^t) \vartriangle (\widetilde{H}^t)$;

b) $(P^t) \vartriangle (\widetilde{P}^t)$;

c) $\overline{\lim}_{t\to\infty} \widetilde{P}^t(z_t > N) = 1$ *for any* $N < \infty$;

d) $\overline{\lim}_{t\to\infty} P^t(z_t < N) = 1$ *for any* $N > 0$;

e) $\underline{\lim}_{t\to\infty} \|P^t \wedge \widetilde{P}^t\| = 0$;

f) $\underline{\lim}_{t\to\infty} H(\varepsilon; \widetilde{P}^t, P^t) = 0$ *for any* $\varepsilon \in (0,1)$;

g) $\overline{\lim}_{t\to\infty} \|\widetilde{P}^t - P^t\| = 2$;

h) $\overline{\lim}_{t\to\infty} \rho(P^t, \widetilde{P}^t) = 1$.

PROOF. *a*) ⇒ *b*). Let $t_n \uparrow \infty$, $n \to \infty$, and let $\delta_{t_n} \in \Sigma^{t_n}$ be such that $\alpha(\delta_{t_n}) \to 0$ and $\beta(\delta_{t_n}) \to 0$. Putting $A_n = \chi(\delta_{t_n} > \gamma)$, where $0 < \gamma < 1$, in view of the Chebyshev inequality we have

$$P^{t_n}(A_n) = P^{t_n}(\delta_{t_n} > \gamma) \leq \gamma^{-1}\alpha(\delta_{t_n}) \to 0.$$

Similarly,

$$\widetilde{P}^{t_n}(\delta_{t_n} \leq \gamma) \leq \widetilde{P}^{t_n}(1 - \delta_{t_n} \geq 1 - \gamma) \leq (1-\gamma)^{-1}\beta(\delta_{t_n}) \to 0,$$

which implies $\widetilde{P}^{t_n}(A_n) \to 1$.

b) ⇒ *d*). Let $t_n \uparrow \infty, n \to \infty$, and let $A_n \in \mathfrak{B}^{t_n}$ be such that

$$P^{t_n}(A_n) \to 0, \qquad \widetilde{P}^{t_n}(A_n) \to 1. \tag{2.2.14}$$

According to the Lebesgue decomposition, for any $A \in \mathfrak{B}^t$ we have

$$P^t(A) \le N\widetilde{P}^t(A) + P^t(\widetilde{z}_t > N), \qquad 0 < N < \infty. \tag{2.2.15}$$

The relations (2.2.14) and estimate (2.2.15) for $A = A_n^c$ and $t = t_n$ imply that $P^{t_n}(\widetilde{z}_{t_n} > N) \to 1$ for any $0 < N < \infty$. From this, taking into account the equality

$$P^t\left(\widetilde{z}_t > N\right) = P^t\left(z_t < N^{-1}\right), \tag{2.2.16}$$

we get assertion d).

$d) \Rightarrow f)$. For any ε, $\delta \in (0,1)$ and $\gamma > 0$ we have

$$\begin{aligned} H(\varepsilon; \widetilde{P}^t, P^t) &= \mathsf{E}_Q^t\, \tilde{\mathfrak{z}}_t^\varepsilon \mathfrak{z}_t^{1-\varepsilon} \chi(\tilde{\mathfrak{z}}_t < \gamma) + \mathsf{E}_Q^t\, \tilde{\mathfrak{z}}_t^\varepsilon \mathfrak{z}_t^{1-\varepsilon} \chi(\tilde{\mathfrak{z}}_t \ge \gamma, \mathfrak{z}_t \le \delta) \\ &\quad + \mathsf{E}_Q^t\, \tilde{\mathfrak{z}}_t^\varepsilon \mathfrak{z}_t^{1-\varepsilon} \chi(\tilde{\mathfrak{z}}_t \ge \gamma, \mathfrak{z}_t > \delta) \\ &\le \gamma^\varepsilon + \delta^{1-\varepsilon} + \mathsf{E}_Q^t\, \mathfrak{z}_t z_t^\varepsilon \chi(\tilde{\mathfrak{z}}_t \ge \gamma, \mathfrak{z}_t > \delta) \\ &\le \gamma^\varepsilon + \delta^{1-\varepsilon} + \left[P^t(\tilde{\mathfrak{z}}_t \ge \gamma)\right]^{1-\varepsilon}. \end{aligned} \tag{2.2.17}$$

Taking into account (2.2.16), from relation d) we get

$$\overline{\lim_{t\to\infty}}\, P^t\left(\widetilde{z}_t > N\right) = 1.$$

From this, in view of the equality $\widetilde{z}_t = 2/\tilde{\mathfrak{z}}_t - 1$, we have

$$\underline{\lim_{t\to\infty}}\, P^t(\tilde{\mathfrak{z}}_t \ge \gamma) = 0 \tag{2.2.18}$$

for every $\gamma > 0$. From (2.2.17) and (2.2.18) we obtain

$$\underline{\lim_{t\to\infty}}\, H(\varepsilon; \widetilde{P}^t, P^t) \le \gamma^\varepsilon + \delta^{1-\varepsilon}.$$

Since γ and δ are arbitrary, we conclude that f) is true.

$f) \Rightarrow h)$. It suffices to apply (2.2.9).

$h) \Rightarrow g)$. This implication follows from the left inequality in (2.2.8) and the estimate $\|\widetilde{P}^t - P^t\| \le 2$.

$g) \Rightarrow e)$. It is enough to use (2.2.11).

$e) \Rightarrow a)$. By virtue of definition (2.2.10) and equality (2.1.16) for $c = 1$, condition e) implies that there exists a sequence $t_n \uparrow \infty$, $n \to \infty$, such that

$$\left\|P^{t_n} \wedge P^{t_n}\right\| = \alpha(\delta_{t_n}^{*,1}) + \beta(\delta_{t_n}^{*,1}) \to 0.$$

This implies a).

$b) \Rightarrow c)$. Let $t_n \to \infty$, $n \to \infty$, and let $A_n \in \mathfrak{B}^{t_n}$ be such that the relations (2.2.14) are valid. In view of the Lebesgue decomposition, for any $A \in \mathfrak{B}^t$ and $N > 0$ we have

$$\widetilde{P}^t(A) \le NP^t(A) + \widetilde{P}^t(z_t > N).$$

This inequality and (2.2.14) imply c).

$c) \Rightarrow b)$. Since c) is valid, there exists a sequence $t_n \uparrow \infty$, $n \to \infty$, such that

$$\widetilde{P}^{t_n}(z_{t_n} > n) \ge 1 - 1/n.$$

Therefore

$$\widetilde{P}^{t_n}(z_{t_n} > n) \to 1, \qquad n \to \infty.$$

On the other hand, we have

$$P^{t_n}(z_{t_n} > n) \le n^{-1}\, \mathsf{E}^{t_n} z_{t_n} \le n^{-1},$$

whence

$$P^{t_n}(z_{t_n} > n) \to 0, \qquad n \to \infty.$$

Thus assertion $b)$ is valid. □

REMARK 2.2.2. The definition of complete asymptotic distinguishability implies the alternative: either $(H^t) \vartriangle (\widetilde{H}^t)$ or $(H^t) \overline{\vartriangle} (\widetilde{H}^t)$. Therefore, the characterization of the notion $(H^t) \overline{\vartriangle} (\widetilde{H}^t)$ can be obtained from Theorem 2.2.1. For example, $(H^t) \overline{\vartriangle} (\widetilde{H}^t) \Leftrightarrow \varliminf_{t\to\infty} \|P^t \wedge \widetilde{P}^t\| > 0$, etc.

Complete asymptotic indistinguishability.

DEFINITION 2.2.6. Families of hypotheses (H^t) and $(\widetilde{H}^t)$ are called *completely asymptotically indistinguishable* (denoted $(H^t) \cong (\widetilde{H}^t)$), if for any $\alpha \in [0,1]$ and any family (δ_t) of tests $\delta_t \in \Sigma^t$ such that the limit $\lim_{t\to\infty} \alpha(\delta_t)$ exists, the implication

$$\lim_{t\to\infty} \alpha(\delta_t) = \alpha \Rightarrow \lim_{t\to\infty} \beta(\delta_t) = 1 - \alpha \tag{2.2.19}$$

is true.

DEFINITION 2.2.7. Families of measures (P^t) and $(\widetilde{P}^t)$ are called *completely asymptotically inseparable* (denoted $(P^t) \cong (\widetilde{P}^t)$) if for any $\alpha \in [0,1]$ and any families (A_t) of sets $A_t \in \mathfrak{B}^t$ such that the limit $\lim_{t\to\infty} P^t(A_t)$ exists the implication

$$\lim_{t\to\infty} P^t(A_t) = \alpha \Rightarrow \lim_{t\to\infty} \widetilde{P}^t(A_t) = \alpha \tag{2.2.20}$$

is valid.

The following theorem yields the characterization of the notion of complete asymptotic indistinguishability among families of hypotheses.

THEOREM 2.2.2. *The following conditions are equivalent:*

$a)$ $(H^t) \cong (\widetilde{H}^t)$;
$b)$ $(P^t) \cong (\widetilde{P}^t)$;
$c)$ $\lim_{t\to\infty} P^t(|\Lambda_t| > \gamma) = 0$ *for any* $\gamma > 0$;
$d)$ $\lim_{t\to\infty} \widetilde{P}^t(|\Lambda_t| > \gamma) = 0$ *for any* $\gamma > 0$;
$e)$ $\lim_{t\to\infty} H(\varepsilon; \widetilde{P}^t, P^t) = 1$ *for any* $\varepsilon \in (0,1)$;
$f)$ $\lim_{t\to\infty} \rho(P^t, \widetilde{P}^t) = 0$;
$g)$ $\lim_{t\to\infty} \|\widetilde{P}^t - P^t\| = 0$;
$h)$ $\lim_{t\to\infty} \|P^t \wedge \widetilde{P}^t\| = 1$.

PROOF. $a) \Rightarrow b)$. Let α be an arbitrary number in $[0,1]$, and let (A_t) be an arbitrary family of sets $A_t \in \mathfrak{B}^t$ such that $P^t(A_t) \to \alpha$ as $t \to \infty$. For the test $\delta_t = \chi(A_t)$ we have $\alpha(\delta_t) = P^t(A_t) \to \alpha$. Then, in view of condition $a)$, we have $\widetilde{P}^t(A_t) = 1 - \beta(\delta_t) \to \alpha$ as $t \to \infty$. Thus the implication (2.2.20) is established.

$b) \Rightarrow c)$. We prove it by contradiction. Assume that condition $b)$ holds and condition $c)$ does not hold. Then there exist $\gamma_0 > 0$ and a sequence (t_n) such that $t_n \to \infty$ and $P^{t_n}(|\Lambda_{t_n}| > \gamma_0) \to a > 0$ as $n \to \infty$. This means that there exist subsequences $(m) \subset (n)$ and $(k) \subset (n)$ such that either $P^{t_m}(\Lambda_{t_m} > \gamma_0) \to b$ $(b > 0)$ as $m \to \infty$ or $P^{t_k}(\Lambda_{t_k} < -\gamma_0) \to c$ $(c > 0)$ as $k \to \infty$.

Let $P^{t_m}(\Lambda_{t_m} > \gamma_0) \to b > 0$ as $m \to \infty$. Then, in view of condition b), we have $\widetilde{P}^{t_m}(\Lambda_{t_m} > \gamma_0) \to b$ as $m \to \infty$. But, according to the Lebesgue decomposition,

$$\widetilde{P}^t(\Lambda_t > \gamma_0) = \int_{(\Lambda_t > \gamma_0)} \exp(\Lambda_t)\, dP^t + \widetilde{P}^t(\Lambda_t = \infty) \geq e^{\gamma_0} P^t(\Lambda_t > \gamma_0).$$

Passing to the limit in this inequality over the sequence (t_m) as $m \to \infty$, we get the inequality $b \geq e^{\gamma_0} b$. Since $b > 0$ and $\gamma_0 > 0$, we have $b > b$. Thus we have arrived at a contradiction. Now, putting $P^{t_k}(\Lambda_{t_k} < -\gamma_0) \to c > 0$ as $k \to \infty$, and arguing in the same manner, we also arrive at a contradiction. These contradictions prove the implication $b) \Rightarrow c)$.

$c) \Rightarrow d)$. Using the Lebesgue decomposition, for any $\gamma > 0$ and $\gamma' \in (0, \gamma)$ we get

$$\widetilde{P}^t(|\Lambda_t| \leq \gamma) \geq \widetilde{P}^t(|\Lambda_t| \leq \gamma') = \int_{(|\Lambda_t| \leq \gamma')} \exp(\Lambda_t)\, dP^t \geq e^{-\gamma'} P^t(|\Lambda_t| \leq \gamma').$$

From this, in view of condition c), we have $\underline{\lim}_{t\to\infty} \widetilde{P}^t(|\Lambda_t| \leq \gamma) \geq e^{-\gamma'}$. Now passing to the limit in this inequality as $\gamma' \to 0$, we obtain $\lim_{t\to\infty} \widetilde{P}^t(|\Lambda_t| \leq \gamma) = 1$ for any $\gamma > 0$, that is, condition d) holds.

$d) \Rightarrow e)$. Let $\varepsilon \in (0,1)$ and $\gamma > 0$. Taking into account that $(|\Lambda_t| \leq \gamma) \subset (\tilde{\mathfrak{z}}_t > 0, \mathfrak{z} > 0)$, we get

$$\begin{aligned} H(\varepsilon; \widetilde{P}^t, P^t) &\geq \mathsf{E}_Q^t\, \tilde{\mathfrak{z}}_t^{\varepsilon} \mathfrak{z}_t^{1-\varepsilon} \chi(|\Lambda_t| \leq \gamma) = \mathsf{E}_Q^t\, \tilde{\mathfrak{z}}_t z_t^{\varepsilon-1} \chi(|\Lambda_t| \leq \gamma) \\ &\geq e^{(\varepsilon-1)\gamma} \widetilde{P}^t(|\Lambda_t| \leq \gamma). \end{aligned}$$

In view of condition d), this implies

$$\underline{\lim}_{t\to\infty} H(\varepsilon; \widetilde{P}^t, P^t) \geq e^{(\varepsilon-1)\gamma} \tag{2.2.21}$$

for any $\varepsilon \in (0,1)$ and $\gamma > 0$. Passing to the limit in (2.2.21) as $\gamma \to 0$ and considering the inequality $H(\varepsilon; \widetilde{P}^t, P^t) \leq 1$, we obtain e).

$e) \Rightarrow f)$. It suffices to use the equality (2.2.9).

$f) \Rightarrow g)$. This implication follows from the inequalities (2.2.8).

$g) \Rightarrow h)$. It is enough to use the equality (2.2.11).

$h) \Rightarrow a)$. Let α be an arbitrary number in $[0,1]$, and let $(\bar{\delta}_t)$ be an arbitrary family of tests $\bar{\delta}_t \in \Sigma^t$ such that $\alpha(\bar{\delta}_t) \to \alpha$ as $t \to \infty$. Since the set $\mathfrak{N}^t$ is centrally symmetric with respect to the point $(\frac{1}{2}, \frac{1}{2})$, that is, along with any point (α, β) it contains also the point $(1-\alpha, 1-\beta)$ [**58**], we have

$$\inf\{\alpha(\delta_t) + \beta(\delta_t) \colon \delta_t \in \Sigma^t\} = 2 - \sup\{\alpha(\delta_t) + \beta(\delta_t) \colon \delta_t \in \Sigma^t\} \tag{2.2.22}$$

and hence, by condition h),

$$\lim_{t\to\infty} \sup\{\alpha(\delta_t) + \beta(\delta_t) \colon \delta_t \in \Sigma^t\} = 1. \tag{2.2.23}$$

Thus, taking into account the inequalities

$$\inf\{\alpha(\delta_t) + \beta(\delta_t) \colon \delta_t \in \Sigma^t\} \leq \alpha(\bar{\delta}_t) + \beta(\bar{\delta}_t) \leq \sup\{\alpha(\delta_t) + \beta(\delta_t) \colon \delta_t \in \Sigma^t\}, \tag{2.2.24}$$

condition h) and relation (2.2.23), we get $\beta(\bar{\delta}_t) \to 1 - \alpha$ as $t \to \infty$. Thus the implication (2.2.19) is proved. □

Definitions 2.2.6 and 2.2.7 are not symmetric by their nature. However the following lemma shows that the implications in (2.2.19) and (2.2.20) can be replaced by equivalence.

LEMMA 2.2.3. *If* $(H^t) \cong (\widetilde{H}^t)$, *then for any* $\beta \in [0,1]$ *and any families* (δ_t) *of tests* $\delta_t \in \Sigma^t$ *such that the limit* $\lim_{t\to\infty} \beta(\delta_t)$ *exists the implication*

$$\lim_{t\to\infty} \beta(\delta_t) = \beta \Rightarrow \lim_{t\to\infty} \alpha(\delta_t) = 1 - \beta \tag{2.2.25}$$

is true.

If $(P^t) \cong (\widetilde{P}^t)$, *then for any* $\beta \in [0,1]$ *and any families* (A_t) *of sets* $A_t \in \mathfrak{B}^t$ *such that the limit* $\lim_{t\to\infty} \widetilde{P}^t(A_t)$ *exists the implication*

$$\lim_{t\to\infty} \widetilde{P}^t(A_t) = \beta \Rightarrow \lim_{t\to\infty} P^t(A_t) = \beta \tag{2.2.26}$$

is true.

PROOF. The implication (2.2.25) follows from the equivalence of conditions *a*) and *h*) of Theorem 2.2.2 and relations (2.2.22) and (2.2.24).

Let us prove the second assertion. Let β be an arbitrary number in $[0,1]$, and let (A_t) be an arbitrary family of sets in $\mathfrak{B}^t$ such that $\widetilde{P}^t(A_t) \to \beta$ as $t \to \infty$. Then for the test $\delta_t = \chi(A_t)$ we have $\beta(\delta_t) = \widetilde{P}^t(A_t^c) \to 1 - \beta$ as $t \to \infty$. Since $\alpha(\delta_t) = P^t(A_t)$, the implication (2.2.26) follows from the equivalence of conditions *a*) and *b*) of Theorem 2.2.2 and implication (2.2.25). □

The contiguity of families of hypotheses.

DEFINITION 2.2.8. *A family of hypotheses* $(\widetilde{H}^t)$ is said to be *contiguous to a family of hypotheses* (H^t) (denoted $(\widetilde{H}^t) \lhd (H^t)$) if for any test $\delta_t \in \Sigma^t$ such that $\alpha(\delta_t) \to 0$, $t \to \infty$, we have $\beta(\delta_t) \to 1$, $t \to \infty$. Otherwise, that is, if there exists a family (δ_t) of tests $\delta_t \in \Sigma^t$ such that

$$\lim_{t\to\infty} \alpha(\delta_t) = 0, \qquad \underline{\lim}_{t\to\infty} \beta(\delta_t) < 1,$$

the *family of hypotheses* $(\widetilde{H}^t)$ is called *noncontiguous to the family of hypotheses* (H_t) (denoted $(\widetilde{H}^t) \overline{\lhd} (H_t)$). If $(\widetilde{H}^t) \lhd (H_t)$ and $(H^t) \lhd (\widetilde{H}_t)$, then the families of hypotheses (H^t) and $(\widetilde{H}^t)$ are called *mutually contiguous* (denoted $(H^t) \lhd\rhd (\widetilde{H}_t)$); if $(\widetilde{H}^t) \overline{\lhd} (H_t)$ and $(H^t) \overline{\lhd} (\widetilde{H}_t)$, then they are called *mutually noncontiguous* (denoted $(H^t) \overline{\lhd}\overline{\rhd} (\widetilde{H}_t)$). If either $(\widetilde{H}^t) \overline{\lhd} (H_t)$ or $(H^t) \overline{\lhd} (\widetilde{H}_t)$, we say that the *families of hypotheses* (H^t) *and* $(\widetilde{H}^t)$ *are not mutually contiguous* (denoted $(H^t) \overline{\lhd\rhd} (\widetilde{H}_t)$).

DEFINITION 2.2.9. *A family of measures* $(\widetilde{P}^t)$ is said to be *contiguous to a family of measures* (P^t) (denoted $(\widetilde{P}^t) \lhd (P^t)$) if for any sets $A_t \in \mathfrak{B}^t$ such that $P^t(A_t) \to 0$, $t \to \infty$, we have $\widetilde{P}^t(A_t) \to 0$, $t \to \infty$. If there is a family (A_t) of sets $A_t \in \mathfrak{B}^t$ such that $\lim_{t\to\infty} P^t(A_t) = 0$ and $\overline{\lim}_{t\to\infty} \widetilde{P}^t(A_t) > 0$, then *the family of measures* $(\widetilde{P}^t)$ is said to be *noncontiguous to the family of measures* (P_t) (denoted $(\widetilde{P}^t) \overline{\lhd} (P_t)$). If $(\widetilde{P}^t) \lhd (P_t)$ and $(P^t) \lhd (\widetilde{P}_t)$, the families of measures (P^t) and $(\widetilde{P}^t)$ are called *mutually contiguous* (denoted $(P^t) \lhd\rhd (\widetilde{P}_t)$); if $(\widetilde{P}^t) \overline{\lhd} (P_t)$ and $(P^t) \overline{\lhd} (\widetilde{P}_t)$ they are called *mutually noncontiguous* (denoted $(P^t) \overline{\lhd}\overline{\rhd} (\widetilde{P}_t)$). If either $(\widetilde{P}^t) \overline{\lhd} (P_t)$ or $(P^t) \overline{\lhd} (\widetilde{P}_t)$, we say that *the families of measures* (P^t) *and* $(\widetilde{P}^t)$ *are not mutually contiguous* (denoted $(P^t) \overline{\lhd\rhd} (\widetilde{P}_t)$).

Let $(X^t, \mathfrak{B}^t)$, $t \in \mathbf{R}_+$, be a family of measurable spaces, S^t a probability measure on $(X^t, \mathfrak{B}^t)$, $t \in \mathbf{R}_+$, and let ζ_t be measurable functions with values in $\overline{\mathbf{R}}$ defined on $(X^t, \mathfrak{B}^t)$, $t \in \mathbf{R}_+$.

DEFINITION 2.2.10. A *family* (ζ_t) is said to be *tight with respect to a family of measures* (S^t) (we write: $(\zeta_t|S^t)$ is tight) if

$$\lim_{N\to\infty} \overline{\lim_{t\to\infty}} S^t(|\zeta_t| > N) = 0.$$

DEFINITION 2.2.11. A *family* (ζ_t) is said to be *uniformly integrable with respect to* (S^t) if

$$\lim_{N\to\infty} \sup_t \int \chi(|\zeta_t| > N)|\zeta_t|\, dS^t = 0.$$

The following theorem provides the characterization of contiguity $(\widetilde{H}^t) \lhd (H^t)$.

THEOREM 2.2.3. *The following conditions are equivalent:*

a) $(\widetilde{H}^t) \lhd (H^t)$;

b) $(\widetilde{P}^t) \lhd (P^t)$;

c) $\lim_{t\to\infty} \widetilde{P}^t(z_t = \infty) = 0$ *and the family* (z_t) *is uniformly integrable with respect to* (P^t);

d) $(z_t|\widetilde{P}^t)$ *is tight;*

e) $(1/\mathfrak{z}_t|\widetilde{P}^t)$ *is tight;*

f) $\lim_{\varepsilon\uparrow 1} \underline{\lim}_{t\to\infty} H(\varepsilon; \widetilde{P}^t, P^t) = 1$.

PROOF. *a)* $\Rightarrow$ *b)*. Let $A_t \in \mathfrak{B}^t$ and $P^t(A_t) \to 0$ for $t \to \infty$. Then $\alpha(\tilde{\delta}_t) \to 0$ as $t \to \infty$ for the test $\tilde{\delta}_t = \chi(A_t)$. Therefore, by condition *a)*, we have $\widetilde{P}^t(A_t) = 1 - \beta(\tilde{\delta}_t) \to 0$ as $t \to \infty$.

b) $\Rightarrow$ *a)*. Let (δ_t) be an arbitrary family of tests such that $\alpha(\delta_t) \to 0$, $t \to \infty$, and let ε be an arbitrary positive number. Put $A_t^\varepsilon = (\delta_t \geq \varepsilon)$. Then the inequality

$$P^t(A_t^\varepsilon) \leq \varepsilon^{-1} \int_{A_t^\varepsilon} \delta_t\, dP^t \leq \varepsilon^{-1}\alpha(\delta_t)$$

implies $P^t(A_t^\varepsilon) \to 0$, $t \to \infty$, for any $\varepsilon > 0$. Then condition *b)* implies $\widetilde{P}^t(A_t^\varepsilon) \to 0$, $t \to \infty$, for any $\varepsilon > 0$. Now, taking into account the estimate

$$1 - \beta(\delta_t) = \int_{A_t^\varepsilon} \delta_t\, d\widetilde{P}^t + \int_{(A_t^\varepsilon)^c} \delta_t\, d\widetilde{P}^t \leq \widetilde{P}^t(A_t^\varepsilon) + \varepsilon$$

and arbitrariness of ε, we obtain $\beta(\delta_t) \to 1$, $t \to \infty$.

b) $\Rightarrow$ *c)*. Since $P^t(z_t = \infty) = 0$, condition *b)* implies that

$$\widetilde{P}^t(z_t = \infty) \to 0, \qquad t \to \infty.$$

Note that the uniform integrability of (z_t) with respect to (P^t) is equivalent to the fulfillment of the following two conditions ([**127**], Lemma 2.6.2):

$$\sup_t \int z_t \, dP^t < \infty, \tag{2.2.27}$$

$$P^t(A_t) \to 0, \qquad A_t \in \mathfrak{B}^t \Rightarrow \int_{A_t} z_t \, dP^t \to 0. \tag{2.2.28}$$

By the Lebesgue decomposition, we have

$$\int_{A_t} z_t \, dP^t \le \widetilde{P}^t(A_t) \le 1.$$

This implies (2.2.27), and condition $b)$ implies (2.2.28).

$c) \Rightarrow d)$. In view of the Lebesgue decomposition, we have

$$\widetilde{P}^t(z_t > N) = \int_{(z_t > N)} z_t \, dP^t + \widetilde{P}^t(z_t = \infty).$$

This equality and condition $c)$ imply condition $d)$.

$d) \Rightarrow b)$. Let $A_t \in \mathfrak{B}^t$ be sets such that $P^t(A_t) \to 0$. In view of the Lebesgue decomposition, we have

$$\begin{aligned} \widetilde{P}^t(A_t) &= \widetilde{P}^t(A_t \cap (z_t \le N)) + \widetilde{P}^t(A_t \cap (z_t > N)) \\ &\le \int_{A_t \cap (z_t \le N)} z_t \, dP^t + \widetilde{P}^t(z_t > N) \le N P^t(A_t) + \widetilde{P}^t(z_t > N). \end{aligned}$$

From this, by condition $d)$, we get $\widetilde{P}^t(A_t) \to 0$, $t \to \infty$.

$d) \Leftrightarrow e)$. It is sufficient to use the equality $z_t = 2/\mathfrak{z}_t - 1$.

$e) \Rightarrow f)$. Let $\gamma > 0$. Then for $\varepsilon \in (0,1)$ we have

$$\begin{aligned} H(\varepsilon; \widetilde{P}^t, P^t) &\ge \mathsf{E}_Q^t \, \tilde{\mathfrak{z}}_t \left(\frac{\mathfrak{z}_t}{\tilde{\mathfrak{z}}_t} \right)^{1-\varepsilon} \chi(\mathfrak{z}_t \ge \gamma) \chi(\tilde{\mathfrak{z}}_t > 0) \\ &= \widetilde{\mathsf{E}}^t (\mathfrak{z}_t / \tilde{\mathfrak{z}}_t)^{1-\varepsilon} \chi(\mathfrak{z}_t \ge \gamma) \ge (\gamma/2)^{1-\varepsilon} \widetilde{P}^t(\mathfrak{z}_t \ge \gamma), \end{aligned}$$

since $\mathfrak{z}_t + \tilde{\mathfrak{z}}_t = 2$. Therefore for $\gamma > 0$ we get

$$\underline{\lim}_{\varepsilon \uparrow 1} \, \underline{\lim}_{t \to \infty} H\left(\varepsilon; \widetilde{P}^t, P^t\right) \ge \underline{\lim}_{\varepsilon \uparrow 1} \left(\frac{\gamma}{2}\right)^{1-\varepsilon} \underline{\lim}_{t \to \infty} \widetilde{P}^t(\mathfrak{z}_t \ge \gamma) \ge \underline{\lim}_{t \to \infty} \widetilde{P}^t(\mathfrak{z}_t \ge \gamma).$$

By $e)$, we have

$$\lim_{\gamma \downarrow 0} \, \underline{\lim}_{t \to \infty} \widetilde{P}^t(\mathfrak{z}_t \ge \gamma) = 1.$$

Now, taking into account that $H(\varepsilon; \widetilde{P}^t, P^t) \le 1$, we get assertion $f)$.

$f) \Rightarrow e)$. Similarly to (2.2.17), for any $\varepsilon, \delta \in (0,1)$ and $\gamma > 0$ we obtain

$$\begin{aligned} H(\varepsilon; \widetilde{P}^t, P^t) &\le \gamma^{1-\varepsilon} + \delta^{\varepsilon} + \widetilde{\mathsf{E}}^t (\mathfrak{z}_t / \tilde{\mathfrak{z}}_t)^{1-\varepsilon} \chi(\mathfrak{z}_t \ge \gamma, \tilde{\mathfrak{z}}_t > \delta) \\ &\le \gamma^{1-\varepsilon} + \delta^{\varepsilon} + (2/\delta)^{1-\varepsilon} \widetilde{P}^t(\mathfrak{z}_t \ge \gamma). \end{aligned}$$

From this for any $\varepsilon, \delta \in (0,1)$ we get

$$\underline{\lim}_{\gamma \downarrow 0} \, \underline{\lim}_{t \to \infty} \widetilde{P}^t(\mathfrak{z}_t \ge \gamma) \ge \left(\frac{\delta}{2}\right)^{1-\varepsilon} \underline{\lim}_{t \to \infty} H\left(\varepsilon; \widetilde{P}^t, P^t\right) - \delta/2^{1-\varepsilon}.$$

Now we pass to the limit as $\varepsilon \uparrow 1$ and use $f)$ and then we pass to the limit as $\delta \downarrow 0$

to obtain

$$\varliminf_{\gamma\downarrow 0}\ \varliminf_{t\to\infty} \widetilde{P}^t(\mathfrak{z}_t \geq \gamma) \geq 1,$$

whence assertion e) follows. □

REMARK 2.2.3. Evidently, it is easy to obtain, from Theorem 2.2.3, the characterizations for $(\widetilde{H}^t)\,\overline{\triangleleft}\,(H^t)$, $(H^t) \triangleleft\triangleright (\widetilde{H}^t)$, $(H^t)\,\overline{\triangleleft}\triangleright\,(\widetilde{H}^t)$ and $(H^t)\,\overline{\triangleleft\triangleright}\,(\widetilde{H}^t)$.

Types of asymptotic distinguishability. Using Definitions 2.2.1, 2.2.6, and 2.2.8, we introduce certain *types of asymptotic distinguishability between two families of hypotheses* (H^t) *and* $(\widetilde{H}^t)$. To this end we consider the following conditions:

a$_0$) $(H^t) \cong (\widetilde{H}^t)$;
a) $(H^t) \triangleleft\triangleright (\widetilde{H}^t)$;
b) $(H^t) \triangleleft (\widetilde{H}^t)$, $(\widetilde{H}^t)\,\overline{\triangleleft}\,(H^t)$;
c) $(H^t)\,\overline{\triangleleft}\,(\widetilde{H}^t)$, $(\widetilde{H}^t) \triangleleft (H^t)$;
d) $(H^t)\,\overline{\triangleleft}\triangleright\,(\widetilde{H}^t)$, $(H^t)\,\overline{\triangle}\,(\widetilde{H}^t)$;
e) $(H^t) \triangle (\widetilde{H}^t)$.

DEFINITION 2.2.12. The *asymptotic distinguishability between two families of hypotheses* (H^t) *and* $(\widetilde{H}^t)$ is said to be *of type* **a**$_0$ (respectively, **a**, **b**, **c**, **d**, and **e**) if condition **a**$_0$) (respectively, **a**), **b**), **c**), **d**), and **e**)) is fulfilled.

It is not hard to see that the types of asymptotic distinguishability **a**, **b**, **c**, **d**, and **e** are mutually exclusive and form a complete group of types of asymptotic distinguishability between families of hypotheses. Since **a**$_0$) $\Rightarrow$ **a**), the type **a**$_0$ is a subtype of the type **a**. Therefore, if we introduce the type **a**$_1$ as the subtype of the type **a** for which condition **a**$_0$) is not fulfilled, then, evidently, the types **a**$_0$, **a**$_1$, **b**, **c**, **d**, and **e** also form a complete group of incompatible types of asymptotic distinguishability between the families of hypotheses (H^t) and $(\widetilde{H}^t)$.

The characterization of the types **e** and **a**$_0$ is given in Theorems 2.2.1 and 2.2.2, respectively. The characterizations of the other types are easily obtained by combining the characterizations from Theorems 2.2.1–2.2.3 and for this reason they are omitted here.

To conclude this section, we observe that to each of the above types of asymptotic distinguishability between the families of hypotheses (H^t) and $(\widetilde{H}^t)$ corresponds a certain type of asymptotic behavior of the sets $\mathfrak{N}^t$ as $t \to \infty$. From the above it is easily seen that this type of asymptotic behavior is described by the asymptotic behavior of the boundary of the set $\mathfrak{N}^t$. In turn, the boundary of the set $\mathfrak{N}^t$ is described by the fundamental Neyman–Pearson lemma and is given by formula (2.1.12).

2.3. Complete asymptotic distinguishability under the conditions of the law of large numbers

Consider the Neyman–Pearson test δ_t^{+,α_t} of level $\alpha_t \in [0,1]$ defined by equality (2.1.13) and assume that the complete asymptotic distinguishability $(H^t) \triangle (\widetilde{H}^t)$ holds. In view of Theorem 2.2.1, we have

$$(H^t) \triangle (\widetilde{H}^t) \Leftrightarrow \lim_{t\to\infty} \left[\alpha\left(\delta_t^{*,c}\right) + \beta\left(\delta_t^{*,c}\right)\right] = 0 \quad \text{for all } c \in (0,\infty), \tag{2.3.1}$$

where $\delta_t^{*,c}$ is the Bayes test of the form (2.1.14).

From this it follows that in the case of $(H^t) \vartriangle (\widetilde{H}^t)$ for the test δ_t^{+,α_t} with $\alpha_t = \alpha(\delta_t^{*,c})$ and any $c \in (0,\infty)$ there exists a sequence (t_n), $t_n \to \infty$, such that $\alpha_{t_n} \to 0$ and $\beta(\delta_t^{+,\alpha_{t_n}}) \to 0$ as $n \to \infty$. In view of the equivalence (2.3.1), in order to obtain finer properties of the function $\beta(\delta_t^{+,\alpha})$, it is necessary to consider a more restrictive condition than $(H^t) \vartriangle (\widetilde{H}^t)$.

The law of large numbers. Consider the following conditions:

$\Lambda 1$. $\lim_{t\to\infty} P^t(|\chi_t^{-1}\Lambda_t + 1| > \gamma) = 0$ for any $\gamma > 0$, where χ_t is a nonrandom positive function such that $\chi_t \to \infty$ as $t \to \infty$;

$\Lambda 1'$. $\lim_{t\to\infty} P^t(\chi_t^{-1}\Lambda_t > a) = 0$ for any $a > -1$;

$\Lambda 1''$. $\lim_{t\to\infty} P^t(\chi_t^{-1}\Lambda_t < a) = 0$ for any $a < -1$.

It is not difficult to see that $\Lambda 1', \Lambda 1'' \Leftrightarrow \Lambda 1$. In addition, Theorem 2.2.1 implies that $\Lambda 1' \Rightarrow (H^t) \vartriangle (\widetilde{H}^t)$, that is, under condition $\Lambda 1'$ (and hence under condition $\Lambda 1$) the complete asymptotic distinguishability holds.

DEFINITION 2.3.1 ([**106, 107**]). The quantity

$$I(P^t|\widetilde{P}^t) = \mathsf{E}_Q^t\, \mathfrak{z}_t \ln(\mathfrak{z}_t/\tilde{\mathfrak{z}}_t), \tag{2.3.2}$$

where $\mathfrak{z}_t \ln(\mathfrak{z}_t/\tilde{\mathfrak{z}}_t)$ is 0 for $\mathfrak{z}_t = 0$ and ∞ for $\tilde{\mathfrak{z}}_t = 0$, is called the *entropy of the measure P^t with respect to the measure $\widetilde{P}^t$.*

The relative entropy $I(P^t|\widetilde{P}^t)$ is often called the *Kullback–Leibler divergence, distance, deviation* [**9, 53, 124**], or *Kullback–Leibler information* [**53, 130**] between the measures P^t and $\widetilde{P}^t$.

The definition of entropy (2.3.2) and the definition of likelihood ratio (2.1.1) imply the equality

$$I(P^t|\widetilde{P}^t) = \mathsf{E}^t \ln \tilde{z}_t = -\,\mathsf{E}^t \ln z_t. \tag{2.3.3}$$

REMARK 2.3.1. Let the entropy $I(P^t|\widetilde{P}^t)$ be finite for every $t \in \mathbf{R}_+$ and

$$I(P^t|\widetilde{P}^t) \to \infty, \qquad t \to \infty.$$

If condition $\Lambda 1$ is fulfilled with $\chi_t = I(P^t|\widetilde{P}^t)$, then, evidently, condition $\Lambda 1$ is the law of large numbers for Λ_t, since by (2.3.3) we have $\mathsf{E}^t\,\Lambda_t = -I(P^t|\widetilde{P}^t)$.

LEMMA 2.3.1. *The following implications hold:*

$$\Lambda 1'' \;\Rightarrow\; \lim_{t\to\infty} \bar{\alpha}_t = 1, \tag{2.3.4}$$

$$\Lambda 1'' \;\Rightarrow\; \varliminf_{t\to\infty} \chi_t^{-1} \ln \bar{\beta}_t \ge -1. \tag{2.3.5}$$

PROOF. The implication (2.3.4) follows from (2.1.4) and the estimate

$$P^t(\Lambda_t < -a\chi_t) \ge P^t(\Lambda_t = -\infty) = P^t(z_t = 0),$$

where $1 < a < \infty$. Using the Lebesgue decomposition (2.1.2) and taking into account equality (2.1.5), for every $a \in (1,\infty)$ we get

$$\begin{aligned} P^t(\Lambda_t \ge -a\chi_t) &\le e^{a\chi_t}\,\mathsf{E}^t\, e^{\Lambda_t}\chi(\Lambda_t \ge -a\chi_t) \\ &\le e^{a\chi_t}\,\mathsf{E}^t\, z_t = e^{a\chi_t}\widetilde{P}^t(z_t < \infty) = e^{a\chi_t}\bar{\beta}_t. \end{aligned}$$

From this, in view of the arbitrariness of a, we obtain the implication (2.3.5). $\square$

REMARK 2.3.2. By virtue of (2.1.12), the implication (2.3.5) is equivalent to the implication $\Lambda 1'' \Rightarrow \underline{\lim}_{t\to\infty} \chi_t^{-1} \ln \beta(\delta_t^{+,0}) \geq -1$.

Now we consider the following conditions:

$\alpha 1$. $\underline{\lim}_{t\to\infty} \alpha_t > 0$;
$\alpha 2$. $\overline{\lim}_{t\to\infty} \alpha_t < 1$;
$d1$. $\overline{\lim}_{t\to\infty} \chi_t^{-1} d_t \leq -1$;
$d2$. $\underline{\lim}_{t\to\infty} \chi_t^{-1} d_t \geq -1$;
$\beta 1$. $\overline{\lim}_{t\to\infty} \chi_t^{-1} \ln \beta(\delta_t^{+,\alpha_t}) \leq -1$;
$\beta 2$. $\underline{\lim}_{t\to\infty} \chi_t^{-1} \ln \beta(\delta_t^{+,\alpha_t}) \geq -1$,

where χ_t is the normalizing function from the law of large numbers $\Lambda 1$.

The following theorem describes the relationship among the behaviors of α_t, d_t, and $\beta(\delta_t^{+,\alpha_t})$ under conditions $\Lambda 1'$ and $\Lambda 1''$.

THEOREM 2.3.1. *The following implications are valid:*

$$\Lambda 1', \alpha 1 \Rightarrow d1 \Rightarrow \beta 1, \tag{2.3.6}$$

$$\Lambda 1'', \alpha 2 \Rightarrow \beta 2 \Rightarrow d2. \tag{2.3.7}$$

PROOF. The implication $\Lambda 1', \alpha 1 \Rightarrow d1$ will be proved by contradiction. From the representation (2.1.13) we have

$$\alpha_t = P^t(Y_t > y_t) + \varepsilon_t P^t(Y_t = y_t), \tag{2.3.8}$$

where $Y_t = \chi_t^{-1} \Lambda_t$ and $y_t = \chi_t^{-1} d_t$.

Let (t_n) be a sequence such that $t_n \to \infty$ and $y_{t_n} \to \bar{y} = \overline{\lim}_{t\to\infty} y_t$ as $n \to \infty$. Let us assume that $\underline{\alpha} = \underline{\lim}_{t\to\infty} \alpha_t > 0$ and condition $\Lambda 1'$ is fulfilled but $\bar{y} > -1$. Then, in view of condition $\Lambda 1'$, we get $P^{t_n}(Y_{t_n} \geq y_{t_n}) \to 0$ as $n \to \infty$. On the other hand, the inequality $\underline{\alpha} > 0$ and equality (2.3.8) imply

$$\lim_{n\to\infty} P^{t_n}(Y_{t_n} \geq y_{t_n}) \geq \underline{\lim}_{n\to\infty} \alpha_{t_n} \geq \underline{\lim}_{t\to\infty} \alpha_t = \underline{\alpha} > 0.$$

The obtained contradiction proves the implication $\Lambda 1', \alpha 1 \Rightarrow d1$.

Using the equality (2.1.8), we get the estimate

$$\begin{aligned} \beta\left(\delta_t^{+,\alpha_t}\right) &= \widetilde{\mathsf{E}}^t\left(1 - \delta_t^{+,\alpha_t}\right) \\ &= \mathsf{E}^t\left(1 - \delta_t^{+,\alpha_t}\right) z_t + \widetilde{\mathsf{E}}^t\left(1 - \delta_t^{+,\alpha_t}\right) \chi(z_t = \infty) \\ &= \mathsf{E}^t\left(1 - \delta_t^{+,\alpha_t}\right) z_t \leq e^{d_t}, \end{aligned} \tag{2.3.9}$$

which gives the implications $d1 \Rightarrow \beta 1$ and $\beta 2 \Rightarrow d2$.

Now we prove the implication $\Lambda 1'', \alpha 2 \Rightarrow \beta 2$. Let $a > 1$ and $\gamma > 0$ be arbitrary numbers, and let $S_t = \left\{\chi_t^{-1} \Lambda_t \geq -a\right\} \subset X^t$. Then, in view of condition $\Lambda 1''$, there exists a number $t_0 = t_0(a, \gamma)$, depending on a and γ, such that $P^t(S_t^c) < \gamma$ for all $t > t_0$. From this we obtain, with due regard for (2.1.8),

$$\begin{aligned} \beta\left(\delta_t^{+,\alpha_t}\right) &= \mathsf{E}^t z_t\left(1 - \delta_t^{+,\alpha_t}\right) \geq \mathsf{E}^t \chi(S_t) z_t\left(1 - \delta_t^{+,\alpha_t}\right) \\ &\geq e^{-a\chi_t} \mathsf{E}^t \chi(S_t)\left(1 - \delta_t^{+,\alpha_t}\right) \\ &\geq e^{-a\chi_t}(1 - \alpha_t - \gamma) \end{aligned} \tag{2.3.10}$$

for $t > t_0$. From this estimate, by condition $\alpha 2$ and arbitrariness of $a > 1$ and $\gamma > 0$, the implication $\Lambda 1'', \alpha 2 \Rightarrow \beta 2$ follows. □

From the implications (2.3.6) and (2.3.7) the following assertion follows.

COROLLARY 2.3.1. *If the law of large numbers $\Lambda 1$ is fulfilled and the level α_t satisfies conditions $\alpha 1$ and $\alpha 2$, then the following relation is true:*

$$\lim_{t\to\infty} \chi_t^{-1} \ln \beta\left(\delta_t^{+,\alpha_t}\right) = -1. \tag{2.3.11}$$

Corollary 2.3.1 contains the following well-known result [109, 185].

COROLLARY 2.3.2. *Let $\xi^t = (\xi_1, \xi_2, \ldots, \xi_t)$, where $\xi_1, \xi_2, \ldots, \xi_t$ are independent identically distributed random variables under the hypothesis H^t and under the hypothesis $\widetilde{H}^t$ with t-independent distributions and finite positive relative entropy $a = -\mathsf{E}^1 \Lambda_1$. Then, if $\alpha_t \to \alpha \in (0,1)$ as $t \to \infty$, we have*

$$\lim_{t\to\infty} \frac{1}{t} \ln \beta\left(\delta_t^{+,\alpha_t}\right) = -a. \tag{2.3.12}$$

PROOF. It is evident that under the hypothesis H^t the quantity Λ_t is the sum of independent identically distributed random variables with finite mathematical expectation $\mathsf{E}^1 \Lambda_1 = -a$. Therefore the law of large numbers $\Lambda 1$ is fulfilled with $\chi_t = at$. Since $\alpha_t \to \alpha \in (0,1)$, the conditions of Corollary 2.3.1 are fulfilled. Hence the relation (2.3.12) follows from the relation (2.3.11). □

If under the conditions of Corollary 2.3.2 $\alpha_t = \alpha \in (0,1)$ for all t, then Corollary 2.3.2 turns into the well-known Stein lemma [**130, 135**].

Large deviations. It is seen from the implications (2.3.6) and (2.3.7) that under conditions $\Lambda 1'$ and $\Lambda 1''$ the relations $\beta 1$ and $\beta 2$ are fulfilled under conditions $\alpha 1$ and $\alpha 2$, which prevent the level α_t from tending to 0 or 1, respectively, as $t \to \infty$. However, under more severe constraints on the likelihood ratio z_t, the relations $\beta 1$ and $\beta 2$ can also be obtained for the level α_t tending to 0 or 1, as $t \to \infty$, in a sufficiently slow manner. To give more exact formulations, we introduce the following conditions:

$\alpha 1'$. $\lim_{t\to\infty} \chi_t^{-1} \ln \alpha_t = 0$;

$\alpha 2'$. $\lim_{t\to\infty} \chi_t^{-1} \ln(1-\alpha_t) = 0$;

$\Lambda 2$. $\overline{\lim}_{\varepsilon\downarrow 0} \overline{\lim}_{t\to\infty} \varepsilon^{-1} \chi_t^{-1} \ln H_t(\varepsilon) \le -1$;

$\Lambda 3$. $\underline{\lim}_{\varepsilon\uparrow 0} \underline{\lim}_{t\to\infty} \varepsilon^{-1} \chi_t^{-1} \ln H_t(\varepsilon) \ge -1$,

where χ_t is the normalizing function from condition $\Lambda 1$ and $H_t(\varepsilon) = H(\varepsilon; \widetilde{P}^t, P^t)$ is the Hellinger integral of order ε for the measures $\widetilde{P}^t$ and P^t.

In this section and subsequent discussion we shall need some properties of the Hellinger integral $H_t(\varepsilon)$. Observe that from the definition of $H_t(\varepsilon)$ we have

$$H_t(\varepsilon) = H(\varepsilon; \widetilde{P}^t, P^t) = \begin{cases} \widetilde{P}^t(\mathfrak{z}_t > 0), & \varepsilon = 1, \\ P^t(\tilde{\mathfrak{z}}_t > 0), & \varepsilon = 0, \\ \mathsf{E}_Q^t\, \tilde{\mathfrak{z}}_t^{\varepsilon} \mathfrak{z}_t^{1-\varepsilon}, & \varepsilon \ne 0,\ \varepsilon \ne 1. \end{cases} \tag{2.3.13}$$

We introduce the following notation:

$$\varepsilon_t^- = \inf\{\varepsilon\colon H_t(\varepsilon) < \infty\}, \qquad \varepsilon_t^+ = \sup\{\varepsilon\colon H_t(\varepsilon) < \infty\}. \tag{2.3.14}$$

LEMMA 2.3.2. *For any* $\varepsilon \in (\varepsilon_t^-, \varepsilon_t^+)$, *different from* 0 *or* 1, *we have*

$$H_t(\varepsilon) = \mathsf{E}^t\, z_t^\varepsilon = \widetilde{\mathsf{E}}^t \widetilde{z}_t^{1-\varepsilon}. \tag{2.3.15}$$

PROOF. First, let $0 < \varepsilon < 1$. Then

$$H_t(\varepsilon) = \mathsf{E}_Q^t\, \tilde{\mathfrak{z}}_t^\varepsilon \mathfrak{z}_t^{1-\varepsilon} \chi(\mathfrak{z}_t > 0) = \mathsf{E}^t\, z_t^\varepsilon \chi(\mathfrak{z}_t > 0) = \mathsf{E}^t\, z_t^\varepsilon,$$
$$H_t(\varepsilon) = \mathsf{E}_Q^t\, \tilde{\mathfrak{z}}_t^\varepsilon \mathfrak{z}_t^{1-\varepsilon} \chi(\tilde{\mathfrak{z}}_t > 0) = \widetilde{\mathsf{E}}^t \widetilde{z}_t^{1-\varepsilon} \chi(\tilde{\mathfrak{z}}_t > 0) = \widetilde{\mathsf{E}}^t \widetilde{z}_t^{1-\varepsilon},$$

and equalities (2.3.15) are established for $0 < \varepsilon < 1$.

Now let $\varepsilon_t^- < 0$ and $\varepsilon \in (\varepsilon_t^-, 0)$. Since, in view of (2.3.14), $H_t(\varepsilon) < \infty$, we have $\mathsf{E}_Q^t\, \tilde{\mathfrak{z}}_t^\varepsilon \mathfrak{z}^{1-\varepsilon} \chi(\tilde{\mathfrak{z}}_t = 0, \mathfrak{z}_t > 0) < \infty$. From this it follows that

$$Q^t(\tilde{\mathfrak{z}}_t = 0, \mathfrak{z}_t > 0) = 0$$

and hence $P^t(\tilde{\mathfrak{z}}_t = 0) = 0$. Taking this fact into account, we get

$$H_t(\varepsilon) = \mathsf{E}_Q^t\, \tilde{\mathfrak{z}}_t^\varepsilon \mathfrak{z}_t^{1-\varepsilon} \chi(\tilde{\mathfrak{z}}_t > 0) = \mathsf{E}^t (\mathfrak{z}_t/\tilde{\mathfrak{z}}_t)^{-\varepsilon} = \mathsf{E}^t\, z_t^\varepsilon,$$
$$H_t(\varepsilon) = \mathsf{E}_Q^t\, \tilde{\mathfrak{z}}_t \tilde{\mathfrak{z}}^{\varepsilon-1} \mathfrak{z}_t^{1-\varepsilon} \chi(\tilde{\mathfrak{z}}_t > 0) = \widetilde{\mathsf{E}}^t \widetilde{z}_t^{1-\varepsilon} \chi(\tilde{\mathfrak{z}}_t > 0) = \widetilde{\mathsf{E}}^t \widetilde{z}_t^{1-\varepsilon}.$$

Therefore, equalities (2.3.15) are proved for $\varepsilon \in (\varepsilon_t^-, 0)$ and $\varepsilon_t^- < 0$.

Finally, let $\varepsilon_t^+ > 1$ and $\varepsilon \in (1, \varepsilon_t^+)$. Since $H_t(\varepsilon) < \infty$, we have

$$\mathsf{E}_Q^t\, \tilde{\mathfrak{z}}_t^\varepsilon \mathfrak{z}_t^{1-\varepsilon} \chi(\tilde{\mathfrak{z}}_t > 0, \mathfrak{z}_t = 0) < \infty.$$

Therefore, $Q^t(\tilde{\mathfrak{z}}_t > 0, \mathfrak{z}_t = 0) = 0$ and $\widetilde{P}^t(\mathfrak{z}_t = 0)$. Hence

$$H_t(\varepsilon) = \widetilde{\mathsf{E}}_Q^t \tilde{\mathfrak{z}}_t^{\varepsilon-1} \mathfrak{z}_t^{1-\varepsilon} \chi(\mathfrak{z}_t > 0) = \widetilde{\mathsf{E}}^t (\tilde{\mathfrak{z}}_t/\mathfrak{z}_t)^{\varepsilon-1} = \widetilde{\mathsf{E}}^t \widetilde{z}_t^{1-\varepsilon},$$
$$H_t(\varepsilon) = \mathsf{E}_Q^t\, \mathfrak{z}_t \tilde{\mathfrak{z}}^\varepsilon \mathfrak{z}_t^{-\varepsilon} \chi(\mathfrak{z}_t > 0) = \mathsf{E}^t\, z_t^\varepsilon \chi(\mathfrak{z}_t > 0) = \mathsf{E}^t\, z_t^\varepsilon,$$

and the equalities (2.3.15) are proved also for $\varepsilon \in (1, \varepsilon_t^+)$ and $\varepsilon_t^+ > 1$. □

REMARK 2.3.3. If $\varepsilon_t^- < 0$, then, as established in the proof of Lemma 2.3.2, $P^t(\tilde{\mathfrak{z}}_t = 0) = 0$. In view of (2.1.7), this means that $P^t \ll \widetilde{P}^t$. Moreover, by (2.3.13), in this case we have $H(0; \widetilde{P}^t, P^t) = 1$. If $\varepsilon_t^+ > 1$, then also in the proof of Lemma 2.3.2 it was established that $\widetilde{P}^t(\mathfrak{z}_t = 0) = 0$. By (2.1.6), this means that $\widetilde{P}^t \ll P^t$. From (2.3.13) it follows that in this case $H(1; \widetilde{P}^t, P^t) = 1$.

REMARK 2.3.4. Condition $\Lambda 3$ implies that there exist $\varepsilon_0 < 0$ and $t_0 < \infty$ such that $H_t(\varepsilon) < \infty$ for all $\varepsilon \in (\varepsilon_0, 0)$ and $t > t_0$. Therefore, in view of Remark 2.3.3,

$$\Lambda 3 \Rightarrow P^t \ll \widetilde{P}^t \quad \text{for all } t > t_0.$$

In view of (2.1.7), from this we have $\Lambda 3 \Rightarrow \overline{\alpha}_t = 1$ for all $t > t_0$.

The following lemma describes the relationship between conditions $\Lambda 1'$, $\Lambda 1''$, $\alpha 1$, and $\alpha 2$ and conditions $\Lambda 2$, $\Lambda 3$, $\alpha 1'$, and $\alpha 2'$.

LEMMA 2.3.3. *The following implications hold:*

$$\alpha 1 \Rightarrow \alpha 1'; \qquad \alpha 2 \Rightarrow \alpha 2'; \tag{2.3.16}$$
$$\Lambda 2 \Rightarrow \Lambda 1'; \qquad \Lambda 3 \Rightarrow \Lambda 1''. \tag{2.3.17}$$

PROOF. The implications (2.3.16) are evident. First we prove the implication $\Lambda 2 \Rightarrow \Lambda 1'$. Let $\gamma > 0$ be an arbitrary number. Then, by condition $\Lambda 2$, there exists a positive number $\varepsilon_0 = \varepsilon_0(\gamma)$ such that for all $\varepsilon \in (0, \varepsilon_0)$

$$\overline{\lim_{t\to\infty}} \chi_t^{-1} \ln H_t(\varepsilon) \le -\varepsilon + \frac{1}{2}\varepsilon\gamma. \tag{2.3.18}$$

Fix some $\varepsilon \in (0, \varepsilon_0)$. Then, in view of (2.3.18), there exists $t_0 = t_0(\varepsilon, \gamma)$ such that for all $t > t_0$

$$\chi_t^{-1} \ln H_t(\varepsilon) \le -\varepsilon + \frac{1}{2}\varepsilon\gamma + \frac{1}{4}\varepsilon\gamma.$$

From this, applying the Chebyshev inequality, we get

$$\begin{aligned} P^t\left(\chi_t^{-1}\Lambda_t > -1 + \gamma\right) &= P^t\left(z_t^\varepsilon > e^{-\varepsilon(1-\gamma)\chi_t}\right) \le e^{\varepsilon(1-\gamma)\chi_t}\, \mathsf{E}^t z_t^\varepsilon \\ &= e^{\varepsilon(1-\gamma)\chi_t} H_t(\varepsilon) \le \exp\left(-\frac{1}{4}\varepsilon\gamma\chi_t\right) \end{aligned} \tag{2.3.19}$$

for $t > t_0$. Here we have made use of the equality (2.3.15) for $\varepsilon \in (0, 1)$. The estimate (3.3.19) justifies condition $\Lambda 1'$, that is, the implication $\Lambda 2 \Rightarrow \Lambda 1'$ is proved. The second implication in (2.3.17) can be proved similarly, by using Remark 2.3.4 and Lemma 2.3.2. □

Conditions $\Lambda 2$ and $\Lambda 3$ are closely related to large deviations for Λ_t. To see this, we introduce the following condition.

Λ^*. There exists an interval $(\varepsilon_-, \varepsilon_+)$ that contains 0 and such that for every $\varepsilon \in (\varepsilon_-, \varepsilon_+)$ there exists a finite limit

$$\lim_{t\to\infty} \widetilde{\chi}_t^{-1} \ln H_t(\varepsilon) = \Psi(\varepsilon),$$

where $\widetilde{\chi}_t \to \infty$ as $t \to \infty$ and $\Psi(\varepsilon)$ is a strictly convex differentiable function on $(\varepsilon_-, \varepsilon_+)$.

We introduce the notation

$$\gamma_0 = \Psi'(0), \qquad \gamma_- = \lim_{\varepsilon\downarrow\varepsilon_-} \Psi'(\varepsilon), \qquad \gamma_+ = \lim_{\varepsilon\uparrow\varepsilon_+} \Psi'(\varepsilon).$$

The following theorem on large deviations for Λ_t is true. This theorem is a "hybrid" of theorems from [**136, 137, 143, 144, 189, 191**], where the reader can find its versions with somewhat weaker conditions.

LEMMA 2.3.4 (Theorem on large deviations). *Let condition Λ^* be fulfilled. Then for any $\gamma \in (\gamma_0, \gamma_+)$,*

$$\lim_{t\to\infty} \widetilde{\chi}_t^{-1} \ln P^t\left(\widetilde{\chi}_t^{-1}\Lambda_t > \gamma\right) = -I(\gamma) \in (-\infty, 0),$$

and for any $\gamma \in (\gamma_-, \gamma_0)$

$$\lim_{t\to\infty} \widetilde{\chi}_t^{-1} \ln P^t\left(\widetilde{\chi}_t^{-1}\Lambda_t < \gamma\right) = -I(\gamma) \in (-\infty, 0),$$

where $I(\gamma) = \gamma\varepsilon_\gamma - \Psi(\varepsilon_\gamma)$ and ε_γ is the unique solution of the equation $\Psi'(\varepsilon) = \gamma$.

Evidently, if condition Λ^* is fulfilled, then conditions $\Lambda 2$ and $\Lambda 3$ with $\chi_t = -\gamma_0\widetilde{\chi}_t$, where $\gamma_0 < 0$, are also fulfilled. Thus conditions $\Lambda 2$ and $\Lambda 3$ are somewhat weaker than condition Λ^*, which ensures the fulfillment of the theorem on large deviations. For this reason we connect conditions $\Lambda 2$ and $\Lambda 3$ with the theorem on large deviations for Λ_t.

The following theorem provides upper and lower estimates for $\beta(\delta_t^{+,\alpha})$ for all $\alpha \in (0,1)$, which makes it possible to obtain $\beta 1$ and $\beta 2$ if conditions $\alpha 1'$ and $\alpha 2'$ are fulfilled.

THEOREM 2.3.2. *For all* $\alpha \in (0,1)$ *and* $t \in \mathbf{R}_+$ *we have*

$$\beta(\delta_t^{+,\alpha}) \geq (1-\alpha)^{\varepsilon/(\varepsilon-1)}(H_t(1-\varepsilon))^{1/(1-\varepsilon)}, \qquad \varepsilon > 1, \tag{2.3.20}$$

$$\beta(\delta_t^{+,\alpha}) \leq (1-\varepsilon)(\varepsilon/\alpha)^{\varepsilon/(1-\varepsilon)}(H_t(1-\varepsilon))^{1/(1-\varepsilon)}, \qquad 0 < \varepsilon < 1. \tag{2.3.21}$$

PROOF. If $H_t(1-\varepsilon) = \infty$ for $\varepsilon > 1$, then the estimate (2.3.20) is trivial. Because of this we assume that $H_t(1-\varepsilon) < \infty$ for $\varepsilon > 1$. In view of (2.1.16), for any $c \in [0,\infty)$ we have

$$c\alpha + \beta\left(\delta_t^{+,\alpha}\right) \geq c\alpha\left(\delta_t^{*,c}\right) + \beta\left(\delta_t^{*,c}\right) = 1 - \mathsf{E}_Q^t\left(\tilde{\mathfrak{z}}_t - c\mathfrak{z}_t\right)^+. \tag{2.3.22}$$

By (2.1.16) and the fundamental Neyman–Pearson lemma, there exists $c \in [0,\infty)$ such that inequality (2.3.32) turns into an equality. Therefore

$$\beta(\delta_t^{*,c}) = 1 - \inf\left\{c\alpha + \mathsf{E}_Q^t(\tilde{\mathfrak{z}}_t - c\mathfrak{z}_t)^+; c \geq 0\right\}. \tag{2.3.23}$$

Condition $H_t(1-\varepsilon) < \infty$ for $\varepsilon > 1$ implies that $Q^t(\tilde{\mathfrak{z}}_t = 0) = 0$. Hence we have

$$\mathsf{E}_Q^t(\tilde{\mathfrak{z}}_t - c\mathfrak{z}_t)^+ = \widetilde{\mathsf{E}}^t(1 \vee cz_t) - c. \tag{2.3.24}$$

Taking into account the inequality

$$1 \vee z \leq az^\varepsilon + 1, \qquad \varepsilon > 1,\ z \geq 0,\ a = \varepsilon^{-\varepsilon}(\varepsilon-1)^{\varepsilon-1},$$

and (2.3.15), from (3.3.23) and (2.3.24) we get

$$\beta\left(\delta_t^{+,\alpha}\right) \geq \sup\bigl\{(1-\alpha)c - ac^\varepsilon H_t(1-\varepsilon); c \geq 0\bigr\} \tag{2.3.25}$$

for $\varepsilon > 1$. On the right-hand side of (2.3.25), the upper bound is attained at

$$c = c^* = (1-\alpha)^{1/(\varepsilon-1)}(a\varepsilon H_t(1-\varepsilon))^{1/(1-\varepsilon)}. \tag{2.3.26}$$

Therefore (2.3.25) and (2.3.26) imply (2.3.20).

Now we establish (2.3.21). Similarly to (2.3.24), we get

$$\mathsf{E}_Q^t(\tilde{\mathfrak{z}}_t - c\mathfrak{z}_t)^+ = 1 - \widetilde{\mathsf{E}}^t(1 \wedge c\tilde{z}_t). \tag{2.3.27}$$

Taking into account the elementary inequality

$$z \wedge 1 \leq z^\varepsilon, \qquad z \geq 0,\ 0 < \varepsilon < 1, \tag{2.3.28}$$

and (2.3.15), from (2.3.23) and (2.3.27) we obtain

$$\beta\left(\delta_t^{+,\alpha}\right) \leq \sup\bigl\{c^\varepsilon H_t(1-\varepsilon) - c\alpha; c \geq 0\bigr\} \tag{2.3.29}$$

for $0 < \varepsilon < 1$. On the right-hand side of (2.3.29), the upper bound is attained at

$$c = c^* = \alpha^{1/(\varepsilon-1)}(\varepsilon H_t(1-\varepsilon))^{1/(1-\varepsilon)}. \tag{2.3.30}$$

Therefore (2.3.29) and (2.3.30) imply the estimate (2.3.21). □

COROLLARY 2.3.3. *For any $\varepsilon, \alpha \in (0,1)$, and $t \in \mathbf{R}_+$*

$$H_t(\varepsilon) \geq \beta^{\varepsilon}\left(\delta_t^{+,\alpha}\right)\alpha^{1-\varepsilon}\exp(H(\varepsilon)),$$

where $H(\varepsilon) = -\varepsilon \ln \varepsilon - (1-\varepsilon)\ln(1-\varepsilon)$ is the Shannon entropy of the distribution of the random variable taking two values with probabilities ε and $1-\varepsilon$ [**106**].

REMARK 2.3.5. If $\overline{\alpha}_t < 1$, then, in view of (2.1.7) and Remark 2.3.3, we have $H_t(\varepsilon) = \infty$ for all $\varepsilon < 0$. Therefore, if $\overline{\alpha}_t < 1$, then the estimate (2.3.20) takes the trivial form $\beta(\delta_t^{+,\alpha}) \geq 0$ for all $\alpha \in (0,1)$. Here it should be observed that, in view of (2.1.12), $\beta(\delta_t^{+,\alpha}) = 0$ for all $\alpha \in [\overline{\alpha}_t, 1]$. As regards the estimate (2.3.21), it should be noted that this estimate is rough for $\alpha \in [\overline{\alpha}_t, 1]$. However, in view of (2.3.4), under condition $\Lambda 1''$ we have $\overline{\alpha}_t \to 1$ as $t \to \infty$. Moreover, in view of Remark 2.3.4, under condition $\Lambda 3$ there exists $t_0 < \infty$ such that $\overline{\alpha}_t = 1$ for all $t > t_0$.

THEOREM 2.3.3. *The following implications are true*:

$$\Lambda 2, \alpha 1' \Rightarrow d1 \Rightarrow \beta 1, \tag{2.3.31}$$

$$\Lambda 3, \alpha 2' \Rightarrow \beta 2 \Rightarrow d2. \tag{2.3.32}$$

PROOF. In view of Theorem 2.3.1, it suffices to prove the left implications in (2.3.31) and (2.3.32). Taking into account the equality in (2.3.22), the equality (2.3.27), and the inequality (2.3.28), for any $c \geq 0$ we have

$$c\alpha\left(\delta_t^{*,c}\right) + \beta\left(\delta_t^{*,c}\right) \leq c^{\varepsilon}\widetilde{\mathsf{E}}^t \widetilde{z}_t^{\varepsilon}, \qquad 0 < \varepsilon < 1.$$

In view of (2.1.15) and (2.3.16), from this we get

$$e^{d_t}\alpha_t + \beta\left(\delta_t^{+,\alpha_t}\right) \leq e^{\varepsilon d_t} H_t(1-\varepsilon)$$

for $0 < \varepsilon < 1$. Therefore for $0 < \varepsilon < 1$ we have

$$d_t \leq (1-\varepsilon)^{-1}\ln H_t(1-\varepsilon) - (1-\varepsilon)^{-1}\ln \alpha_t, \tag{2.3.33}$$

which gives the implication $\Lambda 2, \alpha 1' \Rightarrow d1$.

The implication $\Lambda 3, \alpha 2' \Rightarrow \beta 2$ follows from the estimate (2.3.20). □

REMARK 2.3.6. The implication $\Lambda 2, \alpha 1' \Rightarrow \beta 1$ can be immediately obtained from the estimate (2.3.21).

Theorem 2.3.3 implies the well-known result due to Krafft and Plachky [**164**].

COROLLARY 2.3.4. *If under the conditions of Corollary 2.3.2 conditions $\alpha 1'$ and $\alpha 2'$ are fulfilled with $\chi_t = at$, then (2.3.12) holds.*

2.4. Complete asymptotic distinguishability under conditions of weak convergence

Consider two cases of completely distinguishable families of hypotheses (H^t) and $(\widetilde{H}^t)$:

1) the law of large numbers is not fulfilled for Λ_t;
2) the law of large numbers is fulfilled for Λ_t, and, moreover, $(\Lambda_t - A_t)/\psi_t$ weakly converges for some A_t and ψ_t.

The case where the law of large numbers is not fulfilled. Consider the following condition:

Λ4. $\mathcal{L}(\psi_t^{-1}\Lambda_t|P^t) \xrightarrow{w} L$ as $t \to \infty$, where ψ_t is a positive function such that $\psi_t \to \infty$ as $t \to \infty$ and L is a certain probability distribution on $\mathbf{R}^1$ with distribution function $L(x)$; $\mathcal{L}(\eta_t|P^t)$ is the distribution law of η_t with respect to the measure P^t, and $\xrightarrow{w}$ stands for the weak convergence of distribution laws.

LEMMA 2.4.1. *If condition Λ4 is fulfilled, then* $\overline{\alpha}_t \to 1$ *as* $t \to \infty$ *and* $L(x) = 1$ *for any* $x > 0$. *Moreover, if condition Λ4 is fulfilled, then*

$$(H^t) \vartriangle \left(\widetilde{H}^t\right) \Leftrightarrow L(0) = 1. \tag{2.4.1}$$

PROOF. For any $a \in (-\infty, 0)$ we have

$$P^t\left(\psi_t^{-1}\Lambda_t < a\right) \geq P^t(\Lambda_t = -\infty) = 1 - \overline{\alpha}_t.$$

If a is a continuity point of the function $L(x)$, then, in view of condition Λ4, we have $L(a) \geq \overline{\lim}_{t\to\infty}(1 - \overline{\alpha}_t)$. Now, passing to the limit over continuity points $a \to -\infty$, we get $\overline{\alpha}_t \to 1$ as $t \to \infty$.

Putting $a > 0$ and applying the Chebyshev inequality, we obtain

$$P^t\left(\psi_t^{-1}\Lambda_t > a\right) = P^t\left(z_t > e^{a\psi_t}\right) \leq e^{-a\psi_t}\,\mathsf{E}^t\, z_t \leq e^{-a\psi_t}.$$

Passing to the limit as $t \to \infty$, we have $1 - L(a) = 0$ and hence $L(a) = 1$ for any $a > 0$.

By condition Λ4 we have

$$P^t(\Lambda_t > N) = P^t\left(\frac{\Lambda_t}{\psi_t} > \frac{N}{\psi_t}\right) \to \begin{cases} 1 - L(0), & N < 0, \\ 1 - L(0+), & N > 0 \end{cases}$$

as $t \to \infty$. Since $L(0+) = 1$, by Theorem 2.2.1 we get the required equivalence (2.4.1). □

The following theorem describes the behavior of tests δ_t^{+,α_t} as $t \to \infty$ in the case where condition Λ4 is fulfilled. This theorem is an analog of Theorem 2.3.1.

THEOREM 2.4.1. *If condition Λ4 is fulfilled, then for any* $\alpha \in (0,1)$ *the following implications are valid:*

$$\lim_{t\to\infty} \alpha_t = \alpha \Rightarrow \overline{\lim_{t\to\infty}} \frac{d_t}{\psi_t} \leq \bar{l}_{1-\alpha} \Rightarrow \overline{\lim_{t\to\infty}} \frac{\ln \beta(\delta_t^{+,\alpha_t})}{\psi_t} \leq \bar{l}_{1-\alpha}, \tag{2.4.2}$$

$$\lim_{t\to\infty} \alpha_t = \alpha \Rightarrow \underline{\lim_{t\to\infty}} \frac{\ln \beta(\delta_t^{+,\alpha_t})}{\psi_t} \geq \underline{l}_{1-\alpha} \Rightarrow \underline{\lim_{t\to\infty}} \frac{d_t}{\psi_t} \geq \underline{l}_{1-\alpha}, \tag{2.4.3}$$

where

$$\bar{l}_p = \inf\{u\colon L(u) > p\}, \qquad \underline{l}_p = \sup\{u\colon L(u) < p\}, \qquad p \in (0,1). \tag{2.4.4}$$

PROOF. We have

$$\alpha_t = P^t(Y_t > y_t) + \varepsilon_t P^t(Y_t = y_t), \tag{2.4.5}$$

where $Y_t = \psi_t^{-1}\Lambda_t$ and $y_t = \psi_t^{-1}d_t$.

First we prove the left implication in (2.4.2). Let $\alpha_t \to \alpha$ but $\overline{\lim}_{t\to\infty} y_t = \bar{y} > u$, where u is a continuity point of the function $L(x)$ such that $L(u) > 1-\alpha$. Let t_n be a sequence such that $y_{t_n} \to \bar{y}$ as $n \to \infty$. Then

$$\alpha = \lim_{t\to\infty} \alpha_t \le \overline{\lim_{n\to\infty}}\, P^{t_n}(Y_{t_n} \ge y_{t_n}) \le \overline{\lim_{n\to\infty}}\, P^{t_n}(Y_{t_n} \ge u) = 1 - L(u) < \alpha.$$

Thus we have arrived at the contradiction that proves that $\overline{\lim}_{t\to\infty} y_t \le u$ for any continuity point u. From this the left implication in (2.4.2) follows. The right implications in (2.4.2) and (2.4.3) follow from (2.3.9).

Let $\alpha_t \to \alpha$ as $t \to \infty$, and let $u < 0$ be a continuity point of the function $L(x)$ such that $L(u) < 1-\alpha$. Using the Lebesgue decomposition, we get

$$\begin{aligned}\beta\left(\delta_t^{+,\alpha_t}\right) &\ge \widetilde{\mathsf{E}}^t \chi(Y_t \ge u)\left(1-\delta_t^{+,\alpha_t}\right) \ge \mathsf{E}^t\,\chi(Y_t \ge u)\left(1-\delta_t^{+,\alpha_t}\right) z_t \\ &\ge e^{u\psi_t}\,\mathsf{E}^t\,\chi(Y_t \ge u)\left(1-\delta_t^{+,\alpha_t}\right) \ge e^{u\psi_t}\left[P^t(Y_t \ge u) - \alpha_t\right].\end{aligned}$$

From this it follows that $\underline{\lim}_{t\to\infty} \psi_t^{-1} \ln \beta(\delta_t^{+,\alpha_t}) \ge u$, which yields the left implication in (2.4.3). □

If the distribution function $L(x)$ in condition $\Lambda 4$ is continuous, then the assertion of Theorem 2.4.1 can be made more precise. First, we give the following definition.

DEFINITION 2.4.1. A solution $x_p \in \mathbf{R}^1$ of two inequalities $S(x_p) \le p \le S(x_p+0)$ is called a *quantile* of order $p \in [0,1]$ or, briefly, a *p-quantile* of a probability law S defined on $\mathbf{R}^1$ or $\overline{\mathbf{R}}^1$ with distribution function $S(x)$. If for some $p \in [0,1]$ the inequality $S(x) > p$ is fulfilled for all $x \in \mathbf{R}^1$, then we put $x_p = -\infty$; if $S(x) < p$ for all $x \in \mathbf{R}^1$, then we put $x_p = \infty$.

Note that if L is a probability law from condition $\Lambda 4$, then, evidently, for $p \in (0,1)$ a p-quantile l_p of the law L and quantities $\underline{l}_p$ and $\bar{l}_p$ defined by the equalities (2.4.4) are subject to the relation $\underline{l}_p \le l_p \le \bar{l}_p$.

THEOREM 2.4.2. *Let condition $\Lambda 4$ be fulfilled, in which $L(x)$ is a continuous strictly monotone increasing function on the interval $(\underline{l}, \bar{l})$, where*

$$\underline{l} = \sup\{x\colon L(x) = 0\} \quad \textit{and} \quad \bar{l} = \inf\{x\colon L(x) = 1\}.$$

Then for any $\alpha \in (0,1)$

$$\lim_{t\to\infty} \alpha_t = \alpha \Leftrightarrow \lim_{t\to\infty} \frac{d_t}{\psi_t} = l_{1-\alpha} \Leftrightarrow \lim_{t\to\infty} \frac{\ln \beta\left(\delta_t^{+,\alpha_t}\right)}{\psi_t} = l_{1-\alpha}, \tag{2.4.6}$$

where $l_p = \underline{l}_p = \bar{l}_p$ is a p-quantile of the law L for $p \in (0,1)$ and $\underline{l}_p$ and $\bar{l}_p$ are defined by equalities (2.4.4).

PROOF. Since $L(x)$ is a continuous strictly monotone function on the interval $(\underline{l}, \bar{l})$, we have $\underline{l}_p = \bar{l}_p$ for $p \in (0,1)$. Put $L_t(x) = P^t(Y_t < x)$, where $Y_t = \psi_t^{-1}\Lambda_t$. Then for any $\varepsilon > 0$ and $y_t \in \mathbf{R}^1$

(2.4.7)

$$\begin{aligned} P^t(Y_t = y_t) &\le L_t(y_t + \varepsilon) - L_t(y_t) \\ &= [L_t(y_t+\varepsilon) - L(y_t+\varepsilon)] - [L_t(y_t) - L(y_t)] + [L(y_t+\varepsilon) - L(y_t)]. \end{aligned}$$

By the Pólya lemma ([**109**]), we have

$$\lim_{t\to\infty} \sup_y |L_t(y) - L(y)| = 0. \tag{2.4.8}$$

Taking into account that the continuity of the function $L(x)$ implies its uniform continuity, from (2.4.7) and (2.4.8) we get

$$\lim_{t\to\infty} (Y_t = y_t) = 0. \tag{2.4.9}$$

Thus (2.4.5) and (2.4.9) imply

$$\lim_{t\to\infty} \alpha_t = \alpha \Leftrightarrow \lim_{t\to\infty} L_t(y_t) = 1 - \alpha. \tag{2.4.10}$$

In view of (2.4.8) and (2.4.10), we have

$$\lim_{t\to\infty} \alpha_t = \alpha \Leftrightarrow \lim_{t\to\infty} L(y_t) = 1 - \alpha.$$

Hence

$$\lim_{t\to\infty} \alpha_t = \alpha \Leftrightarrow \lim_{t\to\infty} y_t = l_{1-\alpha}. \tag{2.4.11}$$

By virtue of equality (2.1.8), for any $\varepsilon > 0$ we have

$$\begin{aligned} \beta\left(\delta_t^{+,\alpha_t}\right) &\ge \mathsf{E}^t z_t \left(1 - \delta_t^{+,\alpha_t}\right) \ge \mathsf{E}^t \chi(y_t - \varepsilon \le Y_t) z_t \left(1 - \delta_t^{+,\alpha_t}\right) \\ &\ge P^t(y_t - \varepsilon \le Y_t < y_t) \exp\big((y_t - \varepsilon)\psi_t\big). \end{aligned} \tag{2.4.12}$$

In view of the right implication in (2.4.2) and arbitrariness of ε, from (2.4.12) we get the implication

$$\lim_{t\to\infty} y_t = l_{1-\alpha} \Rightarrow \lim_{t\to\infty} \psi_t^{-1} \ln \beta\left(\delta_t^{+,\alpha_t}\right) = l_{1-\alpha}. \tag{2.4.13}$$

In view of the right implication in (2.4.3), in order to prove the implication inverse to (2.4.13), it suffices to show that

$$\lim_{t\to\infty} \psi_t^{-1} \ln \beta\left(\delta_t^{+,\alpha_t}\right) = l_{1-\alpha} \Rightarrow \overline{\lim_{t\to\infty}}\, y_t \le l_{1-\alpha}. \tag{2.4.14}$$

The assertion (2.4.14) can be proved by contradiction. Let

$$\lim_{t\to\infty} \psi_t^{-1} \ln \beta\left(\delta_t^{+,\alpha_t}\right) = l_{1-\alpha}, \qquad \overline{\lim_{t\to\infty}}\, y_t > l_{1-\alpha}, \tag{2.4.15}$$

and let (t_n) be a sequence such that $t_n \to \infty$ and $y_{t_n} \to \overline{\lim}_{t\to\infty} y_t$ as $n \to \infty$. Then, in view of (2.4.12),

$$\overline{\lim_{t\to\infty}}\, \psi_t^{-1} \ln \beta\left(\delta_t^{+,\alpha_t}\right) \ge \overline{\lim_{n\to\infty}}\, \psi_{t_n}^{-1} \ln \beta\big(\delta_{t_n}^{+,\alpha_{t_n}}\big) > l_{1-\alpha},$$

which contradicts the equality in (2.4.15). This contradiction proves the implication (2.4.14). Combining (2.4.11), (2.4.13), (2.4.14), and (2.4.3), we get (2.4.6). □

REMARK 2.4.1. If in condition $\Lambda 4$ the limit law L is degenerate and concentrated at some point $x_0 \in (-\infty, 0)$, then the law of large numbers $\Lambda 1$ is fulfilled and the assertions of the previous section are applicable. But if under condition $\Lambda 4$ the function $L(x)$ has a discontinuity at some point $x_0 \in (-\infty, 0)$ such that $0 < L(x_0) < L(x_0 + 0) < 1$, then Theorem 2.4.1 implies the following result. If $\alpha_t \to \alpha$ as $t \to \infty$, then for any $\alpha \in (1 - L(x_0 + 0), 1 - L(x_0))$ we have

$$\lim_{t\to\infty} \frac{d_t}{\psi_t} = l_{1-\alpha}, \qquad \lim_{t\to\infty} \frac{\ln \beta\left(\delta_t^{+,\alpha_t}\right)}{\psi_t} = l_{1-\alpha}, \tag{2.4.16}$$

where $l_{1-\alpha} = x_0$.

Moreover, if condition $\Lambda 4$ is fulfilled and $L^{-1}(\alpha) = \inf\{u \colon L(u) > \alpha\}$, where $\alpha \in [0, 1]$, is the inverse of the function $L(x)$, then Theorems 2.4.1 and 2.4.2 imply the following result. If $\alpha_t \to \alpha$ as $t \to \infty$, then for any continuity point $\alpha \in (0, 1)$ of the function $L^{-1}(\alpha)$ the relation (2.4.16) with $l_{1-\alpha} = L^{-1}(\alpha)$ holds.

REMARK 2.4.2. If the conditions of Theorem 2.4.2 are fulfilled and δ_t^{+,α_t} is the Neyman–Pearson test with limiting value $\alpha \in (0, 1)$ for the level α_t, then Theorem 2.4.2 implies

$$d_t = l_{1-\alpha}\psi_t + o(\psi_t), \qquad \ln \beta(\delta_t^{+,\alpha_t}) = l_{1-\alpha}\psi_t + o(\psi_t).$$

Here $l_{1-\alpha}$ strictly monotonically decreases from 0 to $-\infty$ as a increases from 0 to 1. Thus the rates of decrease of d_t and $\beta(\delta_t^{+,\alpha_t})$ depend on α and they are different for different α. This case is substantially different from the one where the law of large numbers is fulfilled and the rate of decrease of d_t and $\beta(\delta_t^{+,\alpha_t})$ does not depend on α (see Section 2.3).

Second order behavior of the Neyman–Pearson test. Consider the case where the law of large numbers is fulfilled for Λ_t. In this case the rate of decrease of the probability of errors of the second kind of the Neyman–Pearson test δ_t^{+,α_t} does not depend on $\alpha = \lim_{t\to\infty} \alpha_t$ if $\alpha \in (0, 1)$. Here we formulate a condition that allows us to find the second term of the asymptotic expansion of $\ln \beta(\delta_t^{+,\alpha_t})$ as $t \to \infty$. First, we introduce the following condition:

$\Lambda 5$. $\mathcal{L}\big(\varphi_t^{-1}(\Lambda_t + \psi_t) | P^t\big) \xrightarrow{w} L$ as $t \to \infty$, where φ_t and ψ_t are positive functions of t such that $\varphi_t \to \infty$ and $\psi_t \to \infty$ as $t \to \infty$ but $\varphi_t = o(\psi_t)$ and L is the distribution on $\mathbf{R}^1$ with distribution function $L(x)$.

It is evident that $\Lambda 5 \Rightarrow \Lambda 1$ with $\chi_t = \psi_t$.

THEOREM 2.4.3. *If condition $\Lambda 5$ is fulfilled, then for any $\alpha \in (0, 1)$ the following implications are true*:

$$\lim_{t\to\infty} \alpha_t = \alpha \Rightarrow \overline{\lim_{t\to\infty}} \frac{d_t + \psi_t}{\varphi_t} \le \bar{l}_{1-\alpha} \Rightarrow \overline{\lim_{t\to\infty}} \frac{\ln \beta\left(\delta_t^{+,\alpha_t}\right) + \psi_t}{\varphi_t} \le \bar{l}_{1-\alpha}, \tag{2.4.17}$$

$$\lim_{t\to\infty} \alpha_t = \alpha \Rightarrow \underline{\lim_{t\to\infty}} \frac{\ln \beta\left(\delta_t^{+,\alpha_t}\right) + \psi_t}{\varphi_t} \ge \underline{l}_{1-\alpha} \; \underline{\lim_{t\to\infty}} \frac{d_t + \psi_t}{\varphi_t} \ge \underline{l}_{1-\alpha}, \tag{2.4.18}$$

where $\underline{l}_p$ and $\bar{l}_p$ are defined in (2.4.4).

PROOF. It suffices to repeat the proof of Theorem 2.4.1 with appropriate modifications. First of all we have the equality (2.4.5), in which $Y_t = \varphi_t^{-1}(\Lambda_t + \psi_t)$ and $y_t = \varphi_t^{-1}(d_t + \psi_t)$. Furthermore, the left implication in (2.4.17) can be proved by repeating literally the proof of the left implication in (2.4.2). The right implications in (2.4.17) and (2.4.18) follow from the estimate (2.3.9). Now let $\alpha_t \to \alpha$ and $u \in \mathbf{R}^1$ be some continuity point of the function $L(x)$ such that $L(u) < 1 - \alpha$. In view of the Lebesgue decomposition, we have the estimate

$$\beta\left(\delta_t^{+,\alpha_t}\right) \geq \mathsf{E}^t\,\chi(Y_t \geq u)\left(1 - \delta_t^{+,\alpha_t}\right) z_t \geq \exp(u\varphi_t - \psi_t)\left[P^t(Y_t \geq u) - \alpha_t\right],$$

which implies the left implication in (2.4.18). □

THEOREM 2.4.4. *Let condition* $\Lambda 5$ *be fulfilled, in which the function* $L(x)$ *is continuous and strictly monotonically increasing on the interval* $(\underline{l}, \bar{l})$. *Then for any* $\alpha \in (0,1)$

$$\lim_{t\to\infty} \alpha_t = \alpha \Leftrightarrow \lim_{t\to\infty} \frac{d_t + \psi_t}{\varphi_t} = l_{1-\alpha} \Leftrightarrow \lim_{t\to\infty} \frac{\ln\beta\left(\delta_t^{+,\alpha_t}\right) + \psi_t}{\varphi_t} = l_{1-\alpha}, \tag{2.4.19}$$

where l_p *is a* p*-quantile of the law* L.

PROOF. It suffices to repeat the proof of Theorem 2.4.2 with appropriate modifications. First it is necessary to use the equality (2.4.5), in which $Y_t = \varphi_t^{-1}(\Lambda_t + \psi_t)$ and $y_t = \varphi_t^{-1}(d_t + \psi_t)$. Repeating word for word the proof of the equivalence (2.4.11), we get the left equivalence in (2.4.19).

Furthermore, for any $\varepsilon > 0$ we have

$$\begin{aligned}\beta\left(\delta_t^{+,\alpha_t}\right) &\geq \mathsf{E}^t\,\chi(y_t - \varepsilon \leq Y_t) z_t \left(1 - \delta_t^{+,\alpha_t}\right)\\ &\geq P^t(y_t - \varepsilon \leq Y_t < y_t)\exp\big((y_t - \varepsilon)\varphi_t - \psi_t\big).\end{aligned} \tag{2.4.20}$$

Taking into account the right implications in (2.4.17) and the fact that ε in (2.4.20) is arbitrary, we get the implication

$$\lim_{t\to\infty} y_t = l_{1-\alpha} \Rightarrow \lim_{t\to\infty} \varphi_t^{-1}\left(\ln\beta\left(\delta_t^{+,\alpha_t}\right) + \psi_t\right) = l_{1-\alpha}. \tag{2.4.21}$$

In view of implication (2.4.18), to prove the implication inverse to (2.4.21), it is enough to show that

$$\lim_{t\to\infty} \varphi_t^{-1}\left(\ln\beta\left(\delta_t^{+,\alpha_t}\right) + \psi_t\right) = l_{1-\alpha} \Rightarrow \varlimsup_{t\to\infty} y_t \leq l_{1-\alpha}. \tag{2.4.22}$$

The implication (2.4.22) is easily proved by contradiction. Indeed, let

$$\lim_{t\to\infty} \varphi_t^{-1}\left(\ln\beta\left(\delta_t^{+,\alpha_t}\right) + \psi_t\right) = l_{1-\alpha}, \qquad \varlimsup_{t\to\infty} y_t > l_{1-\alpha}, \tag{2.4.23}$$

and let (t_n) be a sequence such that $t_n \to \infty$ and $y_{t_n} \to \varlimsup_{t\to\infty} y_t$ as $n \to \infty$. Then, in view of (2.4.20),

$$\varlimsup_{t\to\infty} \varphi_t^{-1}\left(\ln\beta\left(\delta_t^{+,\alpha_t}\right) + \psi_t\right) \geq \varlimsup_{n\to\infty} \varphi_{t_n}^{-1}\big(\ln\beta(\delta_{t_n}^{+,\alpha_{t_n}}) + \psi_{t_n}\big) > l_{1-\alpha},$$

which contradicts the equality in (2.4.23). Therefore the implication (2.4.22) is valid. This completes the proof of the right equivalence in (2.4.19). □

REMARK 2.4.3. If the conditions of Theorem 2.4.4 are fulfilled and $\alpha_t = \alpha + o(1)$, where $\alpha \in (0,1)$, then Theorem 2.4.4 gives the asymptotic decompositions

$$d_t = -\psi_t + l_{1-\alpha}\varphi_t + o(\varphi_t), \tag{2.4.24}$$

$$\ln \beta\left(\delta_t^{+,\alpha_t}\right) = -\psi_t + l_{1-\alpha}\varphi_t + o(\varphi_t) \tag{2.4.25}$$

for the Neyman–Pearson test δ_t^{+,α_t}. The decompositions (2.4.24) and (2.4.25) show that under the given conditions the asymptotic behavior of the test δ_t^{+,α_t} depends on the limiting value α of the level α_t in the second term of the asymptotic expansion.

2.5. Contiguous families of hypotheses

Relative compactness and tightness of the families of probability measures. These notions play the major role in studies of contiguous families of hypotheses. Assume that all measures under consideration are defined on a metric space $(E, \mathcal{E}, \rho)$ with distance ρ and σ-algebra of measurable sets $\mathcal{E}$ generated by ρ. In what follows we often assume that $(E, \mathcal{E}) = (\mathbf{R}^m, \mathcal{B}^m)$ with the Euclidean metric ρ.

DEFINITION 2.5.1. A family of probability measures $Q = (Q_u; u \in \mathfrak{A})$ is said to be *relatively compact* if every sequence of measures in Q contains a subsequence weakly convergent to some probability measure.

Note that in this definition the limit measure is assumed to be a probability measure, though it does not necessary belong to the family Q.

DEFINITION 2.5.2. A family of probability measures $Q = (Q_u; u \in \mathfrak{A})$ is said to be *tight* if for any $\varepsilon > 0$ there exists a compact set $K_\varepsilon \subset E$ such that

$$\sup\{Q_u(E \backslash K_\varepsilon); u \in \mathfrak{A}\} \le \varepsilon.$$

The following theorem plays a fundamental role in the study of the weak convergence of probability measures.

THEOREM 2.5.1 (Prokhorov's theorem). *Let $Q = (Q_u; u \in \mathfrak{A})$ be a family of probability measures defined on a complete separable metric space $(E, \mathcal{E}, \rho)$. The family Q is relatively compact if and only if it is tight.*

PROOF. The proof of this theorem can be found in many textbooks in probability theory [**7, 127**].

Consider a family of probability measures $Q = (Q_t; t \in \mathbf{R}_+)$ on the space $(\mathbf{R}, \mathcal{B})$, where Q_t is the probability measure that defines the distribution of the likelihood ratio z_t with respect to $\widetilde{P}^t$. If a family of hypotheses $(\widetilde{H}^t)$ is contiguous to the family (H^t), then, by Theorem 2.2.3, the family (z_t) is tight with respect to $(\widetilde{P}^t)$ and hence the family of measures Q is tight and, by Prokhorov's theorem, it is relatively compact. Similarly, if $(H^t) \lhd (\widetilde{H}^t)$, then the family of distributions of the likelihood ratio $\widetilde{z}_t$ with respect to P^t is relatively compact. Since the family (z_t) is evidently tight with respect to (P^t), we get that the contiguity of $(H^t) \lhd (\widetilde{H}^t)$ implies the relative compactness of the family of distributions of $\Lambda_t = \ln z_t$ with respect to P^t. Thus every sequence of distributions of Λ_t with respect to P^t contains a weakly convergent subsequence. Thus in this section we assume that the distributions of Λ_t with respect to P^t weakly converge.

The weak convergence of the likelihood ratio. We introduce the following condition:

Λ6. $\mathcal{L}(\Lambda_t|P^t) \xrightarrow{w} L$, where L is some probability law on $\mathbf{R}^1$ with distribution function $L(x)$.

THEOREM 2.5.2. *If condition Λ6 is fulfilled, then*

$$\mathcal{L}(\Lambda_t|\widetilde{P}^t) \xrightarrow{w} \widetilde{L}, \tag{2.5.1}$$

where $\widetilde{L}$ is the probability law on $\overline{\mathbf{R}}^1 = [-\infty, \infty]$ with distribution function

$$\widetilde{L}(x) = \int_{-\infty}^{x} e^y \, dL(y). \tag{2.5.2}$$

Generally, $\widetilde{L}(\infty) \leq 1$ and

$$\lim_{N\to\infty} \lim_{t\to\infty} \widetilde{P}^t(\Lambda_t \geq N) = 1 - \widetilde{L}(\infty). \tag{2.5.3}$$

PROOF. In view of the Lebesgue decomposition, for any $x \in \mathbf{R}^1$ we have

$$\widetilde{P}^t(\Lambda_t < x) = \mathsf{E}^t \chi(\ln z_t < x) z_t = \int_0^{\exp(x)} z \, dG_t(z), \tag{2.5.4}$$

where $G_t(z) = P^t(z_t < z)$, $z \in \mathbf{R}_+$. By the Helly theorem [**127**], condition Λ6 implies that for any $x \in (0, \infty)$ for which $\ln x$ is a continuity point of the function $L(x)$, we have

$$\lim_{t\to\infty} \int_0^x z \, dG_t(z) = \int_0^x z \, dG(z), \tag{2.5.5}$$

where $G(z) = L(\ln z)$, $z \in (0, \infty)$. It is evident that

$$\int_0^{\exp(y)} z \, dG(z) = \int_{-\infty}^{y} e^x \, dL(x). \tag{2.5.6}$$

From (2.5.4)–(2.5.6) it follows that for any $y \in \mathbf{R}^1$ that is a continuity point of the function $L(x)$ we have

$$\lim_{t\to\infty} \widetilde{P}^t(\Lambda_t < y) = \int_{-\infty}^{y} e^x \, dL(x), \tag{2.5.7}$$

whence the relations (2.5.1) and (2.5.2) follow. The inequality $\widetilde{L}(\infty) \leq 1$ and relation (2.5.3) also follow from (2.5.7). □

COROLLARY 2.5.1. *The following implication is valid:*

$$\Lambda 6 \Rightarrow \lim_{t\to\infty} \overline{\alpha}_t = 1, \ \ \varliminf_{t\to\infty} \overline{\beta}_t \geq \widetilde{L}(\infty). \tag{2.5.8}$$

The implication (2.5.8) follows from the relations

$$1 - \overline{\alpha}_t = P^t(\Lambda_t = -\infty) \leq P^t(\Lambda_t < x),$$
$$\overline{\beta}_t = \widetilde{P}^t(\Lambda_t < \infty) \geq \widetilde{P}^t(\Lambda_t < x),$$

valid for any $x \in \mathbf{R}^1$, condition Λ6, and relation (2.5.3).

COROLLARY 2.5.2. *If condition* $\Lambda 6$ *is fulfilled with* $L = \mathcal{N}(a, \sigma^2)$, *where* $a \in (-\infty, \infty)$ *and* $\sigma \in (0, \infty)$, *then* $a \leq -\sigma^2/2$ *and relation* (2.5.1) *holds with distribution function*

$$\widetilde{L}(x) = h\Phi\left(\frac{x - a - \sigma^2}{\sigma}\right),$$

where $h = \exp(a + \sigma^2/2)$.

To prove the corollary, it is enough to note that

$$\int_{-\infty}^{x} e^y \, dL(y) = \int_{-\infty}^{x} e^y \, d\Phi\left(\frac{y - a}{\sigma}\right) = h\Phi\left(\frac{x - a - \sigma^2}{\sigma}\right),$$

and then to apply the relation (2.5.7).

REMARK 2.5.1. If condition $\Lambda 6$ is fulfilled, then, by Theorem 2.5.2, the limit law $\widetilde{L}$ is a mixture of two laws $\widetilde{L}^+$ and $\varepsilon_{\{\infty\}}$, respectively, with weights $\widetilde{L}(\infty)$ and $1 - \widetilde{L}(\infty)$, respectively, that is, $\widetilde{L} = \widetilde{L}(\infty)\widetilde{L}^+ + (1 - \widetilde{L}(\infty))\varepsilon_{\{\infty\}}$, where $\varepsilon_{\{\infty\}}$ is the Dirac measure concentrated at the point ∞ and $\widetilde{L}^+$ is the probability law on $\mathbf{R}^1$ defined by the distribution function $\widetilde{L}^+(x) = \widetilde{L}(x)/\widetilde{L}(\infty)$, $x \in \mathbf{R}^1$.

In particular, if $L = \mathcal{N}(a, \sigma^2)$, then, by virtue of Corollary 2.5.2, $\widetilde{L}(\infty) = h = \exp(a + \sigma^2/2)$ and $\widetilde{L}^+ = \mathcal{N}(a + \sigma^2, \sigma^2)$.

REMARK 2.5.2. Let condition $\Lambda 6$ be fulfilled, in which the law L is a mixture of normal laws $\mathcal{N}(-\sigma^2/2, \sigma^2)$ over the parameter σ with some distribution law K on $(0, \infty)$. Then the distribution function $L(x)$ is continuous and strictly monotonically increasing on $(-\infty, \infty)$. It is evident that in this case the law $\widetilde{L}$ is also a mixture of normal laws $\mathcal{N}(\sigma^2/2, \sigma^2)$ over the parameter σ with the same distribution law K. Therefore $\widetilde{L}(\infty) = 1$.

THEOREM 2.5.3. *The following implication is valid:*

$$\Lambda 6 \Rightarrow (H^t) \lhd (\widetilde{H}^t). \tag{2.5.9}$$

In particular, if condition $\Lambda 6$ *is fulfilled, then*

$$(\widetilde{H}^t) \lhd (H^t) \Leftrightarrow \widetilde{L}(\infty) = 1, \tag{2.5.10}$$

$$(\widetilde{H}^t) \overline{\lhd} (H^t) \Leftrightarrow \widetilde{L}(\infty) < 1. \tag{2.5.11}$$

PROOF. Condition $\Lambda 6$ implies that

$$\lim_{N \to -\infty} \lim_{t \to \infty} P^t(\Lambda_t < N) = 0.$$

Now, applying Theorem 2.2.3 with $\widetilde{H}^t$ instead of H^t and taking into account the equality

$$P^t(z_t < N) = P^t\left(\widetilde{z}_t > N^{-1}\right),$$

we obtain the implication (2.5.9).

Let condition $\Lambda 6$ be fulfilled. Then, by Theorem 2.5.2, $\widetilde{L}(\infty) \leq 1$ and the relation (2.5.3) is true. Now, applying Theorem 2.2.3, we get (2.5.10). The equivalence (2.5.11) follows from the inequality $\widetilde{L}(\infty) \leq 1$, the equivalence (2.5.10), and the following alternative: either $(\widetilde{H}^t) \lhd (H^t)$ or $(\widetilde{H}^t) \overline{\lhd} (H^t)$. □

REMARK 2.5.3. Theorem 2.5.3 implies that under condition Λ6 the following alternative is true: either $(\widetilde{H}^t) \lhd\rhd (H^t)$ (type **a**) or $(\widetilde{H}^t) \overline{\lhd} (H^t)$ (type **b**). Moreover, $(\widetilde{H}^t) \lhd\rhd (H^t) \Leftrightarrow \widetilde{L}(\infty) = 1$ and $(\widetilde{H}^t) \overline{\lhd} (H^t) \Leftrightarrow \widetilde{L}(\infty) < 1$. If in condition Λ6 we have $L = \mathcal{N}(a, \sigma^2)$, then, by Corollary 2.5.2, we have $a \leq -\sigma^2/2$ and, moreover, $\widetilde{L}(\infty) = 1 \Leftrightarrow a = -\sigma^2/2$ and $\widetilde{L}(\infty) < 1 \Leftrightarrow a < -\sigma^2/2$.

Behavior of Neyman–Pearson test. The following theorem describes the relationship between the behavior of α_t and $\beta(\delta_t^{+,\alpha_t})$.

THEOREM 2.5.4. *If condition Λ6 is fulfilled, then for any $\alpha \in (0, 1)$ the following implications hold:*

$$\lim_{t\to\infty} \alpha_t = \alpha \Rightarrow \varlimsup_{t\to\infty} \beta\left(\delta_t^{+,\alpha_t}\right) \leq \widetilde{L}(\bar{l}_{1-\alpha} + 0), \tag{2.5.12}$$

$$\lim_{t\to\infty} \alpha_t = \alpha \Rightarrow \varliminf_{t\to\infty} \beta\left(\delta_t^{+,\alpha_t}\right) \geq \widetilde{L}(\underline{l}_{1-\alpha}), \tag{2.5.13}$$

where $\underline{l}_p$ and $\bar{l}_p$ are defined by (2.4.4).

PROOF. Assume that $\alpha_t \to \alpha$ as $t \to \infty$. First let u be a continuity point of the function $L(x)$ such that $L(u) > 1-\alpha$. Exactly as in the proof of Theorem 2.4.1, we get $\varlimsup_{t\to\infty} d_t \leq u$. From this it follows that

$$\varlimsup_{t\to\infty} \beta\left(\delta_t^{+,\alpha_t}\right) \leq \varlimsup_{t\to\infty} \widetilde{P}^t(\Lambda_t \leq d_t) \leq \lim_{t\to\infty} \widetilde{P}^t(\Lambda_t \leq u) = \widetilde{L}(u). \tag{2.5.14}$$

Here we have made use of the fact that, by (2.5.2), the point u is also a continuity point of the function $\widetilde{L}(x)$.

From (2.5.14) we get

$$\varlimsup_{t\to\infty} \beta\left(\delta_t^{+,\alpha_t}\right) \leq \inf\{\widetilde{L}(u) : L(u) > 1-\alpha\} = \widetilde{L}(\bar{l}_{1-\alpha} + 0),$$

where the lower bound is taken over all u that are continuity points of $L(x)$ such that $L(u) > 1 - \alpha$. Thus the implication (2.5.12) is proved.

Now let u be a continuity point of the function $L(x)$ such that $L(u) < 1 - \alpha$. Then $\varliminf_{t\to\infty} d_t \geq u$, whence we obtain

$$\varliminf_{t\to\infty} \beta\left(\delta_t^{+,\alpha_t}\right) \geq \varliminf_{t\to\infty} \widetilde{P}^t(\Lambda_t < d_t) \geq \lim_{t\to\infty} \widetilde{P}^t(\Lambda_t < u) = \widetilde{L}(u)$$

and then

$$\varliminf_{t\to\infty} \beta\left(\delta_t^{+,\alpha_t}\right) \geq \sup\{\widetilde{L}(u); L(u) < 1-\alpha\} = \widetilde{L}(\underline{l}_{1-\alpha}),$$

where the upper bound is taken over all continuity points u of the function $L(x)$ such that $L(u) < 1 - \alpha$. Thus the implication (2.5.13) is also proved. □

If condition Λ6 is fulfilled, in which the function $L(x)$ is continuous, then the assertions of Theorem 2.5.4 can be sharpened. First, we prove an auxiliary lemma, which is of independent interest.

LEMMA 2.5.1. *Let $(Z^t, \mathfrak{A}^t, S^t)$, $t \in \mathbf{R}_+$, be a family of probability spaces, and let Y_t be a measurable map of the space $(Z^t, \mathfrak{A}^t)$ into the space $(\mathbf{R}^1, \mathcal{B}^1)$ such that*

$$\mathcal{L}(Y_t | S^t) \xrightarrow{w} S, \qquad t \to \infty, \tag{2.5.15}$$

where S is a probability law on $\overline{\mathbf{R}}^1$ with continuous distribution function $S(x)$ for $x \in (-\infty, \infty)$, $S(-\infty) = 0$, and $S(\infty) \leq 1$. Then for any family (y_t) of numbers

$y_t \in \mathbf{R}^1$ *satisfying, in the case* $S(\infty) < 1$, *the additional condition* $\overline{\lim}_{t\to\infty} y_t < \infty$ *the following relation holds*:

$$\lim_{t\to\infty} S^t(Y_t = y_t) = 0. \tag{2.5.16}$$

Besides, let the function $S(x)$ *strictly monotonically increase on the interval* $(\underline{x}, \overline{x})$, *where*

$$\underline{x} = \sup\{x \colon S(x) = 0\}, \qquad \overline{x} = \inf\{x \colon S(x) = S(\infty)\}.$$

Then for any families (y_t) *and* (ε_t) *of numbers* $y_t \in \mathbf{R}^1$ *and* $\varepsilon_t \in [0,1]$, *for which the limit*

$$\lim_{t\to\infty} \left[S^t(Y_t > y_t) + \varepsilon_t S^t(Y_t = y_t)\right] = \beta \tag{2.5.17}$$

exists, the limit $\lim_{t\to\infty} y_t$ *exists for any* $\beta \in (1 - S(\infty), 1)$; *moreover*,

$$\lim_{t\to\infty} y_t = s_{1-\beta}, \tag{2.5.18}$$

and for $\beta = 1$

$$\overline{\lim_{t\to\infty}}\, y_t \le \underline{x}, \tag{2.5.19}$$

whereas for $\beta = 1 - S(\infty)$

$$\underline{\lim_{t\to\infty}}\, y_t \ge \overline{x}, \tag{2.5.20}$$

where s_p *is a p-quantile of the law* S.

PROOF. First we assume that $S(\infty) = 1$ and prove the relations (2.5.16) and (2.5.18). Put $S_t(y) = S^t(Y_t < y)$. Then for any $\varepsilon > 0$

$$\begin{aligned} S^t(Y_t = y_t) &\le S_t(y_t + \varepsilon) - S_t(y_t) \\ &= [S_t(y_t+\varepsilon) - S(y_t+\varepsilon)] - [S_t(y_t) - S(y_t)] + [S(y_t+\varepsilon) - S(y_t)]. \end{aligned}$$

Now, applying the Pólya lemma [**109**] and using the uniform continuity of the function $S(x)$, from condition (2.5.15) we get the relation (2.5.16).

It is evident that from the relation (2.5.16) it follows that

$$(2.5.17) \Leftrightarrow \lim_{t\to\infty} S_t(y_t) = 1 - \beta.$$

Now, applying the Pólya lemma once again, we get

$$(2.5.17) \Leftrightarrow \lim_{t\to\infty} S(y_t) = 1 - \beta. \tag{2.5.21}$$

From properties of the function $S(x)$ and equivalence (2.5.21) we obtain the assertion about the existence of the limit $\lim_{t\to\infty} y_t$ and also the relation (2.5.18).

Now let $S(\infty) < 1$. Condition (2.5.15) and equality $S(-\infty) = 0$ imply

$$\lim_{t\to\infty} S_t(x) = S(x) \tag{2.5.22}$$

uniformly with respect to $x \le N$ for any $N < \infty$. The required relation (2.5.16) is proved similarly to the proof of this relation in the case $S(\infty) = 1$ using, instead of the Pólya lemma, the uniform convergence in (2.5.22) only on the interval $\left(-\infty, \overline{\lim}_{t\to\infty} y_t + \varepsilon\right)$ for $\varepsilon > 0$.

Evidently, $\beta \in [1 - S(\infty), 1]$. First let $\beta \in (1 - S(\infty), 1)$. To prove the existence of the limit $\lim_{t\to\infty} y_t$ and relation (2.5.18), we note that by condition

(2.5.15) and equality $S(-\infty) = 0$, we have $\lim_{N\to-\infty} \lim_{t\to\infty} S_t(N) = 0$. From this and conditions (2.5.15) and (2.5.17) we get $\overline{\lim}_{t\to\infty} y_t < \infty$. Let us prove it by contradiction. Assume that $\overline{\lim}_{t\to\infty} y_t = \infty$. Then there exists a sequence (t_n) such that $y_{t_n} \to \infty$ as $n \to \infty$. Let $\varepsilon > 0$ be arbitrary, and let $N_\varepsilon \in \mathbf{R}^1$ be such that $S(N_\varepsilon) > S(\infty) - \varepsilon$. Furthermore, in view of relation (2.5.22), there exists a positive integer $n_0 = n_0(\varepsilon)$ such that $|S_{t_n}(N_\varepsilon) - S(N_\varepsilon)| < \varepsilon$ for all $n > n_0$. Now let $n_1 = n_1(\varepsilon)$ be a positive integer such that $y_{t_n} > N_\varepsilon$ for all $n > n_1$. Thus for all $n > n_0 \vee n_1$ we have

$$S_{t_n}(y_{t_n}) \geq S_{t_n}(N_\varepsilon) \geq S(N_\varepsilon) - \varepsilon \geq S(\infty) - 2\varepsilon$$

and hence, since ε is arbitrary,

$$\underline{\lim}_{t\to\infty} S^t(Y_t \geq y_t) \leq \underline{\lim}_{n\to\infty} S^{t_n}(Y_{t_n} \geq y_{t_n}) \leq 1 - S(\infty).$$

On the other hand, in view of (2.5.17),

$$\underline{\lim}_{t\to\infty} S^t(Y_t \geq y_t) \geq \lim_{t\to\infty} \left[S^t(Y_t > y_t) + \varepsilon_t S^t(Y_t = y_t)\right] = \beta > 1 - S(\infty).$$

The contradiction obtained proves that $\overline{\lim}_{t\to\infty} y_t < \infty$. Now, applying (2.5.16), we get $S^t(Y_t = y_t) \to 0$ as $t \to \infty$. The equality (2.5.18) is proved similarly to the case $S(\infty) = 1$.

In proving the inequalities (2.5.19) and (2.5.20) we assume that $S(\infty) \leq 1$. In both cases the proofs are carried out by contradiction.

Let $\beta = 1$ but $\overline{\lim}_{t\to\infty} y_t > \underline{x}$. Then for any $N \in (\underline{x}, \overline{y})$, where $\overline{y} = \overline{\lim}_{t\to\infty} y_t$, there exists $t_0 = t_0(N)$ such that $y_t > N$ for all $t > t_0$. Hence for $t > t_0$ the following inequality holds:

$$S^t(Y_t < y_t) + (1 - \varepsilon_t)S^t(Y_t = y_t) \geq S^t(Y_t < y_t) \geq S^t(Y_t < N).$$

From this, taking into account conditions (2.5.15) and (2.5.17), strict monotonicity of $S(x)$, and equality $S(\underline{x}) = 0$, we get

$$\underline{\lim}_{t\to\infty} \left[S^t(Y_t < y_t) + (1 - \varepsilon_t)S^t(Y_t = y_t)\right] > 0.$$

However, this inequality contradicts (2.5.17) for $\beta = 1$. Therefore (2.5.19) hold.

Now let $\beta = 1 - S(\infty)$ but $\underline{y} = \underline{\lim}_{t\to\infty} y_t < \overline{x}$. Then there exists a sequence (t_n) such that $t_n \to \infty$ and $y_{t_n} \to \underline{y} < \overline{x}$ as $n \to \infty$. Taking into account conditions (2.5.15) and (2.5.17), the relation (2.5.16), and strict monotonicity of $S(x)$ on the interval $(\underline{x}, \overline{x})$, we get

$$\lim_{t\to\infty} \left[S^{t_n}(Y_{t_n} > y_{t_n}) + \varepsilon_{t_n} S^{t_n}(Y_{t_n} = y_{t_n})\right] = 1 - S(\underline{y}) > 1 - S(\overline{x}) = 1 - S(\infty).$$

The last equality contradicts (2.5.17), and hence (2.5.20) is valid. □

REMARK 2.5.4. If under the conditions of Lemma 2.5.1 $\underline{x} = -\infty$, then for $\beta = 1$ the limit $\lim_{t\to\infty} y_t$ exists and the equality (2.5.18) is valid for $s_0 = \underline{x} = -\infty$. Similarly, if $\overline{x} = \infty$, then for $\beta = 1 - S(\infty)$ the limit $\lim_{t\to\infty} y_t$ also exists and the equality (2.5.18) is valid for $s_{1-\beta} = s_{S(\infty)} = \overline{x} = \infty$.

REMARK 2.5.5. If $S = \mathcal{N}(a, \sigma^2)$, then in the case $a = -\sigma^2/2$, equivalent to the equality $S(\infty) = 1$, Lemma 2.5.1 implies the well-known result [**110**]. In the case $a < -\sigma^2/2$ Lemma 2.5.1 also implies the known result [**88**].

THEOREM 2.5.5. *Let condition* $\Lambda 6$ *be fulfilled, in which the function* $L(x)$ *is continuous and strictly monotonically increasing on the interval* $(\underline{l}, \bar{l})$, *where*

$$\underline{l} = \sup\{x \colon L(x) = 0\}, \qquad \bar{l} = \inf\{x \colon L(x) = 1\}.$$

Then for any $\alpha \in [0, 1]$

$$\lim_{t\to\infty} \alpha_t = \alpha \Leftrightarrow \lim_{t\to\infty} \beta(\delta_t^{+,\alpha_t}) = \widetilde{L}(l_{1-\alpha}), \tag{2.5.23}$$

where l_p *is a* p*-quantile of the law* L *and* $\widetilde{L}(x)$ *is the distribution function of the probability law* $\widetilde{L}$ *defined by the equality* (2.5.2).

PROOF. Condition $\Lambda 6$ and Theorem 2.5.1 imply the weak convergence (2.5.1), where $\widetilde{L}$ is the probability law on $\overline{\mathbf{R}}^1$ with distribution function $\widetilde{L}(x)$ of the form (2.5.2) and $\widetilde{L}(\infty) \le 1$. From the equality (2.5.2) it is seen that

$$\sup\{x \colon \widetilde{L}(x) = 0\} = \underline{l}, \qquad \inf\{x \colon \widetilde{L}(x) = \widetilde{L}(\infty)\} = \bar{l},$$

and the function $\widetilde{L}(x)$ is continuous on $\mathbf{R}^1$ and strictly monotonically increases on the interval $(\underline{l}, \bar{l})$.

Let $\alpha_t \to \alpha$ as $t \to \infty$. If $0 < \alpha < 1$, then, by Lemma 2.5.1, the limit $\lim_{t\to\infty} d_t = l_{1-\alpha}$ exists and hence $\overline{\lim}_{t\to\infty} d_t < \infty$. Then, once again, in view of Lemma 2.5.1, $\widetilde{P}^t(\Lambda_t = d_t) \to 0$ as $t \to \infty$. Taking into consideration (2.5.1), (2.5.2), the inequality $\overline{\lim}_{t\to\infty} d_t < \infty$, and the uniform convergence of $\widetilde{P}^t(\Lambda_t < y)$ to $\widetilde{L}(y)$ for $y \le N$, for any $N < \infty$ we get

$$\beta\left(\delta_t^{+,\alpha_t}\right) \to \widetilde{L}(l_{1-\alpha}), \qquad t \to \infty. \tag{2.5.24}$$

If $\alpha = 0$, then, by Lemma 2.5.1, we have $\underline{\lim}_{t\to\infty} d_t \ge \bar{l}$. Therefore for any $N \in (-\infty, \bar{l})$ there exists $t' = t'(N)$ such that $d_t > N$ for all $t > t'$. Hence for any $t > t'$ we have

$$\beta\left(\delta_t^{+,\alpha_t}\right) \ge \widetilde{P}^t(\Lambda_t < d_t) \ge \widetilde{P}^t(\Lambda_t < N).$$

In view of the convergence (2.5.1) and arbitrariness of N, we get (2.5.24) for $\alpha = 0$.

If $\alpha = 1$, then, by Lemma 2.5.1, we have $\overline{\lim}_{t\to\infty} d_t \le \bar{l}$. Then for any $N \in (\bar{l}, \infty)$ there exists $t'' = t''(N)$ such that $d_t < N$ for all $t > t''$. Therefore for all $t > t''$

$$\beta\left(\delta_t^{+,\alpha_t}\right) \le \widetilde{P}^t(\Lambda_t < N) + \widetilde{P}^t(\Lambda_t = d_t). \tag{2.5.25}$$

It is evident that $\overline{\lim}_{t\to\infty} d_t < \infty$ and, by Lemma 2.5.1, $\widetilde{P}^t(\Lambda_t = d_t) \to 0$ as $t \to \infty$. Therefore, by virtue of the convergence (2.5.1), the estimate (2.5.25) implies (2.5.24) for $\alpha = 1$. Thus the implication $\Rightarrow$ in (2.5.23) is proved.

Now let $\beta(\delta_t^{+,\alpha_t}) \to \widetilde{L}(l_{1-\alpha})$ as $t \to \infty$. Then $1 - \beta(\delta_t^{+,\alpha_t}) \to \beta$, where $\beta = 1 - \widetilde{L}(l_{1-\alpha}) \in [1 - \widetilde{L}(\infty), 1]$. For $\alpha \in (0,1)$ we have $\beta \in (1 - \widetilde{L}(\infty), 1)$. By Lemma 2.5.1, $d_t \to \tilde{l}_{1-\beta} = \tilde{l}_{\widetilde{L}(l_{1-\alpha})} = l_{1-\alpha}$, where $\tilde{l}_p$ is a p-quantile of the law $\widetilde{L}$.

Taking into account condition $\Lambda 6$ and applying Lemma 2.5.1, we get $\alpha_t \to 1 - L(l_{1-\alpha}) = \alpha$, $t \to \infty$.

If $\alpha = 0$, then $\beta = 1 - \widetilde{L}(\infty)$ and, by Lemma 2.5.1, $\underline{\lim}_{t\to\infty} d_t \geq \bar{l}$. Then for every $N \in (-\infty, \bar{l})$ there exists $t_0 = t_0(N)$ such that $d_t > N$ for all $t > t_0$. From this we obtain

$$\alpha_t \leq P^t(\Lambda_t > N) + \varepsilon_t P^t(\Lambda_t = d_t)$$

for any $t > t_0$. Taking into consideration condition Λ6, Lemma 2.5.1, and arbitrariness of N, from this estimate we conclude that $\alpha_t \to 0$ as $t \to \infty$.

In a similar manner, it can be proved that $\alpha_t \to \alpha$ as $t \to \infty$ in the case $\alpha = 1$. Thus the implication $\Leftarrow$ in (2.5.23) is proved. □

Complete asymptotic indistinguishability. By Definition 2.2.12, the mutual contiguity $(\widetilde{H}^t) \lhd\rhd (H^t)$ is an **a** type distinguishability of families of hypotheses (H^t) and $(\widetilde{H}^t)$. The complete asymptotic indistinguishability $(H^t) \cong (\widetilde{H}^t)$ of families of hypotheses (type $\mathbf{a}_0$) is a subtype of the type **a**. To consider the type $\mathbf{a}_0$, we introduce the condition

Λ7. $\lim_{t\to\infty} P^t(|\Lambda_t| > \gamma) = 0$ for any $\gamma > 0$.

From Theorem 2.2.2 it follows that $\Lambda 7 \Leftrightarrow (H^t) \cong (\widetilde{H}^t)$.

THEOREM 2.5.6. *If condition Λ7 is fulfilled, then for any $\alpha \in [0, 1]$ the following equivalence holds*:

$$\lim_{t\to\infty} \alpha_t = \alpha \Leftrightarrow \lim_{t\to\infty} \beta\left(\delta_t^{+,\alpha_t}\right) = 1 - \alpha. \tag{2.5.26}$$

PROOF. The implication $\Rightarrow$ in (2.5.26) follows from the definition of complete asymptotic indistinguishability and Theorem 2.2.2. The inverse implication follows from Lemma 2.2.3. □

REMARK 2.5.6. It is not hard to see that for any Neyman–Pearson test either the equivalence (2.5.26) or the implication $\Rightarrow$ in (2.5.26) can be used as one more characterization of complete asymptotic indistinguishability.

2.6. The case of asymptotic expansion of the likelihood ratio

Consider the case where Λ_t has asymptotic expansion as $t \to \infty$ given by the following condition:

Λ8. $\Lambda_t = u_t'\eta_t - (1/2)u_t'\varkappa_t^2 u_t$, $u_t \in \mathbf{R}^k$ is a nonrandom k-dimensional vector, and η_t and $\varkappa_t$ are a random k-dimensional vector and a symmetric $(k\times k)$-matrix such that

$$\lim_{N\to\infty} \overline{\lim_{t\to\infty}} P^t(|\eta_t| > N) = 0, \qquad \mathcal{L}\left(\varkappa_t | P^t\right) \xrightarrow{w} \mathcal{L}(\varkappa|\mathsf{P}),$$

where $\varkappa$ is a random symmetric positive definite $(k \times k)$-matrix on a certain probability space $(\Omega, \mathcal{F}, \mathsf{P})$ such that $\lambda'\varkappa\lambda > 0$ (P-a.s.) for any $\lambda \in \mathbf{R}^k$, $\lambda \neq 0$.

THEOREM 2.6.1. *If condition Λ8 is fulfilled, then*

$$(H^t) \,\triangle\, (\widetilde{H}^t) \Leftrightarrow \overline{\lim_{t\to\infty}} |u_t| = \infty, \tag{2.6.1}$$

$$(H^t) \lhd (\widetilde{H}^t) \Leftrightarrow \overline{\lim_{t\to\infty}} |u_t| < \infty. \tag{2.6.2}$$

PROOF. If $\overline{\lim}_{t\to\infty}|u_t| = \infty$, then condition Λ8 implies $\overline{\lim}_{t\to\infty} P^t(\Lambda_t < N) = 1$ for any $N > -\infty$. Then, by Theorem 2.2.1, we have the complete asymptotic distinguishability $(H^t) \bigtriangleup (\widetilde{H}^t)$. Thus the implication $\Leftarrow$ in (2.6.1) is proved. If $\overline{\lim}_{t\to\infty}|u_t| < \infty$, then condition Λ8 implies

$$\lim_{N\to\infty}\overline{\lim_{t\to\infty}}\, P^t(\Lambda_t < N) = 0. \tag{2.6.3}$$

Applying Theorem 2.2.1 with H^t and $\widetilde{H}^t$ interchanged, from (2.6.3) we get the contiguity $(H^t) \lhd (\widetilde{H}^t)$. Thus the implication $\Leftarrow$ in (2.6.2) is established. The inverse implications $\Rightarrow$ in (2.6.1) and (2.6.2) are easily proved by contradiction. □

Consider the following particular cases of condition Λ8:

Λ8′. Condition Λ8 in which

$$\mathcal{L}\left((\eta_t, \varkappa_t^2)\,|P^t\right) \xrightarrow{w} \mathcal{L}\left((\eta'\varkappa, \varkappa^2)\,|\mathsf{P}\right),$$

where η is a random k-dimensional vector, independent of $\varkappa$, defined on the probability space $(\Omega, \mathcal{F}, \mathsf{P})$ and having the normal distribution $\mathcal{N}(0, J)$. Here J is the unit $(k\times k)$-matrix and 0 is the zero k-dimensional vector.

Λ8*. Condition Λ8′ in which $\varkappa = J$.

REMARK 2.6.1. Condition Λ8* means that for Λ_t the decomposition from condition Λ8 with $\varkappa_t^2 = J + p_t$ is valid and, moreover,

$$\mathcal{L}\left(\eta_t|P^t\right) \xrightarrow{w} \mathcal{N}(0, J), \tag{2.6.4}$$

$$\lim_{t\to\infty} P^t(|p_t| \geq \gamma) = 0 \quad \text{for all } \gamma > 0. \tag{2.6.5}$$

The following theorem provides the weak convergence of Λ_t under the hypothesis H^t and alternative $\widetilde{H}^t$, when condition Λ8′ is fulfilled.

THEOREM 2.6.2. *Let condition Λ8′ be fulfilled, in which $u_t \to u \in \mathbf{R}^k$ as $t\to\infty$. Then condition Λ6 is valid, in which the law L is a mixture of normal distributions $\mathcal{N}(-u'\varkappa^2u/2, u'\varkappa^2u)$ with respect to the distribution of the matrix $\varkappa$. Moreover, the relation (2.5.1) holds, in which the law $\widetilde{L}$ is also a mixture of normal distributions $\mathcal{N}(u'\varkappa^2u/2, u'\varkappa^2u)$ with respect to the distribution of the matrix $\varkappa$.*

PROOF. By condition Λ8, we have

$$\begin{aligned}\Lambda_t - u'\eta_t + \frac{1}{2}u'\varkappa_t^2u &= (u_t-u)'\eta_t - \frac{1}{2}u_t'\varkappa_t^2u_t + \frac{1}{2}u'\varkappa_t^2u \\ &= (u_t-u)'\eta_t - \frac{1}{2}(u_t-u)'\varkappa_t^2u_t - \frac{1}{2}u'\varkappa_t^2(u_t-u).\end{aligned}$$

Since $u_t \to u$, from this, by condition Λ8′, we obtain

$$\lim_{t\to\infty} P^t\left(\left|\Lambda_t - u'\eta_t + u'\varkappa_t^2u/2\right| > \gamma\right) = 0 \quad \text{for all } \gamma > 0. \tag{2.6.6}$$

By condition Λ8′,

$$\mathcal{L}\left(u'\eta_t - (1/2)u'\varkappa_t^2u|P^t\right) \xrightarrow{w} \mathcal{L}\left(u'\varkappa\eta - (1/2)u'\varkappa^2u|\mathsf{P}\right). \tag{2.6.7}$$

Putting $\varkappa = k$, we have

$$\mathcal{L}\left(u'k\eta - (1/2)u'k^2u|\mathsf{P}\right) = \mathcal{N}\left(-u'k^2u/2, u'k^2u\right),$$

and hence the limit law in (2.6.7) is a mixture of Gaussian laws

$$\mathcal{N}\left(-(1/2)u'\varkappa^2 u, u'\varkappa^2 u\right)$$

with respect to the distribution of the matrix $\varkappa$. Combining (2.6.6) and (2.6.7), we get condition $\Lambda 6$ with limit law $L = \mathcal{L}(u'\varkappa\eta - (1/2)u'\varkappa^2 u|\mathsf{P})$. The validity of the relation (2.5.1), in which $\widetilde{L}$ is a mixture of normal laws $\mathcal{N}((1/2)u'\varkappa^2 u, u'\varkappa^2 u)$ with respect to the distribution of the matrix $\varkappa$, is obtained by using Remark 2.5.2 with $\sigma^2 = u'\varkappa^2 u$. □

If condition $\Lambda 8^*$ is fulfilled, then the requirement $u_t \to u \in \mathbf{R}^k$ can be replaced by the requirement of convergence of only the norm $|u_t|$ as $t \to \infty$. First, we prove the following lemma.

LEMMA 2.6.1. *Let $(Z^t, \mathfrak{A}^t, S^t)$, $t \in \mathbf{R}_+$, be a family of probability spaces, and let Y_t be measurable maps of the space $(Z^t, \mathfrak{A}^t)$ into the space $(\mathbf{R}^k, \mathcal{B}^k)$ such that*

$$\mathcal{L}\left(Y_t|S^t\right) \xrightarrow{w} \mathcal{N}(0, I), \tag{2.6.8}$$

where I is some positive semidefinite $(k \times k)$-matrix.

Then for any family (v_t) of vectors $v_t \in \mathbf{R}^k$ such that

$$\overline{\lim_{t\to\infty}} |v_t| < \infty, \qquad \lim_{t\to\infty} v_t' I v_t = b \in (0, \infty) \tag{2.6.9}$$

the weak convergence

$$\mathcal{L}\left(v_t' Y_t|S^t\right) \xrightarrow{w} \mathcal{N}(0, b) \tag{2.6.10}$$

takes place.

PROOF. By the continuity of the correspondence among distributions and characteristic functions [23], the convergence (2.6.8) implies

$$\lim_{t\to\infty} \sup_{|v|\le r} \left|\mathsf{E}_S^t \exp(ivY_t) - \exp\left(-v'Iv/2\right)\right| = 0 \tag{2.6.11}$$

for every $r < \infty$, where E_S^t is the mathematical expectation with respect to the measure S^t. For any $\lambda \in \mathbf{R}^1$

$$\begin{aligned}
&\left|\mathsf{E}_S^t \exp(i\lambda v_t' Y_t) - \exp\left(-\lambda^2 b/2\right)\right| \\
&\qquad \le \sup_{|v|\le|\lambda|\cdot|v_t|} \left|\mathsf{E}_S^t \exp(ivY_t) - \exp\left(-v'Iv/2\right)\right| \\
&\qquad\quad + \left|\exp\left(-(1/2)\lambda^2 v_t' I v_t\right) - \exp\left(-\lambda^2 b/2\right)\right|.
\end{aligned}$$

Now, taking into account (2.5.9) and (2.5.11), we get

$$\lim_{t\to\infty} \mathsf{E}_S^t \exp(i\lambda v_t' Y_t) = \exp(-\lambda^2 b/2)$$

for any $\lambda \in \mathbf{R}^1$, whence (2.5.10) follows. □

THEOREM 2.6.3. *If condition $\Lambda 8^*$ is fulfilled, in which $|u_t| \to \sigma \in (0, \infty)$ as $t \to \infty$, then*

$$\mathcal{L}\left(\Lambda_t|P^t\right) \xrightarrow{w} \mathcal{N}\left(-\sigma^2/2, \sigma^2\right), \tag{2.6.12}$$

$$\mathcal{L}\left(\Lambda_t|\widetilde{P}^t\right) \xrightarrow{w} \mathcal{N}\left(\sigma^2/2, \sigma^2\right). \tag{2.6.13}$$

PROOF. It is evident that

$$\Lambda_t = |u_t| v_t' \eta_t - (1/2)|u_t|^2 v_t'(J + p_t) v_t, \tag{2.6.14}$$

where $v_t = u_t/|u_t|$ and $J + p_t = \varkappa_t^2$. Condition $\Lambda 8^*$ implies the convergence (2.6.4). Now, applying Lemma 2.6.1 and taking into consideration that $|v_t| = 1$, $I = J$, and $b = v_t' I v_t = 1$, we obtain

$$\mathcal{L}\left(v_t' \eta_t | P^t\right) \xrightarrow{w} \mathcal{N}(0,1). \tag{2.6.15}$$

By (2.6.5), we have

$$\lim_{t\to\infty} P^t\left(|v_t' p_t v_t| > \gamma\right) = 0 \quad \text{for all } \gamma > 0. \tag{2.6.16}$$

Combining (2.6.14)–(2.6.16), we get (2.6.12). Using Theorem 2.5.2 with

$$L = \mathcal{N}\left(-\sigma^2/2, \sigma^2\right)$$

(see also Corollary 2.5.2), from (2.6.12) we obtain the relation (2.6.13). □

The following theorem gives the characterization of distinguishability between hypotheses under condition $\Lambda 8'$.

THEOREM 2.6.4. *If condition $\Lambda 8'$ is fulfilled, then*

$$(H^t) \vartriangle (\widetilde{H}^t) \Leftrightarrow \overline{\lim_{t\to\infty}}\, |u_t| = \infty, \tag{2.6.17}$$

$$(H^t) \lhd\rhd (\widetilde{H}^t) \Leftrightarrow \overline{\lim_{t\to\infty}}\, |u_t| < \infty, \tag{2.6.18}$$

$$(H^t) \cong (\widetilde{H}^t) \Leftrightarrow \overline{\lim_{t\to\infty}}\, |u_t| = 0. \tag{2.6.19}$$

PROOF. The equivalence (2.6.17) was proved in Theorem 2.6.1. The equivalence (2.6.18) follows from Theorem 2.6.1 and implication

$$\overline{\lim_{t\to\infty}}\, |u_t| < \infty \Rightarrow (\widetilde{H}^t) \lhd (H^t). \tag{2.6.20}$$

In view of Theorem 2.2.3 and Prokhorov's theorem, in order to prove the implication (2.6.20), it is enough to demonstrate the relative compactness of the family of probability laws $\{\mathcal{L}(\Lambda_t | \widetilde{P}^t), t \in \mathbf{R}_+\}$. To put it differently, it is enough to demonstrate that any sequence (t_n) contains a subsequence $(t_{n'})$ such that the sequence $\mathcal{L}(\Lambda_{t_{n'}} | \widetilde{P}^{t_{n'}})$ weakly converges, as $n' \to \infty$, to some probability law.

Let (t_n) be any sequence such that $t_n \to \infty$ as $n \to \infty$. Since $\overline{\lim}_{t\to\infty} |u_t| < \infty$, the sequence (u_{t_n}) is bounded and hence there exists a subsequence $(t_{n'})$ such that $t_{n'} \to \infty$ and $u_{t_{n'}}$ tends to some limit $u \in \mathbf{R}^k$ as $n' \to \infty$. By Theorem 2.6.2, we have $\mathcal{L}(\Lambda_{t_{n'}} | \widetilde{P}^{t_{n'}}) \xrightarrow{w} \widetilde{L}$, $n' \to \infty$, where $\widetilde{L}$ is a probability law that is a mixture of normal laws $\mathcal{N}((1/2)u'\varkappa^2 u, u'\varkappa^2 u)$ with respect to the distribution of the matrix $\varkappa$. Thus the implication (2.6.20) is proved.

Now we prove (2.6.19). First let $\overline{\lim}_{t\to\infty} |u_t| = 0$. By virtue of condition $\Lambda 8'$,

$$\lim_{t\to\infty} P^t(|\Lambda_t| > \gamma) = 0 \quad \text{for all } \gamma > 0.$$

Then, by Theorem 2.2.2, we have the complete asymptotic indistinguishability $(H^t) \cong (\widetilde{H}^t)$. Conversely, let $(H^t) \cong (\widetilde{H}^t)$ but $\overline{\lim}_{t\to\infty} |u_t| > 0$. Then there exists a sequence (t_n) such that $t_n \to \infty$ and $u_{t_n} \to u \in \mathbf{R}^k$ as $n \to \infty$, where $|u| > 0$.

Therefore, by Theorem 2.6.2, $\mathcal{L}(\Lambda_{t_n}|P^{t_n}) \xrightarrow{w} L$, $n \to \infty$, where L is a mixture of normal distributions $\mathcal{N}(-(1/2)u'\varkappa u, u'\varkappa^2 u)$ with respect to the distribution of the matrix $\varkappa$.

Since $|u| > 0$ and $u'\varkappa^2 u > 0$ (P-a.s.), the limit law is not equal to the Dirac measure $\varepsilon_{\{0\}}$. Thus, in view of Theorem 2.2.2, we arrive at the contradiction with condition $(H^t) \cong (\widetilde{H}^t)$. Therefore the equivalence (2.6.19) is proved. □

REMARK 2.6.2. If condition $\Lambda 8'$ is fulfilled, then, by Theorem 2.6.4, the following alternative holds: either $(H^t) \vartriangle (\widetilde{H}^t)$ (type **e**) or $(H^t) \lhd\rhd (\widetilde{H}^t)$ (type **a**). Moreover, the following trichotomy is valid: either type $\mathbf{a}_0$, or type $\mathbf{a}_1$, or type **e**, where

$$\mathbf{a}_0 \Leftrightarrow \overline{\lim_{t\to\infty}}\, |u_t| = 0,$$
$$\mathbf{a}_1 \Leftrightarrow 0 < \overline{\lim_{t\to\infty}}\, |u_t| < \infty,$$
$$\mathbf{e} \Leftrightarrow \overline{\lim_{t\to\infty}}\, |u_t| = \infty.$$

If condition $\Lambda 8'$ is fulfilled and $u_t \to u \in \mathbf{R}^k$, then, by Theorem 2.6.2, Theorem 2.5.5 is applicable: for any $\alpha \in [0,1]$

$$\lim_{t\to\infty} \alpha_t = \alpha \Leftrightarrow \lim_{t\to\infty} \beta\left(\delta_t^{+,\alpha_t}\right) = \widetilde{L}(l_{1-\alpha}), \tag{2.6.21}$$

where l_p is a p-quantile of the law L and the laws L and $\widetilde{L}$ are defined in Theorem 2.6.2.

Consider the particular case of assertion (2.6.21) under condition $\Lambda 8^*$, where the limit $\lim_{t\to\infty} |u_t|$ exists.

THEOREM 2.6.5. *If condition $\Lambda 8^*$ is fulfilled, in which $|u_t| \to \sigma \in (0,\infty)$ as $t \to \infty$, then for any $\alpha \in [0,1]$*

$$\lim_{t\to\infty} \alpha_t = \alpha \Leftrightarrow \lim_{t\to\infty} \beta\left(\delta_t^{+,\alpha_t}\right) = \Phi(z_{1-\alpha} - \sigma), \tag{2.6.22}$$

where z_p is a p-quantile and $\Phi(x)$ is the distribution function of the normal law $\mathcal{N}(0,1)$.

PROOF. By Theorem 2.6.3, the relation (2.6.12) is true, that is, condition $\Lambda 6$ is valid, in which $L = \mathcal{N}(-\sigma^2/2, \sigma^2)$. Now we use Theorem 2.5.5. Since $L = \mathcal{N}(-\sigma^2/2, \sigma^2)$, we have $l_p = \sigma z_p - \sigma^2/2$. In view of relation (2.6.13), the relation $\widetilde{L} = \mathcal{N}(\sigma^2/2, \sigma^2)$ is true. Therefore $\widetilde{L}(x) = \Phi\big(\sigma^{-1}(x - \sigma^2/2)\big)$ and $\widetilde{L}(l_{1-\alpha}) = \Phi(z_{1-\alpha} - \sigma)$. Thus Theorem 2.5.5 implies the equivalence (2.6.22). □

In conclusion, we consider the case where condition $\Lambda 8^*$ is fulfilled, in which $|u_t| \to \infty$ as $t \to \infty$.

THEOREM 2.6.6. *Let condition $\Lambda 8^*$ be fulfilled, in which $|u_t| \to \infty$ as $t \to \infty$ and, moreover,*

$$\lim_{t\to\infty} P^t(|p_t u_t| > \gamma) = 0 \quad \textit{for all } \gamma > 0, \tag{2.6.23}$$

where $p_t = \varkappa_t^2 - J$ is the matrix defined in Remark 2.6.1.

Then for any $\alpha \in (0,1)$

$$\begin{aligned}\lim_{t\to\infty}\alpha_t = \alpha &\Leftrightarrow \lim_{t\to\infty}|u_t|^{-1}\left(d_t + |u_t|^2/2\right) = z_{1-\alpha}\\ &\Leftrightarrow \lim_{t\to\infty}|u_t|^{-1}\left(\ln\beta\left(\delta_t^{+,\alpha_t}\right) + |u_t|^2/2\right) = z_{1-\alpha}.\end{aligned} \tag{2.6.24}$$

PROOF. In view of the asymptotic decomposition, from condition $\Lambda 8^*$ we have

$$Y_t = \left(\Lambda_t + |u_t|^2/2\right)/|u_t| = \left(u_t'\eta_t - u_t'p_t u_t/2\right)/|u_t|.$$

By condition $\Lambda 8^*$, using the assumption (2.6.23) and Lemma 2.6.1, we get

$$\mathcal{L}\left(Y_t|P^t\right) \xrightarrow{w} \mathcal{N}(0,1), \qquad t\to\infty.$$

Hence condition $\Lambda 5$ is fulfilled, in which $L = \mathcal{N}(0,1)$, $\varphi_t = |u_t|$ and $\psi_t = |u_t|^2/2$. Thus Theorem 2.4.4 implies the relation (2.6.24). □

REMARK 2.6.3. Theorem 2.6.6 is also valid in the case $\alpha = 0$ and $\alpha = 1$, though in this case it carries little information. To prove assertion (2.6.24) for $\alpha = 0$ and $\alpha = 1$, it is enough to use the representation

$$\alpha_t = P^t(Y_t > y_t) + \varepsilon_t P^t(Y_t = y_t),$$

where

$$y_t = \left(d_t + (1/2)|u_t|^2\right)/|u_t|, \qquad Y_t = \left(u_t'\eta_t - (1/2)u_t'p_t u_t\right)/|u_t|,$$

and to use Lemma 2.5.1 with $S = \mathcal{N}(0,1)$ with due regard to Remark 2.5.4.

REMARK 2.6.4. Though for $\alpha = 0$ and $\alpha = 1$ the assertion (2.6.24) carries little information, the following remark can be made. If under the conditions of Theorem 2.6.6 we have $\alpha = 0$, then, by the previous remark,

$$d_t = -(1/2)|u_t|^2 + z_{1-\widetilde{\alpha}_t}|u_t|,$$

where $\widetilde{\alpha}_t \to 0$.

If $z_{1-\widetilde{\alpha}_t} = o(|u_t|)$, then $d_t/|u_t|^2 \to -1/2$ and, by Theorem 2.3.1,

$$\lim_{t\to\infty}|u_t|^{-2}\ln\beta\left(\delta_t^{+,\alpha_t}\right) = -1/2.$$

Using the well-known expansion for $\Phi(x)$ for $x > 0$ [**121**], for $\widetilde{\alpha}_t \to 0$ we have

$$\ln\widetilde{\alpha}_t = -(1/2)z^2_{1-\widetilde{\alpha}_t} - \ln\left(\sqrt{2\pi}z_{1-\widetilde{\alpha}_t}\right) + o(1).$$

Therefore the condition $z_{1-\widetilde{\alpha}_t} = o(|u_t|)$ as $\widetilde{\alpha}_t \to 0$ is equivalent to the condition $\lim_{t\to\infty}|u_t|^{-2}\ln\widetilde{\alpha}_t = 0$ (compare with condition $\alpha 1'$). Similar remark concerns also the case $\alpha = 1$.

2.7. Reduction of the problem of testing hypotheses

The conditions on Λ_t considered in the previous sections impose severe constraints on the behavior of $\overline{\alpha}_t$ and $\overline{\beta}_t$ as $t\to\infty$ (see, for example, (2.3.4), (2.3.5), Remark 2.3.4, Lemma 2.4.1, etc.). However, we can reduce the problem of hypotheses testing so as to remove these constraints. In the following we consider two reductions of this kind.

The first reduction. Let $(X_0^t, \mathfrak{B}_0^t, P_0^t, \widetilde{P}_0^t)$, $t \in \mathbf{R}_+$, be the family of statistical experiments generated by an observation ξ_0^t, and let H_0^t and $\widetilde{H}_0^t$ be simple hypotheses consisting in that the distribution of the observation ξ_0^t is given by the measures P_0^t and $\widetilde{P}_0^t$, respectively. We assume that

$$X_0^t = \{\tilde{\mathfrak{z}}_t > 0\}, \qquad \mathfrak{B}_0^t = \mathfrak{B}^t \cap X_0^t, \qquad P_0^t = P^t/\overline{\alpha}_t, \qquad \widetilde{P}_0^t = \widetilde{P}^t.$$

Let a measurable map δ_t from $(X_0^t, \mathfrak{B}_0^t)$ into $([0,1], \mathcal{B}([0,1]))$ be a test for distinguishing hypotheses H_0^t and $\widetilde{H}_0^t$ from observation ξ_0^t. We denote by $\Sigma^{t,0}$ the set of all such tests and for every test $\delta_t \in \Sigma^{t,0}$ we introduce the probabilities of errors of the first and the second kind

$$\alpha_0(\delta_t) = \mathsf{E}_0^t\, \delta_t, \qquad \beta_0(\delta_t) = \widetilde{\mathsf{E}}_0^t(1 - \delta_t),$$

where E_0^t and $\widetilde{\mathsf{E}}_0^t$ are the mathematical expectations with respect to the distributions P_0^t and $\widetilde{P}_0^t$, respectively.

Let $\mathfrak{z}_t^0 = dP_0^t/dQ_0^t$ and $\tilde{\mathfrak{z}}_t^0 = d\widetilde{P}_0^t/dQ_0^t$ be finite versions of the Radon–Nikodym derivatives, where Q_0^t is the restriction of the measure Q^t to the σ-algebra $\mathfrak{B}_0^t$. Evidently, on the set $\{\tilde{\mathfrak{z}}_t > 0\}$ we have $\mathfrak{z}_t^0 = \mathfrak{z}_t/\overline{\alpha}_t$ and $\tilde{\mathfrak{z}}_t^0 = \tilde{\mathfrak{z}}_t$ (Q^t-a.s.). Now we introduce the likelihood ratios $z_t^0 = \tilde{\mathfrak{z}}_t^0/\mathfrak{z}_t^0$ and $\widetilde{z}_t^0 = \mathfrak{z}_t^0/\tilde{\mathfrak{z}}_t^0$ and put $0/0 = 0$. Obviously, on the set $\{\tilde{\mathfrak{z}}_t > 0\}$ we have

$$z_t^0 = \overline{\alpha}_t z_t, \qquad \widetilde{z}_t^0 = \widetilde{z}_t/\overline{\alpha}_t \qquad (Q^t\text{-a.s.}).$$

For testing the hypotheses H_0^t and $\widetilde{H}_0^t$ in the problem under consideration, we introduce the test

$$\delta_{t,0}^{c,\varepsilon} = \chi\left(z_t^0 > c\right) + \varepsilon\chi\left(z_t^0 = c\right), \tag{2.7.1}$$

where $c \in [0, \infty]$ and $\varepsilon \in [0,1]$ are parameters of the test $\delta_{t,0}^{c,\varepsilon}$.

Now, applying Neyman–Pearson lemma, we see that in this problem the Neyman–Pearson test of level α (we denote it by $\delta_{t,0}^{+,\alpha}$) has the form (2.7.1) with parameters $c = c^0(\alpha)$ and $\varepsilon = \varepsilon^0(\alpha)$, where $(c^0(\alpha), \varepsilon^0(\alpha))$ is the solution of the equation $\alpha_0(\delta_{t,0}^{c,\varepsilon}) = \alpha$ with respect to (c, ε); $\varepsilon^0(0) = 1$. It is not hard to show that $c^0(\alpha)$ and $\varepsilon^0(\alpha)$ can be chosen so that for every $\alpha \in [0,1]$

$$c^0(\alpha) = \overline{\alpha}_t c(\alpha\overline{\alpha}_t), \qquad \varepsilon^0(\alpha) = \varepsilon(\alpha\overline{\alpha}_t), \tag{2.7.2}$$

where $c(\alpha)$ and $\varepsilon(\alpha)$ are defined by the equalities (2.1.11).

Moreover, for every $\alpha \in [0,1]$ we have

$$\beta_0\left(\delta_{t,0}^{+,\alpha}\right) = \beta\left(\delta_t^{+,\alpha\overline{\alpha}_t}\right), \tag{2.7.3}$$

where $\delta_t^{+,\alpha}$ is the Neyman–Pearson test of level α in the problem of testing the hypotheses H^t and $\widetilde{H}^t$.

The above problem of testing the hypotheses H_0^t and $\widetilde{H}_0^t$ and its relation to the problem of testing the hypotheses H^t and $\widetilde{H}^t$ allow us to consider the case $\overline{\lim}_{t\to\infty} \overline{\alpha}_t < 1$ and, as a consequence, to improve the results of Sections 2.3–2.6. To give an exact formulation of results, we introduce the following conditions:

$\Lambda 1_0'$. $\lim_{t\to\infty} \overline{\alpha}_t^{-1} P^t(\Lambda_t + \ln\overline{\alpha}_t > -a\chi_t) = 0$ for all $a < 1$;

$\Lambda 1_0''$. $\lim_{t\to\infty} \overline{\alpha}_t^{-1} P^t(-\infty < \Lambda_t + \ln\overline{\alpha}_t < -a\chi_t) = 0$ for all $a > 1$;

$\alpha 1_0$. $\underline{\lim}_{t\to\infty} \alpha_t/\overline{\alpha}_t > 0$;

$\alpha 2_0$. $\overline{\lim}_{t\to\infty} \alpha_t/\overline{\alpha}_t < 1$;
$d1_0$. $\overline{\lim}_{t\to\infty} \chi_t^{-1}(d_t + \ln \overline{\alpha}_t) \le -1$;
$d2_0$. $\underline{\lim}_{t\to\infty} \chi_t^{-1}(d_t + \ln \overline{\alpha}_t) \ge -1$;
$d1_0'$. $\overline{\lim}_{t\to\infty} \chi_t^{-1} \ln c^0(\alpha_t) \le -1$;
$d2_0'$. $\underline{\lim}_{t\to\infty} \chi_t^{-1} \ln c^0(\alpha_t) \ge -1$;
$\beta 1_0$. $\overline{\lim}_{t\to\infty} \chi_t^{-1} \ln \beta_0(\delta_{t,0}^{+,\alpha_t}) \le -1$;
$\beta 2_0$. $\underline{\lim}_{t\to\infty} \chi_t^{-1} \ln \beta_0(\delta_{t,0}^{+,\alpha_t}) \ge -1$,
where χ_t is a normalizing factor such that $\chi_t \to \infty$ as $t \to \infty$.

THEOREM 2.7.1. *The following implications are valid:*

$$\Lambda 1_0', \alpha 1_0 \Rightarrow d1_0 \Rightarrow \beta 1, \tag{2.7.4}$$

$$\Lambda 2_0', \alpha 2_0 \Rightarrow \beta 2 \Rightarrow d2_0. \tag{2.7.5}$$

PROOF. It is evident that

$$\Lambda 1_0' \Leftrightarrow \lim_{t\to\infty} P_0^t\left(\ln z_t^0 > -a\chi_t\right) = 0 \quad \text{for all } a < 1.$$

From this, applying the implications (2.3.6) to the test $\delta_{t,0}^{+,\alpha_t}$, we get

$$\Lambda 1_0', \alpha 1 \Rightarrow d1_0' \Rightarrow \beta 1_0. \tag{2.7.6}$$

Taking into account the equalities (2.7.2) and (2.7.3) and replacing α_t by $\alpha_t/\overline{\alpha}_t$ in the implications (2.7.6), we obtain the implications (2.7.4).

We have

$$\Lambda 1_0'' \Leftrightarrow \lim_{t\to\infty} P_0^t\left(\ln z_t^0 < -a\chi_t\right) = 0 \quad \text{for all } a > 1.$$

Therefore, applying the implication (2.3.7) to $\delta_{t,0}^{+,\alpha_t}$, we get

$$\Lambda 1_0'', \alpha 2 \Rightarrow \beta 2_0 \Rightarrow d2_0'. \tag{2.7.7}$$

Taking into account (2.7.2) and (2.7.3) once again and replacing α_t by $\alpha_t/\overline{\alpha}_t$ in (2.7.7), we obtain (2.7.5). □

The following theorem improves the estimates from Theorem 2.3.2.

THEOREM 2.7.2. *For any $\alpha \in (0, \overline{\alpha}_t)$ and $t \in \mathbf{R}_+$*

$$\beta\left(\delta_t^{+,\alpha}\right) \ge (\overline{\alpha}_t - \alpha)^{\varepsilon/(\varepsilon-1)} \left(\mathsf{E}^t\, z_t^{1-\varepsilon}\chi(\mathfrak{z}_t > 0)\right)^{1/(1-\varepsilon)}, \quad \varepsilon > 1, \tag{2.7.8}$$

$$\beta\left(\delta_t^{+,\alpha}\right) \le (1-\varepsilon)(\varepsilon/\alpha)^{\varepsilon/(1-\varepsilon)} \left(\mathsf{E}^t\, z_t^{1-\varepsilon}\chi(\tilde{\mathfrak{z}}_t > 0)\right)^{1/(1-\varepsilon)}, \quad 0 < \varepsilon < 1. \tag{2.7.9}$$

PROOF. Applying Theorem 2.3.2 to the problem of testing the hypotheses H_0^t and $\widetilde{H}_0^t$, for any $\alpha \in (0,1)$ and $t \in \mathbf{R}_+$ we get

$$\beta\left(\delta_{t,0}^{+,\alpha}\right) \ge (1-\alpha)^{\varepsilon/(\varepsilon-1)} \left(\mathsf{E}_0^t (z_t^0)^{1-\varepsilon}\right)^{1/(1-\varepsilon)}, \quad \varepsilon > 1, \tag{2.7.10}$$

$$\beta\left(\delta_{t,0}^{+,\alpha}\right) \le (1-\varepsilon)(\varepsilon/\alpha)^{\varepsilon/(1-\varepsilon)} \left(\mathsf{E}_0^t (z_t^0)^{1-\varepsilon}\right)^{1/(1-\varepsilon)}, \quad 0 < \varepsilon < 1. \tag{2.7.11}$$

The equality

$$\mathsf{E}_0^t\left(z_t^0\right)^{1-\varepsilon} = \overline{\alpha}_t^{-\varepsilon}\,\mathsf{E}^t\, z_t^{1-\varepsilon}\chi(\tilde{\mathfrak{z}}_t > 0) \tag{2.7.12}$$

is true. Replacing α_t by $\alpha_t/\overline{\alpha}_t$ in the estimates (2.7.10) and (2.7.11) and taking into account (2.7.3) and (2.7.12), we get the estimates (2.7.8) and (2.7.9). □

To formulate the next theorem, we introduce the conditions:

$\Lambda 2_0$. $\overline{\lim}_{\varepsilon\downarrow 0}\overline{\lim}_{t\to\infty}\,\varepsilon^{-1}\chi_t^{-1}\ln(\overline{\alpha}_t^{\varepsilon-1}\,\mathsf{E}^t\, z_t^{\varepsilon}\chi(\tilde{\mathfrak{z}}_t > 0)) \le -1$;

$\Lambda 3_0$. $\underline{\lim}_{\varepsilon\uparrow 0}\underline{\lim}_{t\to\infty}\,\varepsilon^{-1}\chi_t^{-1}\ln(\overline{\alpha}_t^{\varepsilon-1}\,\mathsf{E}^t\, z_t^{\varepsilon}\chi(\tilde{\mathfrak{z}}_t > 0)) \ge -1$;

$\alpha 1_0'$. $\lim_{t\to\infty}\chi_t^{-1}\ln(\alpha_t/\overline{\alpha}_t) = 0$;

$\alpha 2_0'$. $\lim_{t\to\infty}\chi_t^{-1}\ln(1-\alpha_t/\overline{\alpha}_t) = 0$.

THEOREM 2.7.3. *The following implications are valid:*

$$\Lambda 2_0, \alpha 1_0' \Rightarrow d1_0 \Rightarrow \beta 1, \tag{2.7.13}$$

$$\Lambda 3_0, \alpha 2_0' \Rightarrow \beta 2 \Rightarrow d2_0. \tag{2.7.14}$$

PROOF. Repeating the proof of the inequality (2.3.33) for the test $\delta_{t,0}^{+,\alpha_t}$, we get

$$\ln c^0(\alpha_t) \le \varepsilon^{-1}\ln \mathsf{E}_0^t\left(z_t^0\right)^{\varepsilon} - \varepsilon^{-1}\ln\alpha_t, \qquad 0<\varepsilon<1. \tag{2.7.15}$$

Taking into consideration (2.7.2) and (2.7.12) and replacing α_t by $\alpha_t/\overline{\alpha}_t$ in (2.7.15), we get the inequality

$$d_t + \ln\overline{\alpha}_t \le \varepsilon^{-1}\ln\left(\overline{\alpha}_t^{\varepsilon-1}\,\mathsf{E}^t\left(z_t^{\varepsilon}\chi(\tilde{\mathfrak{z}}_t>0)\right)\right) - \varepsilon^{-1}\ln(\alpha_t/\overline{\alpha}_t), \qquad 0<\varepsilon<1.$$

From this the implication $\Lambda 2_0$, $\alpha 1_0' \Rightarrow d1_0$ follows. The implication $\Lambda 3_0$, $\alpha 2_0' \Rightarrow \beta 2$ follows from the estimate (2.7.10). To complete the proof of the implications (2.7.13) and (2.7.14), it is enough to take into account (2.7.4) and (2.7.5). □

REMARK 2.7.1. If $0 < \underline{\lim}_{t\to\infty}\overline{\alpha}_t \le \overline{\lim}_{t\to\infty}\overline{\alpha}_t < 1$, then it is not hard to see that conditions $\Lambda 1_0'$, $\Lambda 1_0''$, $\alpha 1_0$, $\alpha 2_0$, $d1_0$, $d2_0$, and $\alpha 1_0'$ are equivalent to conditions $\Lambda 1'$, $\Lambda 1''$, $\alpha 1$, $\alpha 2$, $d1$, $d2$, and $\alpha 1'$, respectively, while conditions $\Lambda 2_0$, $\Lambda 3_0$, and $\alpha 2_0'$ get the form

$\Lambda 2_0$. $\overline{\lim}_{\varepsilon\downarrow 0}\overline{\lim}_{t\to\infty}\,\varepsilon^{-1}\chi^{-1}\ln\mathsf{E}^t(z_t^{\varepsilon}\chi(\tilde{\mathfrak{z}}_t>0)) \le -1$;

$\Lambda 3_0$. $\underline{\lim}_{\varepsilon\uparrow 0}\underline{\lim}_{t\to\infty}\,\varepsilon^{-1}\chi^{-1}\ln\mathsf{E}^t(z_t^{\varepsilon}\chi(\tilde{\mathfrak{z}}_t>0)) \ge -1$;

$\alpha 2_0'$. $\lim_{t\to\infty}\chi^{-1}\ln(\overline{\alpha}_t-\alpha_t) = 0$.

REMARK 2.7.2. If $\chi_t^{-1}\ln\overline{\alpha}_t \to 0$ as $t\to\infty$, then conditions $d1_0$, $d2_0$, and $\alpha 1_0'$ are equivalent to conditions $d1$, $d2$, and $\alpha 1'$, respectively, while conditions $\Lambda 2_0$, $\Lambda 3_0$, and $\alpha 2_0'$ acquire the form specified in Remark 2.7.1.

Now we consider the following analog of condition $\Lambda 4$:

$\Lambda 4_0$. $\lim_{t\to\infty}\overline{\alpha}_t^{-1}P^t(-\infty < \psi_t^{-1}(\Lambda_t+\ln\overline{\alpha}_t) < x) = L_0(x)$ for any $x\in\mathbf{R}^1$ that is a continuity point of the function $L_0(x)$, where $\psi_t\to\infty$ as $t\to\infty$ and $L_0(x)$ is the distribution function of some probability law L_0 on $\mathbf{R}^1$.

REMARK 2.7.3. If $\overline{\alpha}_t \to 1$ as $t\to\infty$, then condition $\Lambda 4_0$ is equivalent to condition $\Lambda 4$ with $L(x) = L_0(x)$.

REMARK 2.7.4. If $\overline{\alpha}_t \to \overline{\alpha} \in (0,1)$ for $t \to \infty$, then under condition $\Lambda 4_0$ we have $\lim_{t\to\infty} P^t(\psi_t^{-1}\Lambda_t < x) = 1 - \overline{\alpha} + \overline{\alpha} L_0(x)$ for any continuity point of the function $L_0(x)$. Thus, condition $\Lambda 4$ is fulfilled with the degenerate distribution function $L(x) = 1 - \overline{\alpha} + \overline{\alpha} L_0(x)$.

THEOREM 2.7.4. *Let condition $\Lambda 4_0$ be fulfilled, in which the function $L_0(x)$ is continuous and strictly monotonically increases on $(\underline{l}_0, \overline{l}_0)$, where*

$$\underline{l}_0 = \sup\{L_0(x) = 0\}, \qquad \overline{l}_0 = \inf\{L_0(x) = 1\}.$$

Then for any $\alpha \in (0,1)$

$$\lim_{t\to\infty} \frac{\alpha_t}{\overline{\alpha}_t} = \alpha \Leftrightarrow \lim_{t\to\infty} \frac{d_t + \ln \overline{\alpha}_t}{\psi_t} = l^0_{1-\alpha} \Leftrightarrow \lim_{t\to\infty} \frac{\ln \beta\left(\delta_t^{+,\alpha_t}\right)}{\psi_t} = l^0_{1-\alpha}, \tag{2.7.16}$$

where l^0_p is the p-quantile of the law L_0.

PROOF. Applying Theorem 2.4.2 to the test $\delta_{t,0}^{+,\alpha_t}$ in the problem of testing the hypotheses H_0^t and $\widetilde{H}_0^t$, for any $\alpha \in (0,1)$ we get

$$\lim_{t\to\infty} \alpha_t = \alpha \Leftrightarrow \lim_{t\to\infty} \frac{\ln c^0(\alpha_t)}{\psi_t} = l^0_{1-\alpha} \Leftrightarrow \lim_{t\to\infty} \frac{\ln \beta_0\left(\delta_{t,0}^{+,\alpha_t}\right)}{\psi_t} = l^0_{1-\alpha}. \tag{2.7.17}$$

Taking into account the equalities (2.7.2) and (2.7.3) and replacing α_t in (2.7.17) by $\alpha_t/\overline{\alpha}_t$, we obtain (2.7.16). □

The corresponding modification of condition $\Lambda 5$ and the analog of Theorem 2.4.4 can be considered similarly.

Consider the analog of condition $\Lambda 6$:

$\Lambda 6_0$. $\lim_{t\to\infty} \overline{\alpha}_t^{-1} P^t(-\infty < \Lambda_t + \ln \overline{\alpha}_t < x) = L_0(x)$
for any continuity point of the function $L_0(x)$, where $L_0(x)$ is a distribution function on $(-\infty, \infty)$.

REMARK 2.7.5. If $\overline{\alpha}_t \to 1$ as $t \to \infty$, then, evidently, condition $\Lambda 6_0$ is equivalent to condition $\Lambda 6$ with $L(x) = L_0(x)$. If $\overline{\alpha}_t \to \overline{\alpha} \in (0,1)$, then under condition $\Lambda 6_0$ we have

$$\lim_{t\to\infty} P^t(\Lambda_t < x) = 1 - \overline{\alpha} + \overline{\alpha} L_0(x + \ln \overline{\alpha})$$

for all $x \in \mathbf{R}^1$ for which $x + \ln \overline{\alpha}$ is a continuity point of the function $L_0(x)$.

The following theorem can be proved similarly to Theorem 2.5.2.

THEOREM 2.7.5. *If condition $\Lambda 6_0$ is fulfilled and $\overline{\alpha}_t \to \overline{\alpha} \in (0,1)$ as $t \to \infty$, then*

$$\lim_{t\to\infty} \widetilde{P}^t(\Lambda_t < y) = \widetilde{L}_0(y + \ln \overline{\alpha})$$

for any $y \in \mathbf{R}^1$ for which $y + \ln \overline{\alpha}$ is a continuity point of the function $L_0(x)$, where

$$\widetilde{L}_0(y) = \int_{-\infty}^{y} e^x \, dL_0(x). \tag{2.7.18}$$

The following analog of Theorem 2.5.5 is true.

THEOREM 2.7.6. *Let $\overline{\alpha}_t \to \overline{\alpha} \in (0,1)$ as $t \to \infty$, and let condition $\Lambda 6_0$ be fulfilled, in which the function $L_0(x)$ is continuous and strictly monotonically increasing on $(\underline{l}_0, \overline{l}_0)$, where $\underline{l}_0 = \sup\{x: L_0(x) = 0\}$ and $\overline{l}_0 = \inf\{x: L_0(x) = 1\}$. Then for any $\alpha \in [0, \overline{\alpha}]$ we have*

$$\lim_{t\to\infty} \alpha_t = \alpha \Leftrightarrow \lim_{t\to\infty} \beta\left(\delta_t^{+,\alpha_t}\right) \Leftrightarrow \widetilde{L}_0\left(l^0_{1-\alpha/\overline{\alpha}}\right),$$

where l^0_p is the p-quantile of the distribution function $L_0(x)$ and $\widetilde{L}_0(x)$ is the distribution function of the form (2.7.18).

In the general case (without any additional constraints on the behavior of $\overline{\alpha}_t$) the following theorem holds.

THEOREM 2.7.7. *If condition $\Lambda 6_0$ is fulfilled, in which the function $L_0(x)$ is continuous and strictly monotonically increases on $(\underline{l}_0, \overline{l}_0)$, then for any $\alpha \in [0,1]$*

$$\lim_{t\to\infty} \alpha_t/\overline{\alpha}_t = \alpha \Leftrightarrow \lim_{t\to\infty} \beta(\delta_t^{+,\alpha_t}) \Leftrightarrow \widetilde{L}_0(l^0_{1-\alpha}).$$

Theorems 2.7.6 and 2.7.7 can be proved using the same arguments as those used in the proofs of Theorems 2.7.1, 2.7.3, and 2.7.4.

The second reduction. Let $(X_1^t, \mathfrak{B}_1^t, P_1^t, \widetilde{P}_1^t)$, $t \in \mathbf{R}_+$, be the family of statistical experiments generated by an observation ξ_1^t, and let H_1^t and $\widetilde{H}_1^t$ be simple hypotheses consisting in that the distribution of the observation ξ_1^t is defined by the measures P_1^t and $\widetilde{P}_1^t$, respectively. We assume that

$$X_1^t = \{\mathfrak{z}_t > 0, \tilde{\mathfrak{z}}_t > 0\}, \qquad \mathfrak{B}_1^t = \mathfrak{B}^t \cap X_1^t, \qquad P_1^t = P^t/\overline{\alpha}_t, \qquad \widetilde{P}_1^t = \widetilde{P}^t/\overline{\beta}_t.$$

Let a measurable map δ_t from $(X_1^t, \mathfrak{B}_1^t)$ into $([0,1], \mathcal{B}([0,1]))$ be a test for distinguishing between the hypotheses H_1^t and $\widetilde{H}_1^t$ from an observation ξ_1^t. We denote by $\Sigma^{t,1}$ the family of all such tests and for any test $\delta_t \in \Sigma^{t,1}$ we introduce the probabilities of errors of the first and the second kind

$$\alpha_1(\delta_t) = \mathsf{E}_1^t\, \delta_t, \qquad \beta_1(\delta_t) = \widetilde{\mathsf{E}}_1^t(1-\delta_t),$$

where E_1^t and $\widetilde{\mathsf{E}}_1^t$ are the mathematical expectations with respect to the distributions P_1^t and $\widetilde{P}_1^t$, respectively.

Let $\mathfrak{z}_t^1 = dP_1^t/dQ_1^t$ and $\tilde{\mathfrak{z}}_t^1 = d\widetilde{P}_1^t/dQ_1^t$ be finite versions of the Radon–Nikodym derivatives, where Q_1^t is the restriction of the measure Q^t to the σ-algebra $\mathfrak{B}_1^t$. Evidently, on the set $\{\mathfrak{z}_t > 0, \tilde{\mathfrak{z}}_t > 0\}$ we have $\mathfrak{z}_t^1 = \mathfrak{z}_t/\overline{\alpha}_t$ and $\tilde{\mathfrak{z}}_t^1 = \tilde{\mathfrak{z}}_t/\overline{\beta}_t$ (Q^t-a.s.). We introduce the likelihood ratios $z_t^1 = \tilde{\mathfrak{z}}_t^1/\mathfrak{z}_t^1$ and $\widetilde{z}_t^1 = \mathfrak{z}_t^1/\tilde{\mathfrak{z}}_t^1$ and put $0/0 = 0$. Clearly, on the set $\{\mathfrak{z}_t > 0, \tilde{\mathfrak{z}}_t > 0\}$ we have

$$z_t^1 = z_t \overline{\alpha}_t/\overline{\beta}_t, \qquad \widetilde{z}_t^1 = \widetilde{z}_t \overline{\beta}_t/\overline{\alpha}_t \qquad (Q^t\text{-a.s.}).$$

For testing the hypotheses H_1^t and $\widetilde{H}_1^t$ in the problem under consideration, we introduce the test

$$\delta_{t,1}^{c,\varepsilon} = \chi\left(z_t^1 > c\right) + \varepsilon\chi\left(z_t^1 = c\right), \tag{2.7.19}$$

where $c \in [0,\infty]$ and $\varepsilon \in [0,1]$ are parameters of the test $\delta_{t,1}^{c,\varepsilon}$.

Applying the Neyman–Pearson lemma to this problem of testing hypotheses, we see that the Neyman–Pearson test of level α (we denote it by $\delta_{t,1}^{+,\alpha}$) has the form (2.7.19) with parameters $c = c^1(\alpha)$ and $\varepsilon = \varepsilon^1(\alpha)$, where $(c^1(\alpha), \varepsilon^1(\alpha))$ is the solution of the equation $\alpha_1(\delta_{t,1}^{c,\varepsilon}) = \alpha$ with respect to (c, ε), where $\varepsilon^1(0) = 1$. Obviously, $c^1(\alpha)$ and $\varepsilon^1(\alpha)$ can be chosen so that for any $\alpha \in [0,1]$

$$c^1(\alpha) = \overline{\alpha}_t \overline{\beta}_t^{-1} c(\alpha\overline{\alpha}_t), \qquad \varepsilon^1(\alpha) = \varepsilon(\alpha\overline{\alpha}_t), \tag{2.7.20}$$

where $c(\alpha)$ and $\varepsilon(\alpha)$ are defined in (2.1.11). Then it is easy to demonstrate that for any $\alpha \in [0,1]$

$$\beta_1(\delta_{t,1}^{+,\alpha}) = \overline{\beta}_t^{-1} \beta(\delta_{t,1}^{+,\alpha\overline{\alpha}_t}). \tag{2.7.21}$$

As in the case of the first reduction, we introduce the conditions:

$\Lambda 1_1'$. $\lim_{t\to\infty} \overline{\alpha}_t^{-1} P^t(\Lambda_t + \ln(\overline{\alpha}_t/\overline{\beta}_t) > -a\chi_t) = 0$ for all $a < 1$;

$\Lambda 1_1''$. $\lim_{t\to\infty} \overline{\alpha}_t^{-1} P^t(-\infty < \Lambda_t + \ln(\overline{\alpha}_t/\overline{\beta}_t) < -a\chi_t) = 0$ for all $a > 1$;

$d1_1$. $\overline{\lim}_{t\to\infty} \chi_t^{-1}(d_t + \ln(\overline{\alpha}_t/\overline{\beta}_t)) \le -1$;

$d2_1$. $\underline{\lim}_{t\to\infty} \chi_t^{-1}(d_t + \ln(\overline{\alpha}_t/\overline{\beta}_t)) \ge -1$;

$d1_1'$. $\overline{\lim}_{t\to\infty} \chi_t^{-1} \ln c^1(\alpha_t) \le -1$;

$d2_1'$. $\underline{\lim}_{t\to\infty} \chi_t^{-1} \ln c^1(\alpha_t) \ge -1$;

$\beta 1_1$. $\overline{\lim}_{t\to\infty} \chi_t^{-1} \ln \beta_1(\delta_{t,1}^{+,\alpha_t}) \le -1$;

$\beta 2_1$. $\underline{\lim}_{t\to\infty} \chi_t^{-1} \ln \beta_1(\delta_{t,1}^{+,\alpha_t}) \ge -1$,

where $\chi_t \to \infty$ as $t \to \infty$.

THEOREM 2.7.8. *The following implications are valid:*

$$\Lambda 1_1', \alpha 1_0 \Rightarrow d1_1 \Rightarrow \beta 1, \tag{2.7.22}$$

$$\Lambda 1_1'', \alpha 2_0 \Rightarrow \beta 2 \Rightarrow d2_1. \tag{2.7.23}$$

PROOF. If is evident that

$$\Lambda 1_1' \Leftrightarrow \lim_{t\to\infty} P_1^t\left(\ln z_t^1 > -a\chi_t\right) = 0 \text{ for all } a < 1.$$

Now, applying the implication (2.3.6) to the test $\delta_{t,1}^{+,\alpha_t}$, we obtain

$$\Lambda 1_1', \alpha 1 \Rightarrow d1_1' \Rightarrow \beta 1_1. \tag{2.7.24}$$

Taking into account the equalities (2.7.20) and (2.7.21) and replacing α_t by $\alpha_t/\overline{\alpha}_t$ in the implication (2.7.24), we get the implication (2.7.22).

It is evident that

$$\Lambda 1_1'' \Leftrightarrow \lim_{t\to\infty} P_1^t\left(\ln z_t^1 < -a\chi_t\right) = 0 \quad \text{for all } a > 1.$$

Applying the implication (2.3.7) to $\delta_{t,1}^{+,\alpha_t}$, we get

$$\Lambda 1_1'', \alpha 2 \Rightarrow \beta 2_1 \Rightarrow d2_1'. \tag{2.7.25}$$

Taking into consideration once again the equalities (2.7.20) and (2.7.21) and replacing α_t by $\alpha_t/\overline{\alpha}_t$ in the implication (2.7.25), we obtain the implication (2.7.23). □

To formulate the next theorem, we introduce the following conditions:

$\Lambda 2_1$. $\overline{\lim}_{\varepsilon\downarrow 0} \overline{\lim}_{t\to\infty} \varepsilon^{-1}\chi_t^{-1} \ln\left(\overline{\alpha}_t^{\varepsilon-1} \overline{\beta}_t^{-\varepsilon} \mathsf{E}^t z_t^\varepsilon \chi(\mathfrak{z}_t > 0)\right) \le -1$;

$\Lambda 3_1$. $\underline{\lim}_{\varepsilon\uparrow 0} \underline{\lim}_{t\to\infty} \varepsilon^{-1}\chi_t^{-1} \ln\left(\overline{\alpha}_t^{\varepsilon-1} \overline{\beta}_t^{-\varepsilon} \mathsf{E}^t z_t^\varepsilon \chi(\tilde{\mathfrak{z}}_t > 0)\right) \ge -1$.

THEOREM 2.7.9. *The following implications are valid:*

$$\Lambda 2_1, \alpha 1_0' \Rightarrow d1_1 \Rightarrow \beta 1,$$
$$\Lambda 3_1, \alpha 2_0' \Rightarrow \beta 2 \Rightarrow d2_1.$$

PROOF. If we take into account the equality

$$\mathsf{E}_1^t \left(z_t^1\right)^\varepsilon = \overline{\alpha}_t^{\varepsilon-1} \overline{\beta}_t^{-\varepsilon} \mathsf{E}^t z_t^\varepsilon \chi(\tilde{\mathfrak{z}}_t > 0),$$

then the proof of this theorem becomes similar to that of Theorem 2.7.3. □

Now we consider the following analog of condition $\Lambda 4$:

$\Lambda 4_1$. $\lim_{t\to\infty} \overline{\alpha}_t^{-1} P^t\big(-\infty < \psi_t^{-1}(\Lambda_t + \ln(\overline{\alpha}_t/\overline{\beta}_t)) < x\big) = L_1(x)$

for any continuity point of the function $L_1(x)$, where $\psi_t \to \infty$ as $t \to \infty$, and $L_1(x)$ is the distribution function of a probability law on $\mathbf{R}^1$.

The following analog of Theorem 2.4.2 is proved similarly to Theorem 2.7.4.

THEOREM 2.7.10. *Let condition $\Lambda 4_1$ be fulfilled, in which the function $L_1(x)$ is continuous and strictly monotonically increasing on $(\underline{l}_1, \overline{l}_1)$, where*

$$\underline{l}_1 = \sup\{x \colon L_1(x) = 0\}, \qquad \overline{l}_1 = \inf\{x \colon L_1(x) = 1\}.$$

Then for any $\alpha \in (0,1)$

$$\lim_{t\to\infty} \alpha_t/\overline{\alpha}_t = \alpha \Leftrightarrow \lim_{t\to\infty} \psi_t^{-1}(d_t + \ln(\overline{\alpha}_t/\overline{\beta}_t)) = l_{1-\alpha}^1$$
$$\Leftrightarrow \lim_{t\to\infty} \psi_t^{-1} \ln(\beta(\delta_t^{+,\alpha_t})/\overline{\beta}_t) = l_{1-\alpha}^1,$$

where l_p^1 is the p-quantile of the distribution function $L_1(x)$.

The corresponding modification of condition $\Lambda 5$ and an analog of Theorem 2.4.4 can be considered similarly. Moreover, similarly to condition $\Lambda 6_0$, we can introduce a modification of condition $\Lambda 6$ to obtain an analog of Theorem 2.5.5.

CHAPTER 3

Asymptotic Behavior of the Likelihood Ratio in Problems of Distinguishing between Simple Hypotheses for Semimartingales

3.1. Hellinger integrals and Hellinger processes

Let $(\mathbf{D}^t, \mathfrak{B}^t, P^t, \widetilde{P}^t)$, $t \in \mathbf{R}_+$, be a family of binary statistical experiments generated by observations $\xi^t = (\xi_s)_{0\le s\le t}$ of a semimartingale $\xi = (\xi_s)$, which was introduced in Section 1.2. Consider two simple hypotheses H^t and $\widetilde{H}^t$ consisting in that the distribution of the observation ξ^t is given by the measures P^t and $\widetilde{P}^t$, respectively. We assume that the conditions of Theorems 1.2.1 and 1.2.2 ensuring the absolute continuity of the measure $\widetilde{P}^t$ with respect to the measure P^t and giving the form of the corresponding density $z_t = d\widetilde{P}^t/dP^t(\xi^t)$ (P-a.s.) are fulfilled. We consider the problem of distinguishing between the hypotheses H^t and $\widetilde{H}^t$ from an observation ξ^t of the semimartingale ξ and investigate the asymptotic properties of the Neyman–Pearson test δ_t^{+,α_t} of level α_t, introduced in Section 2.1. According to Chapter 2, to do this requires establishing some asymptotic properties of the local density $z = (z_t)$. This chapter is devoted to this problem.

Let ξ be a semimartingale on a space with filtration $(\Omega, \mathcal{F}, \mathbf{F})$ with triples of predictable characteristics $(\alpha, \langle m\rangle, \nu)$ and $(\widetilde{\alpha}, \langle \widetilde{m}\rangle, \widetilde{\nu})$ with respect to measures P and $\widetilde{\mathsf{P}}$, respectively.

According to Definition 2.2.5 of Hellinger integral of order ε for measures $\widetilde{P}^t$ and P^t, we put

$$(3.1.1)\qquad H_t(\varepsilon; \widetilde{\mathsf{P}}, \mathsf{P}) = H(\varepsilon; \widetilde{P}^t, P^t) = \mathsf{E}_Q\, \widetilde{\mathfrak{z}}_t^{\varepsilon} \mathfrak{z}_t^{1-\varepsilon},$$

where $\mathfrak{z}_t = dP^t/d\mathbf{Q}$, $\widetilde{\mathfrak{z}}_t = d\widetilde{P}^t/d\mathbf{Q}$, $\mathbf{Q} = (\mathsf{P} + \widetilde{\mathsf{P}})/2$, and P^t and $\widetilde{P}^t$ are restrictions of measures P and $\widetilde{\mathsf{P}}$ to the σ-algebra $\mathcal{F}_t$. We introduce random sets

$$\Gamma = \{\mathfrak{z}_- > 0\} \cap [\![0]\!], \qquad \widetilde{\Gamma} = \{\widetilde{\mathfrak{z}}_- > 0\} \cap [\![0]\!].$$

THEOREM 3.1.1. *Let $\varepsilon \in (0,1)$ and $Y(\varepsilon) = \widetilde{\mathfrak{z}}^{\varepsilon}\mathfrak{z}^{1-\varepsilon}$. Then there exists a predictable increasing process $h(\varepsilon)$ with $h_0(\varepsilon) = 0$, unique up to $\mathbf{Q}$-indistinguishability, such that:*

1) $h(\varepsilon) = \chi(\Gamma \cap \widetilde{\Gamma}) \circ h(\varepsilon)$;
2) $M(\varepsilon) = Y(\varepsilon) + Y_-(\varepsilon) \circ h(\varepsilon) \in \mathcal{M}(\mathbf{F}, \mathbf{Q})$.

If $H_t(\varepsilon_0; \widetilde{\mathsf{P}}, \mathsf{P}) < \infty$ for some $\varepsilon_0 < 0$ and all $t \in \mathbf{R}_+$, then for each $\varepsilon \in [\varepsilon_0, 0)$ there exists a predictable increasing process $h(\varepsilon)$ with $h_0(\varepsilon) = 0$, unique up to $\mathbf{Q}$-indistinguishability, such that:

1′) $h(\varepsilon) = \chi(\Gamma) \circ h(\varepsilon)$;
2′) $M(\varepsilon) = Y(\varepsilon) - Y_-(\varepsilon) \circ h(\varepsilon) \in \overline{\mathcal{M}}(\mathbf{F}, \mathbf{Q})$.

PROOF. For $\varepsilon \in (0,1)$ the proof can be found in ([**156**], Theorem 4.1.18) and for $\varepsilon < 0$, in ([**49**], Theorem 2.1). □

DEFINITION 3.1.1. a) The increasing P- and $\widetilde{\mathsf{P}}$-unique process $h(\varepsilon) = h(\varepsilon; \widetilde{\mathsf{P}}, \mathsf{P})$ constructed in Theorem 3.1.1 is called a *Hellinger process in the strict sense of order* ε for the measures $\widetilde{\mathsf{P}}$ and P.

b) Any increasing predictable process $h'(\varepsilon)$ such that the processes $\chi_\Delta h'(\varepsilon)$ and $\chi_\Delta h(\varepsilon)$ (or, equivalently, $\chi_\Delta \circ h'(\varepsilon)$ and $\chi_\Delta \circ h(\varepsilon)$) are P- and $\widetilde{\mathsf{P}}$-indistinguishable (here $\Delta = \Gamma \cap \widetilde{\Gamma}$ for $\varepsilon \in (0,1)$ and $\Delta = \Gamma$ for $\varepsilon < 0$) is called a *Hellinger process of order* ε for the measures $\widetilde{\mathsf{P}}$ and P.

REMARK 3.1.1. In Theorem 3.1.1 we do not consider the case $\varepsilon = 0$. This case requires special discussion. If we assume that for some $\varepsilon_0 < 0$ we have $H_t(\varepsilon_0; \widetilde{\mathsf{P}}, \mathsf{P}) < \infty$ for all $t \in \mathbf{R}_+$, then the process $Y(0) = \mathfrak{z}\chi(\tilde{\mathfrak{z}} > 0) = \mathfrak{z}$ ($\mathbf{Q}$-a.s.) is a $\mathbf{Q}$-martingale, and from the decomposition 2) we have $h(0) = 0$. If $H_t(\varepsilon_0; \widetilde{\mathsf{P}}, \mathsf{P})$ is not assumed finite for $\varepsilon_0 < 0$, $h(0)$ can be defined as in [**156**].

The following theorem demonstrates the relationship between Hellinger processes $h(\varepsilon) = h(\varepsilon; \widetilde{\mathsf{P}}, \mathsf{P})$ and Hellinger integrals $H_t(\varepsilon; \widetilde{\mathsf{P}}, \mathsf{P})$.

THEOREM 3.1.2. *Let $h(\varepsilon)$ be a version of a Hellinger process of order $\varepsilon \in [\varepsilon_0, 1)$ and let $H_t(\varepsilon_0; \widetilde{\mathsf{P}}, \mathsf{P}) < \infty$ for all $t \in \mathbf{R}_+$ for some $\varepsilon_0 < 0$. Then for any stopping time τ*

$$H(\varepsilon; \widetilde{P}^\tau, P^\tau) = H(\varepsilon; \widetilde{P}^0, P^0) - \operatorname{sgn}(\varepsilon)\, \mathsf{E}_Q\, Y_-(\varepsilon) \circ h(\varepsilon)_\tau \tag{3.1.2}$$

for all $\varepsilon \in [\varepsilon_0, 1)$.

PROOF. For $\varepsilon \in (0,1)$ the theorem is proved in ([**156**], Assertion 4.1.27) and for $\varepsilon < 0$ in ([**49**], Theorem 2.1). □

Now we present the *multiplicative decomposition* for the process $Y(\varepsilon)$. To do this, we first introduce the *Doléans stochastic exponent* $\mathcal{E}(G)$ for any process of local variation G by putting

$$\mathcal{E}_t(G) = e^{G_t} \prod_{0<s\le t} (1 + \Delta G_s) e^{-\Delta G_s}, \qquad t \in \mathbf{R}_+. \tag{3.1.3}$$

THEOREM 3.1.3. *If $H_t(\varepsilon_0; \widetilde{\mathsf{P}}, \mathsf{P}) < \infty$ for some $\varepsilon_0 < 0$ and all $t \in \mathbf{R}_+$, then for each $\varepsilon \in [\varepsilon_0, 1)$ there exists a process $N(\varepsilon)$ with the following properties:*

1) *for any version of the process $h(\varepsilon)$ we have*

$$Y(\varepsilon) = N(\varepsilon)\mathcal{E}(-\operatorname{sgn}(\varepsilon) h(\varepsilon)); \tag{3.1.4}$$

2) *$N^\tau(\varepsilon)$ is a $\mathbf{Q}$-local martingale if τ is a stopping time such that for $\varepsilon \in (0,1)$ we have $[\![0,\tau]\!] \subset \{\mathcal{E}(-h(\varepsilon)) > 0\}$ and for $\varepsilon \in [\varepsilon_0, 0)$ we have $[\![0,\tau]\!] \subset \{\mathcal{E}(h(\varepsilon)) > 0\}$ for any version of the process $h(\varepsilon)$;*

3) *$N(\varepsilon)$ is a $\mathbf{Q}$-supermartingale.*

PROOF. For $\varepsilon \in [0,1)$ the theorem is proved in [**156**], Assertion 5.4.16, and for $\varepsilon < 0$ in [**49**], Lemma 2.2. □

The following theorem gives an upper bound for $H_t(\varepsilon; \widetilde{\mathsf{P}}, \mathsf{P})$ expressed only in terms of the Hellinger process $h_t(\varepsilon)$.

THEOREM 1.3.4. *Let $H_t(\varepsilon_0; \widetilde{\mathsf{P}}, \mathsf{P}) < \infty$ for all $t \in \mathbf{R}_+$ for some $\varepsilon_0 < 0$. Then for any stopping time τ and any version of the Hellinger process $h(\varepsilon)$ for any $\varepsilon \in [\varepsilon_0, 1)$*

$$H(\varepsilon; \widetilde{P}^\tau, P^\tau) \le H_0^{\varepsilon/\delta}(\delta; \widetilde{\mathsf{P}}, \mathsf{P})\big(\mathsf{E}\,\mathcal{E}_\tau^{\varepsilon/(\delta-\varepsilon)}(-\operatorname{sgn}(\delta)h(\delta))\big)^{(\delta-\varepsilon)/\delta}, \tag{3.1.5}$$

where $0 < \varepsilon < \delta < 1$ for $0 < \varepsilon < 1$ and $\varepsilon_0 \le \delta < \varepsilon$ for $\varepsilon_0 < \varepsilon < 0$.

PROOF. Let $0 < \varepsilon < \delta < 1$ or $\varepsilon_0 \le \delta < \varepsilon < 0$ and $p = \delta/\varepsilon$, $q = \delta/(\delta - \varepsilon)$ so that $1/p + 1/q = 1$. Then the evident equality $\tilde{\mathfrak{z}}^\varepsilon \mathfrak{z}^{1-\varepsilon} = (\tilde{\mathfrak{z}}^\delta \mathfrak{z}^{1-\delta})^{1/p} \mathfrak{z}^{1/q}$ and multiplicative decomposition (3.1.4) imply

$$\tilde{\mathfrak{z}}_\tau^\varepsilon \mathfrak{z}_\tau^{1-\varepsilon} = (N_\tau(\delta))^{1/p} \left(\mathfrak{z}_\tau\, \mathcal{E}_\tau^{q/p}(-\operatorname{sgn}(\delta)h(\delta))\right)^{1/q}.$$

From this, using Hölder's inequality and taking into account that $N(\delta)$ is a **Q**-supermartingale, we get

$$\begin{aligned} H(\varepsilon; \widetilde{P}^\tau, P^\tau) &\le (\mathsf{E}_Q\, N_\tau(\delta))^{1/p} \big(\mathsf{E}_Q\, \mathfrak{z}_\tau \mathcal{E}_\tau^{q/p}(-\operatorname{sgn}(\delta)h(\delta))\big)^{1/q} \\ &\le (\mathsf{E}_Q\, N_0(\delta))^{1/p} \big(\mathsf{E}\,\mathcal{E}_\tau^{q/p}(-\operatorname{sgn}(\delta)h(\delta))\big)^{1/q}. \end{aligned}$$

Since in view of (3.1.4) we have $N_0(\delta) = \tilde{\mathfrak{z}}_0^\delta \mathfrak{z}_0^{1-\delta}$, the last inequality implies the inequality (3.1.5) by virtue of Definition 3.1.1. □

The properties specified in Theorem 3.1.1 characterize the process $h(\varepsilon)$ but do not describe its explicit form. Now we describe a technique for computing $h(\varepsilon)$ in terms of the triples $T = (\alpha, \langle m \rangle, \nu)$ and $\widetilde{T} = (\widetilde{\alpha}, \langle \widetilde{m} \rangle, \widetilde{\nu})$.

Denoting by $\operatorname{Var}_t(\nu - \widetilde{\nu})$ the variation of the difference $\nu - \widetilde{\nu}$ on the interval $[0, t]$, we put

$$\begin{gathered} V_t = \operatorname{Var}_t(\widetilde{\alpha} - \alpha) + |x|\,\chi(|x| \le 1) * \operatorname{Var}(\widetilde{\nu} - \nu)_t, \\ \tau = \inf\{t\colon V_t = \infty\}, \end{gathered}$$

and define a predictable process $B = (B_t)_{t \in \mathbf{R}_+}$, where

$$B_t = \begin{cases} 0, & \tau = 0, \\ \widetilde{\alpha}_t - \alpha_t - x\chi(|x| \le 1) * (\widetilde{\nu} - \nu)_t, & 0 \le t < \tau, \\ \overline{\lim}_{s \uparrow \tau} B_s, & t \ge \tau > 0. \end{cases}$$

Note that $\Delta B_t = 0$ on the set $\{0 \le t < \tau\}$, which means that the process B has continuous trajectories. Put

$$h_t^0(\varepsilon; \widetilde{T}, T) = \operatorname{sgn}(\varepsilon)\left\{\frac{\varepsilon(1-\varepsilon)}{2}\left(\frac{dB}{d\langle m \rangle}\right)^2 \circ \langle m \rangle_t + \varphi_\varepsilon(d\widetilde{\nu}, d\nu)_t + \sum_{0 < s \le t} \varphi_\varepsilon(1 - \widetilde{a}_s, 1 - a_s)\right\}, \tag{3.1.6}$$

where $dB/d\langle m \rangle$ is the Radon–Nikodym derivative of the absolutely continuous part of B with respect to $\langle m \rangle$, which can be chosen $\mathcal{P}$-measurable, the expression $\varphi_\varepsilon(d\widetilde{\nu}, d\nu)_t$ is understood as the integral

$$\varphi_\varepsilon(d\widetilde{\nu}, d\nu)_t = \varphi_\varepsilon\left(\frac{d\widetilde{\nu}}{d\overline{\nu}}, \frac{d\nu}{d\overline{\nu}}\right) * \overline{\nu}_t, \qquad \overline{\nu} = \nu + \widetilde{\nu},$$

$a_s = \nu(\{s\}, R_0)$, $\widetilde{a}_s = \widetilde{\nu}(\{s\}, R_0)$, and the function $\varphi_\varepsilon(u,v)$ is defined by the formula

$$\varphi_\varepsilon(u,v) = \varepsilon u + (1-\varepsilon)v - u^\varepsilon v^{1-\varepsilon}, \qquad \varphi_0(u,v) = 0. \tag{3.1.7}$$

THEOREM 3.1.5. *Let ξ be a semimartingale with triples of predictable characteristics $T = (\alpha, \langle m\rangle, \nu)$ and $\widetilde{T} = (\widetilde{\alpha}, \langle \widetilde{m}\rangle, \widetilde{\nu})$ with respect to the measures P and $\widetilde{\mathsf{P}}$, respectively. Assume that the processes $\langle m\rangle$ and $\langle\widetilde{m}\rangle$ are $\mathbf{Q}$-indistinguishable, at least one of the measures P and $\widetilde{\mathsf{P}}$ possesses the property of local uniqueness, and, in addition, $H_t(\varepsilon_0; \widetilde{\mathsf{P}}, \mathsf{P}) < \infty$ for all $t \in \mathbf{R}_+$ for some $\varepsilon_0 < 0$. Then the process $h^0(\varepsilon; \widetilde{T}, T)$ defined by formula* (3.1.6) *is a version of a Hellinger process of order ε for $\varepsilon \in [\varepsilon_0, 1)$.*

PROOF. For $\varepsilon \in (0,1)$ the theorem is proved in ([**156**], Section 4.3) and for $\varepsilon < 0$ in ([**49**], Assertion 1.1). □

Let the semimartingale ξ have determinate triples of predictable characteristics T and $\widetilde{T}$. Since the measure corresponding to a process with independent increments possesses the property of local uniqueness ([**153**], Proposition 13.2), under the condition $\langle m\rangle = \langle\widetilde{m}\rangle$ Theorem 3.1.5 yields determinate versions of processes $h(\varepsilon; \widetilde{T}, T)$, $\varepsilon \in [\varepsilon_0, 1)$. Moreover, in the case under consideration it is possible to compute exactly the Hellinger integral $H_t(\varepsilon; \widetilde{\mathsf{P}}, \mathsf{P})$ of order $\varepsilon \in [\varepsilon_0, 1)$.

THEOREM 3.1.6. *Let the measures P and $\widetilde{\mathsf{P}}$ be probability distributions of semimartingales with independent increments with triples of predictable characteristics $T = (\alpha, \langle m\rangle, \nu)$, $\widetilde{T} = (\widetilde{\alpha}, \langle\widetilde{m}\rangle, \widetilde{\nu})$, and let $h(\varepsilon) = h(\varepsilon; \widetilde{\mathsf{P}}, \mathsf{P})$ be determinate versions of Hellinger processes of order ε for the measures $\widetilde{\mathsf{P}}$ and P. If $H_t(\varepsilon_0; \widetilde{\mathsf{P}}, \mathsf{P}) < \infty$ for all $t \in \mathbf{R}_+$ for some $\varepsilon_0 < 0$, then for any $\varepsilon \in [\varepsilon_0, 1)$ and $t \in \mathbf{R}_+$*

$$H_t(\varepsilon; \widetilde{\mathsf{P}}, \mathsf{P}) = H_0(\varepsilon; \widetilde{\mathsf{P}}, \mathsf{P}) \mathcal{E}_t(-\operatorname{sgn}(\varepsilon) h(\varepsilon)). \tag{3.1.8}$$

Theorem 3.1.6 follows from the multiplicative decomposition (3.1.14) and determinate nature of $h(\varepsilon)$.

Now we describe the form of the Hellinger integral $h(\varepsilon)$ under the above assumptions concerning the characteristics of the semimartingale ξ.

THEOREM 3.1.7. *Let conditions* (1.2.2) *and the conditions of Theorem* 1.2.1 *be fulfilled and, moreover, let $H_t(\varepsilon_0; \widetilde{\mathsf{P}}, \mathsf{P}) < \infty$ for some $\varepsilon_0 < 0$ and all $t \in \mathbf{R}_+$. Then the process $h(\varepsilon)$ of the form*

$$h_t(\varepsilon) = \operatorname{sgn}(\varepsilon) \left\{ \frac{\varepsilon(1-\varepsilon)}{2} \gamma^2 \circ \langle m\rangle_t + \big(\varepsilon(\lambda - 1) - (\lambda^\varepsilon - 1)\big) * \nu_t \right\}, \qquad t \in \mathbf{R}_+, \tag{3.1.9}$$

is a version of a Hellinger process of order ε for $\varepsilon \in [\varepsilon_0, 1)$.

PROOF. It suffices to use Theorem 3.1.5. In view of condition V) of Theorem 1.2.1, we have $B = \gamma \circ \langle m\rangle$ and $dB/d\langle m\rangle = \gamma$. By condition III) of Theorem 1.2.1 we have ($\mathbf{Q}$-a.s.)

$$\begin{aligned}
\varphi_\varepsilon(d\widetilde{\nu}, d\nu)_t &= \varphi_\varepsilon(\widetilde{n}, n) * \overline{\nu}_t = \varphi_\varepsilon\left(\frac{\widetilde{n}}{n}, 1\right) * \nu_t \\
&= \left(\varepsilon \frac{\widetilde{n}}{n} + (1-\varepsilon) - \left(\frac{\widetilde{n}}{n}\right)^\varepsilon\right) * \nu_t = (\varepsilon(\lambda - 1) - (\lambda^\varepsilon - 1)) * \nu_t,
\end{aligned}$$

where $n = d\nu/d\overline{\nu}$, $\widetilde{n} = d\widetilde{\nu}/d\overline{\nu}$, and $\lambda = \widetilde{n}/n$.

Taking into account condition (1.2.2), from formula (3.1.16) we get that the process (3.1.8) is a version of a Hellinger process of order ε. □

REMARK 3.1.2. In the case $\varepsilon \in (0,1)$ the assertions of Theorems 3.1.2–3.1.7 are apparently valid without the condition of finiteness of the Hellinger integral $H_t(\varepsilon_0; \widetilde{\mathsf{P}}, \mathsf{P})$ for $\varepsilon_0 < 0$.

3.2. Limit theorems for the likelihood ratio

Let ξ be a semimartingale with triples of predictable characteristics $(\alpha, \langle m\rangle, \nu)$ and $(\widetilde{\alpha}, \langle\widetilde{m}\rangle, \widetilde{\nu})$ with respect to $(\mathbf{F}, \mathsf{P})$ and $(\widetilde{\mathbf{F}}, \widetilde{\mathsf{P}})$, respectively (see Section 1.2). Theorem 1.2.1 describes the form of the local density z, which yields a representation of the process $\Lambda = \ln z$.

Law of large numbers. The following theorem is a law of large numbers for the process Λ.

THEOREM 3.2.1. *Let $s \in (0,\infty)$ be fixed, let ψ_t be some positive nonrandom function such that $\psi_t \to \infty$ as $t \to \infty$, and let the following conditions be fulfilled:*

1) $\mathsf{P}\text{-}\lim_{t\to\infty} \psi_t^{-1} V_{st} = a_s$, *where a_s is a nonrandom positive function satisfying the condition $a_1 = 1$;*
2) *for any $\varepsilon \in (0,1]$*

$$\mathsf{P}\text{-}\lim_{t\to\infty} \{\psi_t^{-2}[\gamma^2 \circ \langle m\rangle_{st} + \chi(|\ln\lambda| \le \varepsilon\psi_t)\ln^2\lambda * \nu_{st}] + \psi_t^{-1}\chi(|\ln\lambda| > \varepsilon\psi_t)\,|\ln\lambda| * \nu_{st}\} = 0.$$

Then the relation

$$\mathsf{P}\text{-}\lim_{t\to\infty} \psi_t^{-1}\Lambda_{st} = -a_s \tag{3.2.1}$$

is valid and, in particular,

$$\mathsf{P}\text{-}\lim_{t\to\infty} \psi_t^{-1}\Lambda_t = -1. \tag{3.2.2}$$

PROOF. In view of Theorem 1.2.1 we have $\Lambda = M - V$, where the processes M and V are defined by relations (1.2.4) and (1.2.5). By (1.2.4) we have

$$\psi_t^{-1}M = \psi_t^{-1}\gamma \cdot m + \psi_t^{-1}\ln\lambda * (\mu - \nu) \in \mathcal{M}_{\mathrm{loc}}(\mathbf{F}, \mathsf{P}).$$

Now we apply Theorem 1.3.1, putting $Y^t = \psi_t^{-1}M$, $M^t = \psi_t^{-1}\gamma{\cdot}m$, $\mu^t = \mu$, $\nu^t = \nu$, $U^t = \psi_t^{-1}\ln\lambda$, and $c(s) = 0$. The conditions of Theorem 1.3.1 follow from condition 2). From Theorem 1.3.1 we see that

$$\mathsf{P}\text{-}\lim_{t\to\infty} \psi_t^{-1}M_{st} = 0. \tag{3.2.3}$$

Combining (3.2.3) and condition 1), we get (3.2.1). Relation (3.2.2) is (3.2.1) for $s = 1$. □

COROLLARY 3.2.1. *Suppose that instead of condition* VIII) *of Theorem* 1.2.1 *condition* VIII′) *is fulfilled, and instead of condition* 2) *of Theorem* 3.2.1 *the following condition is satisfied:*

2′) $\mathsf{P}\text{-}\lim_{t\to\infty} \psi_t^{-2}[\gamma^2 \circ \langle m\rangle_{st} + \ln^2\lambda * \nu_{st}] = 0$.

Then the law of large numbers (3.2.1) *holds.*

Large deviations. The following theorem is a form of the theorem on large deviations for Λ_t as $t \to \infty$. The suffficient conditions are formulated in terms of the Hellinger process $h(\varepsilon) = h(\varepsilon; \widetilde{\mathsf{P}}, \mathsf{P})$.

THEOREM 3.2.2. 1) *If for some positive function ψ'_t such that $\psi'_t \to \infty$ as $t \to \infty$ the Hellinger integral $h(\varepsilon; \widetilde{\mathsf{P}}, \mathsf{P})$ satisfies the condition*

$$\overline{\lim_{\varepsilon \downarrow 0}}\ \overline{\lim_{t \to \infty}}\ (\varepsilon \psi'_t)^{-1} \ln \mathsf{E}\, \mathcal{E}_t(-h(\varepsilon; \widetilde{\mathsf{P}}, \mathsf{P})) \leq -1, \tag{3.2.4}$$

then the Hellinger integral $H_t(\varepsilon; \widetilde{\mathsf{P}}, \mathsf{P})$ obeys the relation

$$\overline{\lim_{\varepsilon \downarrow 0}}\ \overline{\lim_{t \to \infty}}\ (\varepsilon \psi'_t)^{-1} \ln H_t(\varepsilon; \widetilde{\mathsf{P}}, \mathsf{P}) \leq -1. \tag{3.2.5}$$

2) *If $H_t(\varepsilon_0; \widetilde{\mathsf{P}}, \mathsf{P}) < \infty$ for all $t \in \mathbf{R}_+$ for some $\varepsilon_0 < 0$ and for some positive function ψ''_t such that $\psi''_t \to \infty$ as $t \to \infty$ the Hellinger process $h(\varepsilon; \widetilde{\mathsf{P}}, \mathsf{P})$ satisfies the condition*

$$\underline{\lim_{\varepsilon \uparrow 0}}\ \underline{\lim_{t \to \infty}}\ (\varepsilon \psi''_t)^{-1} \ln \mathsf{E}\, \mathcal{E}_t(-h(\varepsilon; \widetilde{\mathsf{P}}, \mathsf{P})) \geq -1, \tag{3.2.6}$$

then the Hellinger integral $H_t(\varepsilon; \widetilde{\mathsf{P}}, \mathsf{P})$ satisfies the relation

$$\underline{\lim_{\varepsilon \uparrow 0}}\ \underline{\lim_{t \to \infty}}\ (\varepsilon \psi''_t)^{-1} \ln H_t(\varepsilon; \widetilde{\mathsf{P}}, \mathsf{P}) \geq -1. \tag{3.2.7}$$

PROOF. We use the estimate (3.1.5) for $0 < \varepsilon < \delta < 1$. As we are interested in small ε, we can assume that $0 < \varepsilon < \delta/2$. Then $\varepsilon/(\delta - \varepsilon) < 1$. Applying the Jensen inequality to the right side of (3.1.5), in this case we get

$$H(\varepsilon; \widetilde{P}^\tau, P^\tau) \leq H_0^{\varepsilon/\delta}(\delta; \widetilde{\mathsf{P}}, \mathsf{P}) \big(\mathsf{E}\, \mathcal{E}_\tau(-h(\delta))\big)^{\varepsilon/\delta}.$$

Since this inequality is true for any stopping time τ, from this we get

$$\overline{\lim_{\varepsilon \downarrow 0}}\ \overline{\lim_{t \to \infty}}\ (\varepsilon \psi'_t)^{-1} \ln H_t(\varepsilon; \widetilde{\mathsf{P}}, \mathsf{P}) \leq \overline{\lim_{\delta \downarrow 0}}\ \overline{\lim_{t \to \infty}}\ (\delta \psi'_t)^{-1} \ln \mathsf{E}\, \mathcal{E}_t(-h(\delta)) \leq -1,$$

where the last inequality follows from condition (3.2.4).

Now we use the estimate (3.1.5) for $\varepsilon_0 \leq \delta < \varepsilon < 0$. Since in the desired expression $\varepsilon \to 0$, we may assume that $\delta/2 < \varepsilon < 0$. Then $\varepsilon(\delta - \varepsilon) < 1$. Applying the Jensen inequality and considering condition (3.2.6), we get

$$\underline{\lim_{\varepsilon \uparrow 0}}\ \underline{\lim_{t \to \infty}}\ (\varepsilon \psi''_t)^{-1} \ln H_t(\varepsilon; \widetilde{\mathsf{P}}, \mathsf{P}) \geq \overline{\lim_{\delta \uparrow 0}}\ \overline{\lim_{t \to \infty}}\ (\delta \psi''_t)^{-1} \ln \mathsf{E}\, \mathcal{E}_t(h(\delta)) \geq -1. \qquad \square$$

REMARK 3.2.1. If the semimartingale ξ is a process with independent increments with triples of predictable characteristics

$$T = (\alpha, \langle m \rangle, \nu) \quad \text{and} \quad \widetilde{T} = (\widetilde{\alpha}, \langle \widetilde{m} \rangle, \widetilde{\nu})$$

with respect to P and $\widetilde{\mathsf{P}}$, respectively, then in view of Theorem 3.1.6 condition (3.2.4) is necessary and sufficient for (3.2.5) to be true and relation (3.2.6) is necessary and sufficient for (3.2.7) to be true.

For studying some Hellinger processes $h(\varepsilon)$ the following theorem can be useful.

THEOREM 3.2.3. 1) *Suppose for all* $\varepsilon \in (0,1)$ *and* $t \in \mathbf{R}_+$ *there exist sets* $D_{t,\varepsilon} \in \mathcal{F}_t$ *such that*

$$\overline{\lim_{\varepsilon\downarrow 0}}\, \overline{\lim_{t\to\infty}}\, (\varepsilon\psi_t')^{-1} \ln \mathsf{P}(D_{t,\varepsilon}^c) \le -1, \tag{3.2.8}$$

$$\underline{\lim_{\varepsilon\downarrow 0}}\, \underline{\lim_{t\to\infty}}\, (\varepsilon\psi_t')^{-1} \inf_{\omega\in D_{t,\varepsilon}} h_t(\varepsilon) \ge 1, \tag{3.2.9}$$

where $D_{t,\varepsilon}^c = \Omega \setminus D_{t,\varepsilon}$ *and* $\psi_t' \to \infty$ *as* $t \to \infty$. *Then the relation* (3.2.5) *holds.*

2) *Let* $H_t(\varepsilon_0; \widetilde{\mathsf{P}}, \mathsf{P}) < \infty$ *for all* $t \in \mathbf{R}_+$ *for some* $\varepsilon_0 < 0$, *and suppose for all* $\varepsilon \in [\varepsilon_0, 0)$ *and* $t \in \mathbf{R}_+$ *there exist sets* $D_{t,\varepsilon} \in \mathcal{F}_t$ *such that*

$$\underline{\lim_{\varepsilon\uparrow 0}}\, \underline{\lim_{t\to\infty}}\, (\varepsilon\psi_t'')^{-1} \ln \widetilde{E}^{t,\varepsilon} \chi(D_{t,\varepsilon}^c) \exp\{h_t(\varepsilon)\} \ge -1, \tag{3.2.10}$$

$$\underline{\lim_{\varepsilon\uparrow 0}}\, \underline{\lim_{t\to\infty}}\, (\varepsilon\psi_t'')^{-1} \sup_{\omega\in D_{t,\varepsilon}} h_t(\varepsilon) \ge -1, \tag{3.2.11}$$

where $\psi_t'' \to \infty$ *as* $t \to \infty$ *and* $\widetilde{E}^{t,\varepsilon}(\cdot)$ *is integration with respect to the measure* $\widetilde{P}^{t,\varepsilon}$, $d\widetilde{P}^{t,\varepsilon} = N_t(\varepsilon)\, d\mathbf{Q}$, *and* $N(\varepsilon)$ *is the process from the multiplicative decomposition* (3.1.4). *Then* (3.2.7) *holds.*

PROOF. 1) Using the multiplicative decomposition (3.1.4) for $\varepsilon \in (0,1)$, we have

$$\begin{aligned}
H_t(\varepsilon; \widetilde{\mathsf{P}}, \mathsf{P}) &= \mathsf{E}_Q\, \chi(D_{t,\varepsilon}) Y_t(\varepsilon) + \mathsf{E}_Q\, \chi(D_{t,\varepsilon}^c) Y_t(\varepsilon) \\
&= \mathsf{E}_Q\, \chi(D_{t,\varepsilon}) N_t(\varepsilon) e^{-h_t(\varepsilon)} + \mathsf{E}\, \chi(D_{t,\varepsilon}^c) z_t^\varepsilon \\
&\le \exp\left\{-\inf_{D_{t,\varepsilon}} h_t(\varepsilon)\right\} \mathsf{E}_Q\, N_t(\varepsilon) + (\mathsf{E}\, \chi(D_{t,\varepsilon}^c))^{1-\varepsilon} (\mathsf{E}\, z_t)^\varepsilon \\
&\le \exp\left\{-\inf_{D_{t,\varepsilon}} h_t(\varepsilon)\right\} + (\mathsf{P}(D_{t,\varepsilon}^c))^{1-\varepsilon}.
\end{aligned}$$

To derive the last inequality, we used the fact that $N(\varepsilon)$ is a $\mathbf{Q}$-supermartingale. Using the elementary inequality

$$\ln(x+y) \le \ln 2 + (\ln x) \vee (\ln y), \qquad x > 0,\ y > 0, \tag{3.2.12}$$

and conditions (3.2.8) and (3.2.9), we obtain (3.2.5).

2) Now let $H_t(\varepsilon_0; \widetilde{\mathsf{P}}, \mathsf{P}) < \infty$ for all $t \in \mathbf{R}_+$ for some $\varepsilon_0 < 0$. Using the multiplicative decomposition (3.1.4) for $\varepsilon \in [\varepsilon_0, 0)$, we have

$$\begin{aligned}
H_t(\varepsilon_0; \widetilde{\mathsf{P}}, \mathsf{P}) &= \mathsf{E}_Q\, \chi(D_{t,\varepsilon}) N_t(\varepsilon) e^{h_t(\varepsilon)} + \mathsf{E}_Q\, \chi(D_{t,\varepsilon}^c) N_t(\varepsilon) e^{h_t(\varepsilon)} \\
&\le \exp\left\{\sup_{\omega\in D_{t,\varepsilon}} h_t(\varepsilon)\right\} + \widetilde{E}^{t,\varepsilon} \chi(D_{t,\varepsilon}^c) \exp\{h_t(\varepsilon)\}.
\end{aligned}$$

Using the inequality (3.2.12) once again and considering conditions (3.2.10) and (3.2.11), we derive the relation (3.2.7). $\square$

Theorems on weak convergence. Consider a family of stochastic processes $Y^t = (Y^t_s)_{s\in\mathbf{R}_+}$, where $Y^t_s = \psi_t^{-1}(\Lambda_{st} + \varphi_{st})$, φ_s is a continuous function and $\psi_t \to \infty$ as $t \to \infty$. By Theorem 1.2.1, $Y^t \in \mathcal{S}(\mathbf{F}^t, \mathsf{P})$, where $\mathbf{F}^t = (\mathcal{F}^t_s)_{s\in\mathbf{R}_+}$, $\mathcal{F}^t_s = \mathcal{F}_{st}$ and for every $t \in \mathbf{R}_+$ the semimartingale Y^t has the canonical representation

$$Y^t = B^t + M^t + x\chi(|x| \le 1) * (\mu^t - \nu^t) + x\chi(|x| > 1) * \mu^t,$$

where $B^t = (B^t_s)_{s\in\mathbf{R}_+}$, $M^t = (M^t_s)_{s\in\mathbf{R}_+}$, μ^t is the measure of jumps of the semimartingale Y^t, ν^t is its compensator, and

$$\begin{aligned} B^t_s &= \psi_t^{-1}\big[\varphi_{st} - V_{st} - \chi\left(\left|\psi_t^{-1}\ln\lambda\right| > 1\right)\ln\lambda * \nu_{st}\big], \\ M^t_s &= \psi_t^{-1}\gamma \cdot m_{st}, \\ \mu^t((0,s],C) &= \chi\left(\psi_t^{-1}\ln\lambda \in C\right) * \mu_{st}, \\ \nu^t((0,s],C) &= \chi\left(\psi_t^{-1}\ln\lambda \in C\right) * \nu_{st}, \qquad s \in \mathbf{R}_+,\ C \in \mathcal{B}_0. \end{aligned}$$

Here the process V is given by formula (1.2.5).

In order to formulate the next theorem, we introduce, on the stochastic basis $(\Omega, \mathcal{F}, \mathbf{F}, \mathsf{P})$, a left quasicontinuous semimartingale $Y \in \mathcal{S}(\mathbf{F}, \mathsf{P})$, $Y_0 = 0$, with independent increments and a triple of predictable characteristics (B^Y, C^Y, ν^Y), which is obviously determinate and continuous in time.

Moreover, for any $\delta \in (0,1]$ and $t \in \mathbf{R}_+$ we define a local square-integrable martingale $M^{t,\delta} \in \mathcal{M}^2_{\mathrm{loc}}(\mathbf{F}^t, \mathsf{P})$ of the form

$$M^{t,\delta}_s = \psi_t^{-1}\big[\gamma \cdot m_{st} + \chi\left(\left|\psi_t^{-1}\ln\lambda\right| \le \delta\right)\ln\lambda * (\mu - \nu)_{st}\big],$$

whose quadratic characteristic is defined by the formula

$$\left\langle M^{t,\delta}\right\rangle_s = \psi_t^{-2}\big[\gamma \circ \langle m\rangle_{st} + \chi\left(\left|\psi_t^{-1}\ln\lambda\right| \le \delta\right)\ln^2\lambda * \nu_{st}\big].$$

THEOREM 3.2.4. *Let S be a nonempty subset in $\mathbf{R}_+$, and let the following conditions be fulfilled:*

1) *for any bounded function $f = f(x)$, continuous on $R_0 \setminus (\{-1\} \cup \{1\})$ and such that $\chi(|x| > \varepsilon)f(x) = f(x)$ for some $\varepsilon > 0$,*

$$\mathsf{P}\text{-}\lim_{t\to\infty} f\left(\psi_t^{-1}\ln\lambda\right) * \nu_{st} = f * \nu^Y_s, \qquad s \in S;$$

2) $\mathsf{P}\text{-}\lim_{t\to\infty} B^t_s = B^Y_s$, $s \in S$;

3) *for any $a > 0$ and $s \in S$ we have*

$$\lim_{\delta\to 0}\overline{\lim_{t\to\infty}}\ \mathsf{P}\left\{\left|\left\langle M^{t,\delta}\right\rangle_s - C^Y_s\right| > a\right\} = 0.$$

Then $Y_t \xrightarrow{w_f(S)} Y$ as $t \to \infty$.

To prove the theorem, it is enough to apply Theorem 1.3.7.

REMARK 3.2.2. Theorems 3.2.1–3.2.4 are also valid in the so-called triangular array scheme. Namely, let the triple $\widetilde{T} = (\widetilde{\alpha}, \langle\widetilde{m}\rangle, \widetilde{\nu})$ depend on t so that $\widetilde{T} = T^t = (\alpha^t, \langle m^t\rangle, \nu^t)$, and let the triple $T = (\alpha, \langle m\rangle, \nu)$ be independent of t. In this case the functions λ and γ from conditions III) and V) of Theorem 1.2.1 depend on t so that $\lambda = \lambda^t$ and $\gamma = \gamma^t$. Then Theorems 3.2.1–3.2.4 can be easily reformulated by replacing λ and γ with λ^t and γ^t, respectively. We do not give here the corresponding formulations, restricting ourselves to the analog of Theorem 3.2.4 in the triangular array scheme for $\psi_t = 1$ and $\varphi_t = 0$.

Consider a family of stochastic processes $Y^t = (Y^t_s)_{s\in\mathbf{R}_+}$, where $Y^t_s = \Lambda^t_{st}$. By Theorem 1.2.1, we have $Y^t \in \mathcal{S}(\mathbf{F}^t, \mathsf{P})$, where $\mathbf{F}^t = (\mathcal{F}^t_s)_{s\in\mathbf{R}_+}$, $\mathcal{F}^t_s = \mathcal{F}_{st}$, and for every $t \in \mathbf{R}_+$ the semimartingale Y^t has the canonical representation

$$Y^t = B^t + M^t + x\chi(|x| \le 1) * \left(\mu^t - \nu^t\right) + x\chi(|x| > 1) * \mu^t,$$

where μ^t is the measure of jumps of the semimartingale Y^t, ν^t is its compensator, $B^t = (B^t_s)_{s\in\mathbf{R}_+}$, $M^t = (M^t_s)_{s\in\mathbf{R}_+}$, and

$$B^t_s = -V_{st} - \chi\left(\left|\ln\lambda^t\right| > 1\right)\ln\lambda^t * \nu_{st}, \qquad M^t_s = \gamma^t \cdot m_{st},$$
$$\mu^t((0,s],C) = \chi\left(\ln\lambda^t \in C\right) * \mu_{st},$$
$$\nu^t((0,s],C) = \chi\left(\ln\lambda^t \in C\right) * \nu_{st}, \qquad C \in \mathcal{B}_0.$$

Thus we obtain the process Y^t under consideration from the process Y^t in Theorem 3.2.4 by replacing λ with λ^t and γ with γ^t, and putting $\psi_t = 1$, $\varphi_t = 0$. As in Theorem 3.2.4, the process Y is a left quasicontinuous semimartingale with independent increments. As above, for any $\delta \in (0,1]$ and $t \in \mathbf{R}_+$ we construct a local square-integrable martingale $M^{t,\delta} \in \mathcal{M}^2_{\mathrm{loc}}(\mathbf{F}^t, \mathsf{P})$ by putting

$$M^{t,\delta}_s = \gamma^t \cdot m_{st} + \chi\left(\left|\ln\lambda^t\right| \le \delta\right)\ln\lambda^t * (\mu - \nu)_{st},$$

the quadratic characteristic of which is defined by the formula

$$\left\langle M^{t,\delta}\right\rangle_s = \left(\gamma^t\right)^2 \cdot \langle m\rangle_{st} + \chi\left(\left|\ln\lambda^t\right| \le \delta\right)\ln^2\lambda^t * \nu_{st}.$$

THEOREM 3.2.5. *Let S be a nonempty subset in $\mathbf{R}_+$, and let the following conditions be fulfilled:*

1) *for any bounded function $f = f(x)$, continuous on $R_0 \setminus (\{-1\} \cup \{1\})$ and such that $f(x) = \chi(|x| > \varepsilon)f(x)$ for some $\varepsilon > 0$,*

$$\mathsf{P}\text{-}\lim_{t\to\infty} f\left(\ln\lambda^t\right) * \nu^Y_{st}, \qquad s \in S;$$

2) $\mathsf{P}\text{-}\lim_{t\to\infty} B^t_s = B^Y_s$, $s \in S$;

3) *for any $a > 0$ and $s \in S$*

$$\lim_{\delta\to 0}\varlimsup_{t\to\infty} \mathsf{P}\left\{\left|\left\langle M^{t,\delta}\right\rangle_s - C^Y_s\right| > a\right\} = 0.$$

Then $Y^t \xrightarrow{w_f(S)} Y$ as $t \to \infty$.

REMARK 3.2.3. Theorem 3.2.5 can also be used in the case where λ^t and γ^t are independent of t, that is, in the case where the technique of the triangular array scheme cannot be applied.

Statistical applications. 1. Let δ^{+,α_t}_t be the Neyman–Pearson test of level α_t for distinguishing between the hypotheses H^t and $\widetilde{H}^t$ from an observation ξ^t of a semimartingale ξ (see the definition in Section 2.1). If the conditions of Theorem 3.2.1 (or Corollary 3.2.1) are fulfilled for $s = 1$, then the law of large numbers (3.2.2) holds, that is, condition $\Lambda 1$ with $\chi_t = \psi_t$ is fulfilled (see Section 2.3). If, in addition, the level α_t satisfies conditions $\alpha 1$ and $\alpha 2$, then in view of Theorem 2.3.1 the probability of error of the second kind $\beta(\delta^{+,\alpha_t}_t)$ behaves as follows:

$$\beta\left(\delta^{+,\alpha_t}_t\right) = \exp\{-\psi_t(1 + o(1))\}, \qquad t \to \infty. \tag{3.2.13}$$

2. If the conditions of Theorem 3.2.2 (or Theorem 3.2.3) are fulfilled, then conditions $\Lambda 2$ and $\Lambda 3$ with $\chi_t = \psi_t$ are fulfilled (see Section 2.3). If, in addition, the level α_t satisfies conditions $\alpha 1'$ and $\alpha 2'$ with $\chi_t = \psi_t$, then by Theorem 2.3.3 the probability of error of the second kind $\beta(\delta_t^{+,\alpha_t})$ also has the representation (3.2.13).

3. Now let the conditions of Theorem 3.2.4 be fulfilled with $\varphi_s = 0$, $S = \{1\}$, and $Y_1 \leq 0$ (P-a.s.). Then condition $\Lambda 4$ (Section 2.4) is fulfilled with $L(x) = \mathsf{P}\{Y_1 < x\}$. If, in addition, the function $L(x)$ is continuous and strictly monotonically increasing on the interval $(\underline{l}, \bar{l})$ and $\alpha_t \to \alpha \in (0,1)$ as $t \to \infty$, then the probability of error of the second kind behaves as follows:

$$\beta(\delta_t^{+,\alpha_t}) = \exp\{l_{1-\alpha}\psi_t(1+o(1))\}, \qquad t \to \infty, \tag{3.2.14}$$

where l_p is a p-quantile of the distribution function $L(x)$.

Comparing (3.2.14) with (3.2.13), we see that, contrary to the above two cases, in this case the logarithmic rate of decrease in the probability of error of the second kind depends on the limit behavior of the level α_t.

4. Suppose that the conditions of Theorem 3.2.4 with $S = \{1\}$ are fulfilled but $\varphi_t \to \infty$ as $t \to \infty$ and $\psi_t = o(\varphi_t)$. Then condition $\Lambda 5$ (Section 2.4) is fulfilled with $L(x) = \mathsf{P}\{Y_1 < x\}$. Assuming the function $L(x)$ to be continuous and strictly monotonically increasing on the interval $(\underline{l}, \bar{l})$, from Theorem 2.4.4 we get that if $\alpha_t \to \alpha \in (0,1)$ as $t \to \infty$, then the probability of error of the second kind behaves as follows:

$$\beta\left(\delta_t^{+,\alpha_t}\right) = \exp\{-\varphi_t + l_{1-\alpha}\psi_t + o(\psi_t)\}, \qquad t \to \infty. \tag{3.2.15}$$

Comparing (3.2.15) with (3.2.14), we see that, contrary to the previous case, the dependence on the limit behavior of the level α_t appears in the second term of the asymptotic expansion of $\ln \beta(\delta_t^{+,\alpha_t})$. Note that in this case the law of large numbers holds for Λ_t, and hence the expansion (3.2.15) is a refinement of expansion (3.2.13), valid under the conditions of the law of large numbers.

5. Let the conditions of Theorem 3.2.5 be fulfilled with $S = \{1\}$. Then condition $\Lambda 6$ (Section 2.5) is true with distribution function $L(x) = \mathsf{P}\{Y_1 < x\}$. If the function $L(x)$ is continuous and strictly monotonically increasing on the interval $(\underline{l}, \bar{l})$, then in view of Theorem 2.5.5 for any $\alpha \in [0,1]$ the test δ_t^{+,α_t} obeys the equivalence

$$\lim_{t\to\infty} \alpha_t = \alpha \Leftrightarrow \lim_{t\to\infty} \beta\left(\delta_t^{+,\alpha_t}\right) = \widetilde{L}(l_{1-\alpha}),$$

where l_p is a p-quantile of the distribution function $L(x)$ and $\widetilde{L}(x)$ is the distribution function defined by formula (2.5.2).

If $Y_1 = 0$ (P-a.s.), then condition $\Lambda 7$ (Section 2.5) is fulfilled and hence the complete asymptotic indistinguishability of the families of hypotheses (H^t) and $(\widetilde{H}^t)$ takes place. In this case by virtue of Theorem 2.5.6 for any $\alpha \in [0,1]$ the test δ_t^{+,α_t} obeys the equivalence

$$\lim_{t\to\infty} \alpha_t = \alpha \Leftrightarrow \lim_{t\to\infty} \beta\left(\delta_t^{+,\alpha_t}\right) = 1 - \alpha.$$

3.3. Asymptotic decomposition of the likelihood ratio in parametric formulation

On a measurable space with filtration $(\Omega, \mathcal{G}, (\mathcal{G}_t))$ we consider a parametric family of probability measures $(\mathsf{P}_\theta, \theta \in \Theta)$, where $\Theta \subset \mathbf{R}^k$ is an open set,

$k \geq 1$. Furthermore, for every $\theta \in \Theta$ on the stochastic basis $(\Omega, \mathcal{F}, \mathbf{F}, \mathsf{P}_\theta)$ we consider a semimartingale $\xi \in \mathcal{S}(\mathbf{F}, \mathsf{P}_\theta)$ with a triple of predictable characteristics $(\alpha(\theta), \langle m(\theta)\rangle, \nu(\theta))$ (see Section 1.2). For any fixed θ, $y \in \Theta$ we chose two measures P_θ and P_y and, putting $\mathsf{P} = \mathsf{P}_\theta$, $\widetilde{\mathsf{P}} = \mathsf{P}_y$, we assume that the conditions of Theorems 1.2.1 and 1.2.2 are fulfilled, which yield a process of local density $z(y, \theta) = (z_t(y, \theta))_{t \in \mathbf{R}_+}$ of the measure P_y with respect to the measure P_θ:

$$z(y, \theta) = \exp\{M(y, \theta) - V(y, \theta)\}, \tag{3.3.1}$$

where

$$M(y, \theta) = \gamma(y, \theta) \cdot m(\theta) + \ln \lambda(y, \theta) * (\mu - \nu(\theta)) \in \mathcal{M}_{\mathrm{loc}}(\mathbf{F}, \mathsf{P}_\theta), \tag{3.3.2}$$

and

$$\begin{aligned} V(y, \theta) = {} & \frac{1}{2}\gamma^2(y, \theta) \circ \langle m(\theta)\rangle \\ & + \big(\lambda(y, \theta) - 1 - \ln \lambda(y, \theta)\big) * \nu(\theta) \in \mathcal{V}_{\mathrm{loc}}(\mathbf{F}, \mathsf{P}_\theta). \end{aligned} \tag{3.3.3}$$

Here $\gamma(y, \theta)$ is the predictable process from condition V) and $\lambda(y, \theta)$ is the predictable function from condition III) of Theorem 1.2.1 for $\mathsf{P} = \mathsf{P}_\theta$ and $\widetilde{\mathsf{P}} = \mathsf{P}_y$.

In this section we assume that θ is independent of t, and $y = y_t$ depends on t so that $y_t \to \theta$ as $t \to \infty$, and investigate the asymptotic behavior of $z_t(y_t, \theta)$ as $t \to \infty$ for various rates of convergence of y_t to θ as $t \to \infty$.

Asymptotic decomposition. We introduce the space $\mathbf{C}_1^0(\Theta)$ of functions $f = f(\theta)$ that are defined on $\Theta \subset \mathbf{R}^k$, are almost everywhere differentiable in $\theta \in \Theta$ with respect to Lebesgue measure, and possess the property that for all θ and $\Delta \in \mathbf{R}^k$ such that $\theta + v\Delta \in \Theta$ for all $v \in [0, 1]$ the function $f(\theta + v\Delta)$ is absolutely continuous in v. Evidently, a function $f \in \mathbf{C}_1^0(\Theta)$ satisfies the equality

$$f(\theta + \Delta) - f(\theta) = \int_0^1 \Delta' \nabla_\theta f(\theta + v\Delta)\, dv, \tag{3.3.4}$$

where $\theta = (\theta^1, \theta^2, \ldots, \theta^k)'$, $\theta, \Delta \in \mathbf{R}^k$ are such that $\theta + v\Delta \in \Theta$ for all $v \in [0, 1]$, and

$$\nabla_\theta f(\theta) = \big(\nabla_{\theta^1} f(\theta), \nabla_{\theta^2} f(\theta), \ldots, \nabla_{\theta^k} f(\theta)\big)', \qquad \nabla_{\theta^i} f(\theta) = \frac{\partial f(\theta)}{\partial \theta^i}.$$

To formulate the next theorem, we introduce the following notation (under the condition of existence of the corresponding derivatives):

$$\begin{gathered} g(y, \theta) = \nabla_y \gamma(y, \theta), \qquad l(y, \theta) = \nabla_y \lambda(y, \theta), \\ f(y, \theta) = l(y, \theta)/\lambda(y, \theta), \\ g_v^t = g(\theta + v\Delta_t, \theta), \qquad l_v^t = l(\theta + v\Delta_t, \theta), \qquad f_v^t = f(\theta + v\Delta_t, \theta), \\ g_0 = g_0^t, \qquad l_0 = l_0^t, \qquad f_0 = f_0^t, \qquad \Delta_t = y_t - \theta, \qquad v \in [0, 1]. \end{gathered}$$

Observe that $f_0 = l_0$, since $\lambda(\theta, \theta) = 1$.

THEOREM 3.3.1. *Let $y_t \to \theta$ as $t \to \infty$, and let the following conditions be fulfilled:*

1) the functions $\gamma(y,\theta)$ and $\ln \lambda(y,\theta)$ belong to $\mathbf{C}_1^0(\Theta)$ as functions of $y \in \Theta$, the other variables being fixed, and for all y in some neighborhood of θ,

$$g(y,\theta) \in \mathcal{L}^2_{\mathrm{loc}}(m(\theta),\mathbf{F},\mathsf{P}_\theta), \qquad f(y,\theta) \in \mathcal{G}_{\mathrm{loc}}(\nu(\theta),\mathbf{F},\mathsf{P}_\theta),$$
$$\int_0^1 f_v^t \int_0^v (l_z^t)' dz\, dv \in \mathcal{G}^1_{\mathrm{loc}}(\nu(\theta),\mathbf{F},\mathsf{P}_\theta);$$

2) there exists a positive definite symmetric matrix $\varphi_t(\theta)$ such that $|\varphi_t(\theta)| \to 0$ as $t \to \infty$ and for any $\varepsilon \in (0,1]$ we have

$$\mathsf{P}_\theta\text{-}\lim_{t\to\infty} \varphi_t(\theta)\big[g_0 g_0' \circ \langle m(\theta)\rangle_t + \chi(|\varphi_t(\theta) l_0| \le \varepsilon) l_0 l_0' * \nu(\theta)_t\big]\varphi_t(\theta) = I_k;$$

3) for any $\varepsilon \in (0,1]$,

$$\mathsf{P}_\theta\text{-}\lim_{t\to\infty} \chi(|\varphi_t(\theta) l_0| > \varepsilon)|\varphi_t(\theta) l_0| * \nu(\theta)_t = 0;$$

4) for any $\varepsilon \in (0,1]$,

$$\mathsf{P}_\theta\text{-}\lim_{t\to\infty} \Big[\left|\bar g^t\right|^2 \circ \langle m(\theta)\rangle_t + \chi\left(\left|\bar f^{\,t}\right| > \varepsilon\right)\left|\bar f^{\,t}\right| * \nu(\theta)_t + \chi\left(\left|\bar f^{\,t}\right| \le \varepsilon\right)\left|\bar f^{\,t}\right|^2 * \nu(\theta)_t\Big] = 0,$$

where

$$\bar g^t = \varphi_t(\theta)\left(\int_0^1 g_v^t\, dv - g_0\right), \qquad \bar f^{\,t} = \varphi_t(\theta)\left(\int_0^1 f_v^t\, dv - f_0\right);$$

5) for any $\varepsilon \in (0,1]$,

$$\mathsf{P}_\theta\text{-}\lim_{t\to\infty} \varphi_t(\theta)\int_0^1\int_0^v \left[f_v^t\left(l_u^t\right)' - \chi(|\varphi_t(\theta) l_0| \le \varepsilon) l_0 l_0'\right] du\, dv * \nu(\theta)_t \varphi_t(\theta) = 0.$$

Then the following decomposition is valid:

$$\Lambda(y_t,\theta) = u_t'\left(\eta^t + q^t\right) - (1/2)u_t'\left(I_k + p^t\right)u_t, \tag{3.3.5}$$

where

$$\Lambda(y,\theta) = \ln z(y,\theta), \qquad u_t = \varphi_t^{-1}(\theta)\Delta_t,$$
$$\eta^t = \varphi_t(\theta)\left[g_0 \circ \langle m(\theta)\rangle + l_0 * (\mu - \nu(\theta))\right] \in \mathcal{M}_{\mathrm{loc}}(\mathbf{F},\mathsf{P}_\theta), \tag{3.3.6}$$
$$\mathcal{L}\left(\eta_t^t | \mathsf{P}_\theta\right) \xrightarrow{w} \mathcal{N}(0,I_k), \qquad t\to\infty, \tag{3.3.7}$$

and q^t and p^t are vector and matrix processes, respectively, such that

$$\mathsf{P}_\theta\text{-}\lim_{t\to\infty}\left(|q_t^t| + |p_t^t|\right) = 0. \tag{3.3.8}$$

PROOF. Since $\gamma(\theta,\theta)=0$ and $\lambda(\theta,\theta)=1$, by condition 1) and relation (3.3.4) we have

$$\gamma(y_t,\theta)=\Delta_t'\int_0^1 g_v^t\,dv, \tag{3.3.9}$$

$$\ln\lambda(y_t,\theta)=\Delta_t'\int_0^1 f_v^t\,dv. \tag{3.3.10}$$

Taking into consideration condition 1) and relations (1.1.4) and (1.1.4′), from the relations (3.3.2), (3.3.9) and (3.3.10) we get

$$\begin{aligned} M(y_t,\theta) &= \left(\Delta_t'\int_0^1 g_v^t\,dv\right)\cdot m(\theta)+\left(\Delta_t'\int_0^1 f_v^t\,dv\right)*(\mu-\nu(\theta)) \\ &= u_t'\varphi_t(\theta)\left[\int_0^1 g_v^t\,dv\cdot m(\theta)+\int_0^1 f_v^t\,dv*(\mu-\nu(\theta))\right] \\ &= u_t'\left(\eta^t+q^t\right), \end{aligned} \tag{3.3.11}$$

where

$$\eta^t=\varphi_t(\theta)\big[g_0\cdot m(\theta)+l_0*(\mu-\nu(\theta))\big], \tag{3.3.12}$$

$$q^t=\bar g^t\cdot m(\theta)+\bar f^{\,t}*(\mu-\nu(\theta)). \tag{3.3.13}$$

In view of condition 1) we have $f_v^t\in\mathcal{G}_{\mathrm{loc}}(\nu(\theta),\mathbf{F},\mathsf{P}_\theta)$ for all $v\in[0,1]$. Since the function $x^2/(1+|x|)$, $x\in\mathbf{R}^1$, is convex downwards, we have

$$\int_0^1 f_v^t\,dv\in\mathcal{G}_{\mathrm{loc}}(\nu(\theta),\mathbf{F},\mathsf{P}_\theta), \tag{3.3.14}$$

and by Theorem 1.1.8 the stochastic integral $\int_0^1 f_v^t\,dv*(\mu-\nu(\theta))$ is well defined.

Condition 1) and relations (1.1.4) and (1.1.4′) imply that $\eta^t\in\mathcal{M}_{\mathrm{loc}}(\mathbf{F},\mathsf{P}_\theta)$. Now we apply Theorem 1.3.2 putting $Y^t=\eta^t$, $M^t=\varphi_t(\theta)g_0\cdot m(\theta)$, $U^t=\varphi_t(\theta)l_0$, $\mu^t=\mu$, $\nu^t=\nu(\theta)$, $P^t=\mathsf{P}_\theta$, $s=1$, $I(s)=I_k$. Then in view of conditions 2) and 3) we obtain the relation (3.3.7).

The inclusion (3.3.14), condition 1), and relations (1.1.4) and (1.1.4′) imply

$$\bar g^t\in\mathcal{L}^2_{\mathrm{loc}}(m(\theta),\mathbf{F},\mathsf{P}_\theta),\qquad \bar f^{\,t}\in\mathcal{G}_{\mathrm{loc}}(\nu(\theta),\mathbf{F},\mathsf{P}_\theta),$$

and hence $q^t\in\mathcal{M}_{\mathrm{loc}}(\mathbf{F},\mathsf{P}_\theta)$. Now we apply Theorem 1.3.2, putting $Y^t=q^t$, $M^t=\bar g^t\cdot m(\theta)$, $U^t=\bar f^{\,t}$, $\mu^t=\mu$, $\nu^t=\nu(\theta)$, $P^t=\mathsf{P}_\theta$, $s=1$, $I(s)=\mathcal{O}$, where $\mathcal{O}$ is a matrix with zero elements (condition 4) implies that the conditions of Theorem 1.3.2 are fulfilled in this case). Therefore

$$\mathsf{P}_\theta\text{-}\lim_{t\to\infty}|q_t^t|=0. \tag{3.3.15}$$

Thus we have established the representation (3.3.11)–(3.3.13), in which $u_t=\varphi_t^{-1}(\theta)\Delta_t$, processes η^t satisfy relations (3.3.6) and (3.3.7), and processes q^t satisfy relation (3.3.15). Hence in order to complete the proof of representation (3.3.5)–(3.3.8), it suffices to demonstrate that

$$V(y_t,\theta)=(1/2)u_t'\left(I_k+p^t\right)u_t, \tag{3.3.16}$$

$$\mathsf{P}_\theta\text{-}\lim_{t\to\infty}\left|p_t^t\right|=0. \tag{3.3.17}$$

From relations (3.3.3), (3.3.9), and (3.3.10) we get

$$V(y_t,\theta)=\frac{1}{2}\left(\Delta_t'\int_0^1 g_v^t\,dv\right)^2\circ\langle m(\theta)\rangle+\left(\Delta_t'\int_0^1 f_v^t\int_0^v \left(l_u^t\right)'\,du\,dv\,\Delta_t\right)*\nu(\theta).$$

From this, in view of condition 1), after simple calculations we obtain the representation (3.3.16), in which the matrix p^t is of the form

$$p^t=\bar{p}^{\,t}+\bar{\bar{p}}^{\,t}+\tilde{p}^{\,t}, \tag{3.3.18}$$

where

$$\bar{p}^{\,t}=\varphi_t(\theta)\left[g_0g_0'\circ\langle m(\theta)\rangle+\chi(|\varphi_t(\theta)l_0|\le\varepsilon)l_0l_0'*\nu(\theta)\right]\varphi_t(\theta)-I_k,$$

$$\bar{\bar{p}}^{\,t}=2\varphi_t(\theta)\int_0^1\int_0^v\left[f_v^t\left(l_u^t\right)'-\chi(|\varphi_t(\theta)l_0|\le\varepsilon)l_0l_0'\right]du\,dv*\nu(\theta)\varphi_t(\theta),$$

$$\tilde{p}^{\,t}=\varphi_t(\theta)\int_0^1\int_0^1\left[g_v^t\left(g_u^t\right)'-g_0g_0'\right]du\,dv\circ\langle m(\theta)\rangle\,\varphi_t(\theta),\qquad \varepsilon\in(0,1].$$

By conditions 2) and 5) we have

$$\mathsf{P}_\theta\text{-}\lim_{t\to\infty}\left(\left|\bar{p}_t^t\right|+\left|\bar{\bar{p}}_t^t\right|\right)=0. \tag{3.3.19}$$

Putting $G^t=\varphi_t(\theta)g_0$, we get

$$\tilde{p}^{\,t}=\tilde{p}^{\,t,1}+\tilde{p}^{\,t,2}+\tilde{p}^{\,t,3},$$

where

$$\tilde{p}^{\,t,1}=\bar{g}^{\,t}\left(\bar{g}^{\,t}\right)'\circ\langle m(\theta)\rangle,\qquad \tilde{p}^{\,t,2}=\bar{g}^{\,t}\left(G^t\right)'\circ\langle m(\theta)\rangle,\qquad \tilde{p}^{\,t,3}=G^t(\bar{g}^{\,t})'\circ\langle m(\theta)\rangle.$$

Taking into account the estimates

$$\left|\tilde{p}^{\,t,1}\right|^2\le\left(\left|\bar{g}^{\,t}\right|^2\circ\langle m(\theta)\rangle\right)^2,$$

$$\left|\tilde{p}^{\,t,j}\right|^2\le\left|\bar{g}^{\,t}\right|^2\circ\langle m(\theta)\rangle\left|G^t\right|^2\circ\langle m(\theta)\rangle,\qquad j=2,3,$$

from conditions 2) and 4) we get

$$\mathsf{P}_\theta\text{-}\lim_{t\to\infty}\left|\tilde{p}_t^t\right|=0. \tag{3.3.20}$$

The relations (3.3.18)–(3.3.20) imply the relation (3.3.17) and hence the representation (3.3.5)–(3.3.8) is proved. □

The following theorem gives more restrictive but somewhat simpler conditions for the validity of the representation (3.3.5)–(3.3.8) in which $\eta^t\in\mathcal{M}^2_{\mathrm{loc}}(\mathbf{F},\mathsf{P}_\theta)$.

THEOREM 3.3.2. *Let $y_t\to\theta$ as $t\to\infty$, and let the following conditions be fulfilled:*

1) $\gamma(y,\theta)\in\mathbf{C}_1^0(\Theta)$ and $\ln\lambda(y,\theta)\in\mathbf{C}_1^0(\Theta)$ as functions of y, all the other variables being fixed, and for all y in a neighborhood of the point θ

$$g(y,\theta)\in\mathcal{L}^2_{\mathrm{loc}}(m(\theta),\mathbf{F},\mathsf{P}_\theta),\qquad l(y,\theta),\ f(y,\theta)\in\mathcal{G}^2_{\mathrm{loc}}(\nu(\theta),\mathbf{F},\mathsf{P}_\theta); \tag{3.3.21}$$

2) there exists a positive definite symmetric matrix $\varphi_t(\theta)$ such that $|\varphi_t(\theta)|\to 0$ as $t\to\infty$ and

$$\mathsf{P}_\theta\text{-}\lim_{t\to\infty}\varphi_t(\theta)\left[g_0g_0'\circ\langle m(\theta)\rangle_t+l_0l_0'*\nu(\theta)_t\right]\varphi_t(\theta)=I_k;$$

3) *for any* $\varepsilon \in (0,1]$,

$$\mathsf{P}_\theta\text{-}\lim_{t\to\infty} \chi(|\varphi_t(\theta)l_0| > \varepsilon)|\varphi_t(\theta)l_0|^2 * \nu(\theta)_t = 0;$$

4) $\mathsf{P}_\theta\text{-}\lim_{t\to\infty}\left[|\bar g^{\,t}|^2 \circ \langle m(\theta)\rangle_t + \int_0^1 (|\bar l_v^t|^2 + |\bar f_v^t|^2)\, dv * \nu(\theta)_t\right] = 0$, *where*

$$\bar l_v^t = \varphi_t(\theta)\left(l_v^t - l_0\right), \qquad \bar f_v^t = \varphi_t(\theta)\left(f_v^t - f_0\right)$$

and $\bar g^{\,t}$ *is defined in condition* 4) *of Theorem* 3.3.1.

Then the representation (3.3.5)–(3.3.8) *holds, in which*

$$\eta^t = \varphi_t(\theta)\big[g_0 \circ m(\theta) + l_0 * (\mu - \nu(\theta))\big] \in \mathcal{M}^2_{\mathrm{loc}}(\mathbf{F}, \mathsf{P}_\theta). \tag{3.3.22}$$

PROOF. The theorem is proved by the same scheme as Theorem 3.3.1, with some changes specified below.

In view of condition 1) we get the relations (3.3.9)–(3.3.13), where according to (3.3.21) the relation (3.3.22) holds and, moreover, $q^t \in \mathcal{M}^2_{\mathrm{loc}}(\mathbf{F}, \mathsf{P}_\theta)$. Applying Theorem 1.3.3 to the process η^t, in view of conditions 2) and 3) we get the relation (3.3.7). Then applying Theorem 1.3.3 to the process q^t based on condition 4) we get the relation (3.3.15).

Now, to complete the proof, it suffices to establish the relations (3.3.16) and (3.3.17).

As when proving Theorem 3.3.1, we get the equality (3.3.16), in which the matrix p^t has the form (3.3.18), where

$$\bar p^{\,t} = \varphi_t(\theta)\left[g_0 g_0' \circ \langle m(\theta)\rangle + l_0 l_0' * \nu(\theta)\right]\varphi_t(\theta) - I_k,$$

$$\bar{\bar p}^{\,t} = 2\varphi_t(\theta)\int_0^1\int_0^v \left[f_v^t\left(l_u^t\right)' - f_0 l_0'\right] du\, dv * \nu(\theta)\varphi_t(\theta),$$

$$\tilde p^{\,t} = \varphi_t(\theta)\int_0^1\int_0^1 \left[g_v^t\left(g_u^t\right)' - g_0 g_0'\right] du\, dv \circ \langle m(\theta)\rangle\,\varphi_t(\theta).$$

From conditions 2) and 4), taking into account (3.3.20), we get

$$\mathsf{P}_\theta\text{-}\lim_{t\to\infty}\left(|\bar p_t^t| + |\tilde p_t^t|\right) = 0.$$

Putting $\bar l^{\,t} = \varphi_t(\theta)l_0$ (it is evident that $\varphi_t(\theta)f_0 = \bar l^{\,t}$), we get $\bar{\bar p}^{\,t} = \bar{\bar p}^{\,t,1} + \bar{\bar p}^{\,t,2} + \bar{\bar p}^{\,t,3}$, where

$$\bar{\bar p}^{\,t,1} = \int_0^1\int_0^v \bar f_v^t\left(\bar l_u^t\right)' du\, dv * \nu(\theta), \qquad \bar{\bar p}^{\,t,2} = \int_0^1\int_0^v \bar f_v^t\left(\bar l^{\,t}\right)' du\, dv * \nu(\theta),$$

$$\bar{\bar p}^{\,t,3} = \int_0^1\int_0^v \bar l^{\,t}\left(\bar l_u^t\right)' du\, dv * \nu(\theta).$$

Taking into account the estimates

$$|\bar{\bar p}^{\,t,1}|^2 \le \left(\int_0^1 |\bar f_v^t|^2\, dv * \nu(\theta)\right)\left(\int_0^1 |\bar l_u^t|^2\, du * \nu(\theta)\right),$$

$$|\bar{\bar p}^{\,t,2}|^2 \le \left(\int_0^1 |\bar f_v^t|^2\, dv * \nu(\theta)\right)\left(|\bar l^{\,t}|^2 * \nu(\theta)\right),$$

$$|\bar{\bar p}^{\,t,3}|^2 \le \left(|\bar l^{\,t}|^2 * \nu(\theta)\right)\left(\int_0^1 |\bar l_u^t|^2\, du * \nu(\theta)\right),$$

from conditions 2) and 4) we get $\mathsf{P}_\theta\text{-}\lim_{t\to\infty} |\bar{\bar{p}}_t^t| = 0$. Thus the relations (3.3.16) and (3.3.17) are proved. □

The following two theorems establish the expansion (3.3.5), in which a vector $\eta_t^t + q_t^t$ is bounded only in probability P_θ.

THEOREM 3.3.3. *Let conditions* 1), 2), 4), *and* 5) *of Theorem* 3.3.1 *be fulfilled, but instead of condition* 3) *assume the following*:

3′) *for any* $\varepsilon \in (0,1]$

$$\lim_{N\to\infty} \overline{\lim_{t\to\infty}} \mathsf{P}_\theta\{\chi(|\varphi_t(\theta)l_0| > \varepsilon)|\varphi_t(\theta)l_0| * \nu(\theta)_t > N\} = 0.$$

Then

$$\Lambda(y_t, \theta) = u_t'\widetilde{\eta}_t - (1/2)u_t'\left(I_k + p^t\right)u_t, \tag{3.3.23}$$

where $u_t = \varphi_t^{-1}(\theta)\Delta_t$ *and the processes* $\widetilde{\eta}_t$ *and* p^t *are such that*

$$\lim_{N\to\infty} \overline{\lim_{t\to\infty}} \mathsf{P}_\theta\left\{|\widetilde{\eta}_t^t| > N\right\} = 0, \tag{3.3.24}$$

$$\mathsf{P}_\theta\text{-}\lim_{t\to\infty} |p_t^t| = 0. \tag{3.3.25}$$

PROOF. Theorem 3.3.1 gives representation (3.3.23) in which $\widetilde{\eta}^t = \eta^t + q^t$ and the processes η^t and q^t are defined by formulas (3.3.12) and (3.3.13). The relation (3.3.25) for the process p^t was established when proving Theorem 3.3.1, and for the process q^t the relation (3.3.15) was proved. From condition 1) of Theorem 3.3.1, using relations (1.1.4) and (1.1.4′) and Theorem 1.1.8, for any $\varepsilon \in (0,1]$ we get

$$\chi(|\varphi_t(\theta)l_0| \le \varepsilon)\varphi_t(\theta)l_0 \in \mathcal{G}^2_{\text{loc}}(\nu(\theta), \mathbf{F}, \mathsf{P}_\theta),$$
$$\chi(|\varphi_t(\theta)l_0| > \varepsilon)\varphi_t(\theta)l_0 \in \mathcal{G}^1_{\text{loc}}(\nu(\theta), \mathbf{F}, \mathsf{P}_\theta).$$

Hence for any $\varepsilon \in (0,1]$ the process η^t can be represented in the form

$$\eta^t = \eta^{t,\varepsilon} + \widetilde{\eta}^{t,\varepsilon}, \tag{3.3.26}$$

where $\eta^{t,\varepsilon} \in \mathcal{M}^2_{\text{loc}}(\mathbf{F}, \mathsf{P}_\theta)$, $\widetilde{\eta}^{t,\varepsilon} \in \mathcal{M}^d_{\text{loc}}(\mathbf{F}, \mathsf{P}_\theta)$, and

$$\begin{aligned} \eta^{t,\varepsilon} &= \varphi_t(\theta)[g_0 \cdot m(\theta) + \chi(|\varphi_t(\theta)l_0| \le \varepsilon)l_0 * (\mu - \nu(\theta))], \\ \widetilde{\eta}^{t,\varepsilon} &= \chi(|\varphi_t(\theta)l_0| > \varepsilon)\varphi_t(\theta)l_0 * (\mu - \nu(\theta)) \\ &= \chi(|\varphi_t(\theta)l_0| > \varepsilon)\varphi_t(\theta)l_0 * \mu - \chi(|\varphi_t(\theta)l_0| > \varepsilon)\varphi_t(\theta)l_0 * \nu(\theta)). \end{aligned} \tag{3.3.27}$$

Now we apply the Lenglart inequality, putting $\xi = |\eta^{t,\varepsilon}|^2$ and $\eta = \operatorname{Tr}\langle \eta^{t,\varepsilon}, \eta^{t,\varepsilon}\rangle$ in Theorem 1.1.3. Then for any $N > 0$, $\delta > 0$, and $s \in \mathbf{R}_+$ we have

$$\mathsf{P}_\theta\left\{|\eta_s^{t,\varepsilon}| > N\right\} \le \delta/N^2 + \mathsf{P}_\theta\left\{\operatorname{Tr}\langle \eta^{t,\varepsilon}, \eta^{t,\varepsilon}\rangle_s > \delta\right\}.$$

But since

$$\langle \eta^{t,\varepsilon}, \eta^{t,\varepsilon}\rangle = \varphi_t(\theta)\left[g_0 g_0' \circ \langle m(\theta)\rangle + \chi(|\varphi_t(\theta)l_0| \le \varepsilon)l_0 l_0' * \nu(\theta)\right]\varphi_t(\theta),$$

in view of condition 2) of Theorem 3.3.1 we have

$$\mathsf{P}_\theta\text{-}\lim_{t\to\infty} \operatorname{Tr}\langle \eta^{t,\varepsilon}, \eta^{t,\varepsilon}\rangle_t = \operatorname{Tr} I_k = k.$$

Therefore

$$\lim_{N\to\infty}\overline{\lim_{t\to\infty}}\,\mathsf{P}_\theta\left\{|\eta_t^{t,\varepsilon}|>N\right\}=0. \tag{3.3.28}$$

Let us apply the Lenglart inequality once again, putting

$$\xi=\left|\widetilde{\eta}^{t,\varepsilon}\right|,\qquad \eta=2\sqrt{k}\,\chi(|\varphi_t(\theta)l_0|>\varepsilon)|\varphi_t(\theta)l_0|*\nu(\theta)$$

in Theorem 1.1.3. Taking into account the inequality $|x|\le|x|_1\le\sqrt{k}\,|x|$, $x\in\mathbf{R}^k$, from the relation (3.3.27) for any stopping time $\tau\in\mathfrak{M}(\mathbf{F})$ we have

$$\begin{aligned}\mathsf{E}_\theta\left|\widetilde{\eta}_\tau^{t,\varepsilon}\right|\le\mathsf{E}_\theta\left|\widetilde{\eta}_\tau^{t,\varepsilon}\right|_1&\le2\,\mathsf{E}_\theta\,\chi(|\varphi_t(\theta)l_0|>\varepsilon)|\varphi_t(\theta)l_0|_1*\nu(\theta)_\tau\\&\le2\sqrt{k}\;\mathsf{E}_\theta\,\chi(|\varphi_t(\theta)l_0|>\varepsilon)|\varphi_t(\theta)l_0|*\nu(\theta)_\tau.\end{aligned}$$

Hence for any $N>0$, $\delta>0$, and $s\in\mathbf{R}_+$ we have

$$\mathsf{P}_\theta\left\{\left|\widetilde{\eta}_s^{t,\varepsilon}\right|>N\right\}\le\delta/N+\mathsf{P}_\theta\left\{2\sqrt{k}\,\chi(|\varphi_t(\theta)l_0|>\varepsilon)|\varphi_t(\theta)l_0|*\nu(\theta)_s>\delta\right\},$$

whence in view of condition 3$'$ we get

$$\lim_{N\to\infty}\overline{\lim_{t\to\infty}}\,\mathsf{P}_\theta\left\{\left|\widetilde{\eta}_t^{t,\varepsilon}\right|>N\right\}=0. \tag{3.3.29}$$

Combining the relations (3.3.15), (3.3.26), (3.3.28), and (3.3.29) and taking into account the equality $\widetilde{\eta}^t=\eta^t+q^t$, we get (3.3.24). □

THEOREM 3.3.4. *If conditions* 1), 2), *and* 4) *of Theorem* 3.3.2 *are fulfilled, then the representation* (3.3.23)–(3.3.25) *is valid.*

PROOF. It is enough to repeat the arguments of the proof of Theorem 3.3.3, using Theorem 3.3.2 instead of Theorem 3.3.1 and taking into consideration that both the process η^t in Theorem 3.3.2 defined by formula (3.3.22), and the process $\eta^{t,\varepsilon}$ in Theorem 3.3.3 satisfy the relation

$$\lim_{N\to\infty}\overline{\lim_{t\to\infty}}\,\mathsf{P}_\theta\left\{|\eta_t^t|>N\right\}=0$$

in view of Lenglart inequality and condition 2) of Theorem 3.3.2 . □

The following theorem establishes the relation

$$\mathsf{P}_\theta\text{-}\lim_{t\to\infty}\left|p_t^tu_t\right|=0 \tag{3.3.30}$$

for the matrix process p^t from the representation (3.3.5)–(3.3.8).

THEOREM 3.3.5. *Let the conditions of Theorem* 3.3.2 *be fulfilled, and in addition,*

1) $\mathsf{P}_\theta\text{-}\lim_{t\to\infty}[\varphi_t(\theta)(g_0g_0'\circ\langle m(\theta)\rangle_t+l_0l_0'*\nu(\theta)_t)\varphi_t(\theta)-I_k]u_t=0$;

2)

$$\mathsf{P}_\theta\text{-}\lim_{t\to\infty}\Bigg[\left(\Delta_t'\tilde{g}^t\right)^2\circ\langle m(\theta)\rangle_t+\int_0^1\left(\Delta_t'\tilde{f}_v^t\right)^2dv*\nu(\theta)_t+\int_0^1\left(\Delta_t'\tilde{l}_v^t\right)^2dv*\nu(\theta)_t\Bigg]=0,$$

where

$$\tilde{g}^{\,t} = \int_0^1 g_v^t\,dv - g_0, \qquad \tilde{f}_v^{\,t} = f_v^t - f_0, \qquad \tilde{l}_v^{\,t} = l_v^t - l_0.$$

Then the representation (3.3.5)–(3.3.8) *takes place, in which* $\eta^t \in \mathcal{M}^2_{\mathrm{loc}}(\mathbf{F}, \mathsf{P}_\theta)$ *and the process* p^t *satisfies the relation* (3.3.30).

PROOF. The representation (3.3.5)–(3.3.8) and inclusion $\eta^t \in \mathcal{M}^2_{\mathrm{loc}}(\mathbf{F}, \mathsf{P}_\theta)$ follow from Theorem 3.3.2. It suffices to prove the relation (3.3.30).

From the representations of the matrix p^t appearing in the proofs of Theorems 3.3.1 and 3.3.2 we have

$$p^t = \bar{p}^{\,t} + \hat{p}^{\,t,1} + \hat{p}^{\,t,2} + \hat{p}^{\,t,3}, \tag{3.3.31}$$

where

$$\bar{p}^{\,t} = \varphi_t(\theta)[g_0 g_0' \circ \langle m(\theta)\rangle + l_0 l_0' * \nu(\theta)]\varphi_t(\theta) - I_k,$$
$$\hat{p}^{\,t,1} = \varphi_t(\theta)\left[\tilde{g}^{\,t}\left(\tilde{g}^{\,t}\right)' \circ \langle m(\theta)\rangle + 2\int_0^1\int_0^v \tilde{f}_v^{\,t}\left(\tilde{l}_u^{\,t}\right)'\,du\,dv * \nu(\theta)\right]\varphi_t(\theta),$$
$$\hat{p}^{\,t,2} = \varphi_t(\theta)\left[\tilde{g}^{\,t} g_0' \circ \langle m(\theta)\rangle + 2\int_0^1\int_0^v \tilde{f}_v^{\,t} l_0'\,du\,dv * \nu(\theta)\right]\varphi_t(\theta),$$
$$\hat{p}^{\,t,3} = \varphi_t(\theta)\left[g_0\left(\tilde{g}^{\,t}\right)' \circ \langle m(\theta)\rangle + 2\int_0^1\int_0^v l_0\left(\tilde{l}_u^{\,t}\right)'\,du\,dv * \nu(\theta)\right]\varphi_t(\theta).$$

Since $u_t'Au_t = u_t'A'u_t$, for any square matrix A in the representation (3.3.5) we have $u_t'(I_k + p^t)u^t = u_t'(I_k + \check{p}^{\,t})u_t$, where

$$\check{p}^{\,t} = \bar{p}^{\,t} + \hat{p}^{\,t,1} + \left(\hat{p}^{\,t,2}\right)' + \hat{p}^{\,t,3}. \tag{3.3.32}$$

For this reason, in order to prove the theorem, it is enough to show that

$$\mathsf{P}_\theta\text{-}\lim_{t\to\infty} \left|\check{p}_t^{\,t} u_t\right| = 0. \tag{3.3.33}$$

It is not hard to obtain the following estimates:

$$\begin{aligned}
\left|\hat{p}^{\,t,1}u_t\right|^2 &\le 2\left|\bar{g}^{\,t}\Delta_t'\tilde{g}^{\,t} \circ \langle m(\theta)\rangle\right|^2 + 8\left|\int_0^1\int_0^v \bar{f}_v^{\,t}\Delta_t'\tilde{l}_u^{\,t}\,du\,dv * \nu(\theta)\right|^2 \\
&\le 2\left(\left|\bar{g}^{\,t}\right|^2 \circ \langle m(\theta)\rangle\right)\left(\left(\Delta_t'\tilde{g}^{\,t}\right)^2 \circ \langle m(\theta)\rangle\right) \\
&\quad + 8\left(\int_0^1 \left|\bar{f}_v^{\,t}\right|^2 dv * \nu(\theta)\right)\left(\int_0^1 \left(\Delta_t'\tilde{l}_u^{\,t}\right)^2 du * \nu(\theta)\right), \\
\left|\left(\hat{p}^{\,t,2}\right)' u_t\right|^2 &\le 2\left|G^t\Delta_t'\tilde{g}^{\,t} \circ \langle m(\theta)\rangle\right|^2 + 8\left|\int_0^1\int_0^v \bar{l}^{\,t}\Delta_t'\tilde{f}_v^{\,t}\,du\,dv * \nu(\theta)\right|^2 \\
&\le 2\left(\left|G^t\right|^2 \circ \langle m(\theta)\rangle\right)\left(\left(\Delta_t'\tilde{g}^{\,t}\right)^2 \circ \langle m(\theta)\rangle\right) \\
&\quad + 8\left(\left|\bar{l}^{\,t}\right|^2 * \nu(\theta)\right)\left(\int_0^1 \left(\Delta_t'\tilde{f}_v^{\,t}\right)^2 dv * \nu(\theta)\right),
\end{aligned}$$

$$
\begin{aligned}
\left|\hat{p}^{t,3}u_t\right|^2 &\le 2\left|G^t\Delta_t'\tilde{g}^t\circ\langle m(\theta)\rangle\right|^2+8\left|\int_0^1\int_0^v \bar{l}^t\Delta_t'\tilde{l}_u^t\,du\,dv*\nu(\theta)\right|^2\\
&\le 2\left(\left|G^t\right|^2\circ\langle m(\theta)\rangle\right)\left(\left(\Delta_t'\tilde{g}^t\right)^2\circ\langle m(\theta)\rangle\right)\\
&\quad+8\left(\left|\bar{l}^t\right|^2*\nu(\theta)\right)\left(\int_0^1\left(\Delta_t'\tilde{l}_u^t\right)^2du*\nu(\theta)\right),
\end{aligned}
$$

where, according to the above notation,

$$
G^t=\varphi_t(\theta)g_0,\qquad \bar{l}^t=\varphi_t(\theta)l_0,\qquad \bar{g}^t=\varphi_t(\theta)\tilde{g}^t,\qquad \bar{f}_v^t=\varphi_t(\theta)\tilde{f}_v^t.
$$

From these estimates and equality (3.3.32), taking into account conditions 2) and 4) of Theorem 3.3.2 and conditions 1) and 2), we obtain the required relation (3.3.33). □

Statistical applications. 1. If the conditions of Theorems 3.3.3 and 3.3.4 are fulfilled, then condition $\Lambda 8$ (Section 2.6) with $\varkappa=I_k$ is fulfilled. Hence in view of Theorem 2.6.1 we can claim that the following dichotomy takes place: either $(H^t)\vartriangle(\widetilde{H}^t)$ or $(H^t)\lhd(\widetilde{H}^t)$ and

$$
\begin{aligned}
(H^t)\vartriangle\left(\widetilde{H}^t\right)&\Leftrightarrow\varlimsup_{t\to\infty}|u_t|=\infty,\\
(H^t)\lhd\left(\widetilde{H}^t\right)&\Leftrightarrow\varlimsup_{t\to\infty}|u_t|<\infty.
\end{aligned}
$$

2. Now let the conditions of Theorem 3.3.1 or Theorem 3.3.2 be fulfilled. Then condition $\Lambda 8^*$ (Section 2.6) is fulfilled and in view of Theorem 2.6.4 the following dichotomy takes place: either $(H^t)\vartriangle(\widetilde{H}^t)$ or $(H^t)\lhd\rhd(\widetilde{H}^t)$ and

$$
\begin{aligned}
(H^t)\vartriangle\left(\widetilde{H}^t\right)&\Leftrightarrow\varlimsup_{t\to\infty}|u_t|=\infty,\\
(H^t)\lhd\rhd\left(\widetilde{H}^t\right)&\Leftrightarrow\varlimsup_{t\to\infty}|u_t|<\infty.
\end{aligned}
$$

Moreover, $(H^t)\cong(\widetilde{H}^t)\Leftrightarrow\varlimsup_{t\to\infty}|u_t|=0$. This means that for the types of asymptotic distinguishability between the families of hypotheses (H^t) and $(\widetilde{H}^t)$ the following trichotomy takes place: either type $\mathbf{a}_0$, or type $\mathbf{a}_1$, or type $\mathbf{e}$. Moreover, if $|u_t|\to\sigma\in(0,\infty)$ as $t\to\infty$, then in view of Theorem 2.6.5 for the test δ_t^{+,α_t}

$$
\lim_{t\to\infty}\alpha_t=\alpha\Leftrightarrow\lim_{t\to\infty}\beta\left(\delta_t^{+,\alpha_t}\right)=\Phi(z_{1-\alpha}-\sigma)
$$

for any $\alpha\in[0,1]$, where z_p and $\Phi(x)$ are the p-quantile and the distribution function of the law $\mathcal{N}(0,1)$, respectively.

3. Let the conditions of Theorem 3.3.5 be fulfilled, and let $|u_t|\to\infty$ as $t\to\infty$. Then by Theorem 2.6.6 for the test δ_t^{+,α_t} with $\alpha_t\to\alpha\in(0,1)$ as $t\to\infty$ the probability of error of the second kind behaves as follows (compare with (3.2.15)):

$$
\beta(\delta_t^{+,\alpha_t})=-(1/2)|u_t|^2+z_{1-\alpha}|u_t|+o(|u_t|),\qquad t\to\infty.
$$

Moreover, the limit value α for the level α_t lies in the interval $(0,1)$. Thus behavior of the probability of error of the second kind $\beta(\delta_t^{+,\alpha_t})$ is ensured by the following choice of the parameter d_t of the test δ_t^{+,α_t}:

$$
d_t=-(1/2)|u_t|^2+z_{1-\alpha}|u_t|+o(|u_t|).
$$

3.4. Observations of diffusion-type processes

General theorems. Let ξ be a diffusion-type process with coefficients a and b with respect to the measure P, and $\widetilde{a}$ and b with respect to the measure $\widetilde{\mathsf{P}}$ (see Section 1.2). We assume that the conditions of Theorem 1.2.3 are fulfilled, which ensure the local equivalence of the measures P and $\widetilde{\mathsf{P}}$. From Theorems 1.2.3 and 1.2.4 we get the following formula for the local density z of the measure $\widetilde{\mathsf{P}}$ with respect to the measure P:

$$z = \exp\left\{\gamma \cdot w - (1/2)\gamma^2 \circ I\right\} \quad (\mathsf{P}\text{-a.s.}), \tag{3.4.1}$$

where w is the standard Wiener process and

$$\gamma = b^{\oplus}(\widetilde{a} - a) \quad (\mathsf{P}\text{-a.s.}). \tag{3.4.2}$$

THEOREM 3.4.1. *Let $s \in (0, \infty)$, let ψ_t be some positive nonrandom function such that $\psi_t \to \infty$ as $t \to \infty$, and let the following condition be fulfilled:*

$$\mathsf{P}\text{-}\lim_{t\to\infty} \psi_t^{-1}\gamma^2 \circ I_{st} = c_s, \tag{3.4.3}$$

where c_s is a positive nonrandom function satisfying the condition $c_1 = 1$. Then

$$\mathsf{P}\text{-}\lim_{t\to\infty} \psi_t^{-1}\Lambda_{st} = -\frac{1}{2}c_s, \tag{3.4.4}$$

where $\Lambda = \ln z$.

It is enough to apply Theorem 3.2.1 and to observe that condition (3.4.3) coincides with condition 1) of Theorem 3.2.1 and, moreover, implies condition 2) of Theorem 3.2.1.

In view of Theorem 3.1.7, for $\varepsilon \in (0,1)$ the Hellinger process $h(\varepsilon) = h(\varepsilon; \widetilde{\mathsf{P}}, \mathsf{P})$ of order ε for the measures $\widetilde{\mathsf{P}}$ and P is of the form

$$h(\varepsilon) = \frac{\varepsilon(1-\varepsilon)}{2}\gamma^2 \circ I, \tag{3.4.5}$$

and if $H_t(\varepsilon_0; \widetilde{\mathsf{P}}, \mathsf{P}) < \infty$ for all $t \in \mathbf{R}_+$ for some $\varepsilon_0 < 0$, then for $\varepsilon \in [\varepsilon_0, 0)$ the process $h(\varepsilon)$ is of the form

$$h(\varepsilon) = -\frac{\varepsilon(1-\varepsilon)}{2}\gamma^2 \circ I. \tag{3.4.6}$$

In the case of diffusion-type processes the statements of Theorems 3.2.2 and 3.2.3 do not change except that the Hellinger process $h(\varepsilon)$ is replaced by its expression given by formulas (3.4.5) and (3.4.6); therefore, they are omitted.

Now we consider a family of stochastic processes $Y^t = (Y_s^t)_{s\in\mathbf{R}_+}$, where $Y_s^t = \psi_t^{-1}(\Lambda_{st} + \varphi_{st})$, φ_s is a continuous function, $\Lambda = \ln z$, and z is defined by equalities (3.4.1) and (3.4.2). It is obvious that $Y^t \in \mathcal{S}(\mathbf{F}^t, \mathsf{P})$, where $\mathbf{F}^t = (\mathcal{F}_{st})_{s\in\mathbf{R}_+}$ with a triple of predictable characteristics $(B^t, C^t, 0)$ and

$$B_s^t = \psi_t^{-1}\left(\varphi_{st} - (1/2)\gamma^2 \circ I_{st}\right), \qquad C_s^t = \psi_t^{-2}\gamma^2 \circ I_{st}, \quad s \in \mathbf{R}_+.$$

To formulate the next theorem, on the stochastic basis $(\Omega, \mathcal{F}, \mathbf{F}, \mathsf{P})$ we introduce a left quasicontinuous semimartingale $Y \in \mathcal{S}(\mathbf{F}, \mathsf{P})$, $Y_0 = 0$, with independent increments and a triple of predictable characteristics $(B^Y, C^Y, 0)$. Obviously, the process Y is a Gaussian continuous process.

THEOREM 3.4.2. *Let S be a nonempty subset in $\mathbf{R}_+$ and let the following conditions be fulfilled:*

1) $\mathsf{P}\text{-}\lim_{t\to\infty} B_s^t = B_s^Y$, $s \in S$;
2) $\mathsf{P}\text{-}\lim_{t\to\infty} C_s^t = C_s^Y$, $s \in S$.

Then $Y^t \xrightarrow{w_f(S)} Y$ *as* $t \to \infty$.

PROOF. The proof follows from Theorems 3.2.4 and 3.2.5. □

Theorem 3.4.2 is also true in the case of the triangular array scheme (see Remark 3.2.2).

Suppose that there is given a family of probability measures $(P_\theta, \theta \in \Theta)$, where Θ is a subset in $\mathbf{R}^k$, $k \geq 1$, and ξ is a diffusion-type process with coefficients $a(\theta)$ and b with respect to the measure P_θ for every $\theta \in \Theta$ (see Section 1.2). We assume that the conditions of Theorem 1.2.3 with $\mathsf{P} = \mathsf{P}_\theta$ and $\widetilde{\mathsf{P}} = \mathsf{P}_y$ are fulfilled for any $\theta, y \in \Theta$, which ensure the local equivalence of the measures P_y and P_θ. From Theorems 1.2.3 and 1.2.4 we get the following form for the process of local density $z(y,\theta)$ of the measure P_y with respect to P_θ:

$$z(y,\theta) = \exp\{\gamma(y,\theta)\cdot w - (1/2)\gamma^2(y,\theta)\circ I\} \quad (\mathsf{P}_\theta\text{-a.s.}), \tag{3.4.7}$$

where

$$\gamma(y,\theta) = b^{\oplus}(a(y) - a(\theta)) \quad (\mathsf{P}_\theta\text{-a.s.}). \tag{3.4.8}$$

As in Section 3.3, we assume that θ is independent of t, while $y = y_t$ depends on t so that $y_t \to \theta$ as $t \to \infty$. In the notation of Section 3.3, from Theorem 3.3.2 and representation (3.4.7)–(3.4.8) we derive the following result.

THEOREM 3.4.3. *Let $y_t \to \theta$ as $t \to \infty$, and let the following conditions be fulfilled:*

1) *the function $\gamma(y,\theta)$ belongs to $\mathbf{C}_1^0(\Theta)$ as a function of the variable y, all the other variables being fixed, and $g(y,\theta) \in \mathcal{L}^2_{\mathrm{loc}}(w, \mathbf{F}, \mathsf{P}_\theta)$ for all y in a neighborhood of θ;*

2) $\mathsf{P}_\theta\text{-}\lim_{t\to\infty} \varphi_t(\theta) g_0 g_0' \circ I_t \varphi_t(\theta) = I_k$, *where $\varphi_t(\theta)$ is a positive definite symmetric nonrandom matrix such that $|\varphi_t(\theta)| \to 0$ as $t \to \infty$;*

3) $\mathsf{P}_\theta\text{-}\lim_{t\to\infty} |\bar{g}^{\,t}|^2 \circ I_t = 0$.

Then the following representation is true:

$$\Lambda(y_t,\theta) = u_t'\left(\eta^t + q^t\right) - (1/2)u_t'(I_k + p^t)u_t, \tag{3.4.9}$$

where

$$u_t = \varphi_t^{-1}(\theta)\Delta_t, \qquad \eta^t = \varphi_t(\theta) g_0 \cdot w \in \mathcal{M}^c_{\mathrm{loc}}(\mathbf{F}, \mathsf{P}_\theta), \tag{3.4.10}$$

$$\mathcal{L}\left(\eta_t^t | \mathsf{P}_\theta\right) \xrightarrow{w} \mathcal{N}(0, I_k), \qquad t \to \infty, \tag{3.4.11}$$

$$\mathsf{P}_\theta\text{-}\lim_{t\to\infty}\left(|g_t^t| + |p_t^t|\right) = 0. \tag{3.4.12}$$

The next theorem follows from Theorem 3.3.5.

THEOREM 3.4.4. *Let conditions of Theorem 3.4.3 be fulfilled and in addition,*

1) $\mathsf{P}_\theta\text{-}\lim_{t\to\infty}(\varphi_t(\theta) g_0 g_0' \circ I_t \varphi_t(\theta) - I_k)u_t = 0$;
2) $\mathsf{P}_\theta\text{-}\lim_{t\to\infty}(\Delta_t' \tilde{g}^{\,t})^2 \circ I_t = 0$.

Then the representation (3.4.9)–(3.4.12) *is valid, in which*

$$\mathsf{P}_\theta\text{-}\lim_{t\to\infty} |p_t^t u_t| = 0. \tag{3.4.13}$$

The case $a(\theta) = h + s(\theta)b$ with determinate function $s(\theta)$. We assume that the function $s(\theta) = (s_t(\theta))_{t\in\mathbf{R}_+}$ depends only on $\theta \in \Theta$ and $t \in \mathbf{R}_+$ and is independent of ξ. Besides the conditions of Theorem 1.2.3 we suppose that $b_t \neq 0$ (P_θ-a.s.) and $s^2(\theta)\circ I_t < \infty$ for all $\theta \in \Theta$ and $t \in \mathbf{R}_+$. Then, evidently, $\gamma(y,\theta) = s(y) - s(\theta)$. Put $v_t = (\gamma^2(y_t,\theta)\circ I_t)^{1/2}$, where θ is independent of t, while y_t is dependent of t in some way. We assume that $v_t > 0$ for all $t > 0$. Under the above assumptions $\mathsf{P}_y \overset{\text{loc}}{\sim} \mathsf{P}_\theta$ for any $y, \theta \in \Theta$. Then from Theorems 1.2.3 and 1.2.4 we get

$$\Lambda_t(y_t,\theta) = v_t\zeta_t - v_t^2/2 \quad (\mathsf{P}_\theta\text{-a.s.}), \tag{3.4.14}$$

$$\Lambda_t(y_t,\theta) = v_t\widetilde{\zeta}_t + v_t^2/2 \quad (\mathsf{P}_{y_t}\text{-a.s.}), \tag{3.4.15}$$

where

$$\mathcal{L}(\zeta_t|\mathsf{P}_\theta) = \mathcal{L}(\widetilde{\zeta}_t|\mathsf{P}_{y_t}) = \mathcal{N}(0,1). \tag{3.4.16}$$

The relations (3.4.14) and (3.4.16) imply that for the likelihood ratio, condition $\Lambda 8^*$ is fulfilled for $k = 1$, $u_t = v_t$, $\eta_t = \zeta_t$, $p_t = 0$. Then in view of Theorem 2.6.4 we get the following characteristic relation for the families of hypotheses (H^t) and $(\widetilde{H}^t)$ (here H^t and $\widetilde{H}^t$ are the hypotheses consisting in that the distribution of observation ξ^t of the process ξ is given by the measures P_θ and P_{y_t}, respectively):

$$(H^t) \vartriangle (\widetilde{H}^t) \Leftrightarrow \overline{\lim_{t\to\infty}}\, v_t = \infty, \tag{3.4.17}$$

$$(H^t) \vartriangleleft\vartriangleright (\widetilde{H}^t) \Leftrightarrow \overline{\lim_{t\to\infty}}\, v_t < \infty, \tag{3.4.18}$$

$$(H^t) \cong (\widetilde{H}^t) \Leftrightarrow \lim_{t\to\infty} v_t = 0. \tag{3.4.19}$$

The relations (3.4.17) and (3.4.18) imply that for the types of asymptotic distinguishability of the families of hypotheses (H^t) and $(\widetilde{H}^t)$ we have the alternative: either **a** or **e**. Moreover, the relations (3.4.17)–(3.4.19) imply the following trichotomy: either type $\mathbf{a}_0$, or type $\mathbf{a}_1$, or type **e**, and

$$\mathbf{a}_0 \Leftrightarrow \overline{\lim_{t\to\infty}}\, v_t = 0, \tag{3.4.20}$$

$$\mathbf{a}_1 \Leftrightarrow 0 < \overline{\lim_{t\to\infty}}\, v_t < \infty, \tag{3.4.21}$$

$$\mathbf{e} \Leftrightarrow \overline{\lim_{t\to\infty}}\, v_t = \infty. \tag{3.4.22}$$

The relations (3.4.14) and (3.4.16) imply that the parameter d_t of the test δ_t^{+,α_t} for distinguishing between the hypotheses H^t and $\widetilde{H}^t$ is of the form

$$d_t = -(1/2)v_t^2 + z_{1-\alpha_t}v_t, \tag{3.4.23}$$

and the parameter ε_t is any number in the interval $[0,1]$. From the relations (3.4.15) and (3.4.16) we get that the probability of error of the second kind of the test δ_t^{+,α_t} is given by the formula

$$\beta\left(\delta_t^{+,\alpha_t}\right) = \Phi(z_{1-\alpha_t} - v_t). \tag{3.4.24}$$

This equality allows us to investigate the relationship among the behavior of the level α_t, function v_t, and the probability of error of the second kind $\beta(\delta_t^{+,\alpha_t})$ as $t \to \infty$ for various types of asymptotic distinguishability of two families of hypotheses. For simplicity, in order not to deal with subfamilies or hypotheses, we assume that the limit $\lim_{t\to\infty} v_t$ exists.

First, let $\lim_{t\to\infty} v_t = \infty$, that is, in view of (3.4.22) the asymptotic distinguishability of type **e** takes place. In this case for $s = 1$ condition (3.4.3) of Theorem 3.4.1 is fulfilled with $\psi_t = v_t^2$ and hence condition $\Lambda 1$ (the law of large numbers for $\Lambda_t(y_t, \theta)$) from Section 2.3 is fulfilled with $\chi_t = v_t^2/2$. Therefore, in view of Theorem 2.3.1 for the test δ_t^{+,α_t}, whose level α_t satisfies conditions $\alpha 1$ and $\alpha 2$, we have

$$(3.4.25) \qquad \beta(\delta_t^{+,\alpha_t}) = \exp\{-(1/2)v_t^2(1+o(1))\}, \qquad t \to \infty.$$

Note that under the given conditions formula (3.4.24) allows us to obtain, due to a well-known asymptotic expansion of $1 - \Phi(x)$ for $x > 0$ [**121**], a more exact asymptotic estimate:

$$(3.4.26) \qquad \beta\left(\delta_t^{+,\alpha_t}\right) = \left(\sqrt{2\pi}\, v_t\right)^{-1} \exp\left\{-(1/2)v_t^2 + O(v_t)\right\}, \qquad t \to \infty.$$

From the relations (3.4.14) and (3.4.16) for any $t \in \mathbf{R}_+$ and $\varepsilon \in \mathbf{R}^1$ we get

$$(3.4.27) \qquad H_t(\varepsilon; \widetilde{\mathsf{P}}, \mathsf{P}) = \exp\left\{-\frac{\varepsilon(1-\varepsilon)}{2} v_t^2\right\}.$$

Hence condition Λ^* (Section 2.3) is fulfilled for $\varepsilon_- = -\infty$, $\varepsilon_+ = \infty$, $\widetilde{\chi}_t = (1/2)v_t^2$, and $\Psi(\varepsilon) = -\varepsilon(1-\varepsilon)$. It means that the theorem on large deviations (Lemma 2.3.4) is valid with $\gamma_0 = -1$, $\gamma_- = -\infty$, $\gamma_+ = \infty$, and $I(\gamma) = (1+\gamma)^2/4$. Also, formula (3.4.27) implies

$$\lim_{\varepsilon\to 0} \lim_{t\to\infty} (\varepsilon\chi_t)^{-1} \ln H_t(\varepsilon; \widetilde{\mathsf{P}}, \mathsf{P}) = -1$$

for $\chi_t = (1/2)v_t^2$, that is, conditions $\Lambda 2$ and $\Lambda 3$ of Section 2.3 are fulfilled. By virtue of Theorem 2.3.3, for the test δ_t^{+,α_t} of the level α_t satisfying conditions $\alpha 1'$ and $\alpha 2'$ the relation (3.4.25) is also valid. Note that under the given conditions the more exact asymptotic estimate (3.4.26) does not hold any more. This can be easily seen from the asymptotic expansion of $1 - \Phi(x)$ as $x \to \infty$ and the following theorem, which is of interest in its own right.

THEOREM 3.4.5. *If* $\lim_{t\to\infty} v_t = \infty$, *then the following assertions are valid:*
1) *if* $\alpha_t \to 0$ *as* $t \to \infty$, *then*

$$(3.4.28) \qquad \alpha 1' \Leftrightarrow d1 \Leftrightarrow \beta 1 \Leftrightarrow z_{1-\alpha_t} = o(v_t);$$

2) *if* $\alpha_t \to 1$ *as* $t \to \infty$, *then*

$$(3.4.29) \qquad \alpha 2' \Leftrightarrow d2 \Leftrightarrow \beta 2 \Leftrightarrow z_{1-\alpha_t} = o(v_t)$$

(*in* (3.4.28) *and* (3.4.29) $\chi_t = v_t^2/2$).

PROOF. Let $\alpha_t \to 0$ as $t \to \infty$. Then by the relation between α_t and $z_{1-\alpha_t}$ (see Remark 2.6.4) we get

$$\ln \alpha_t = -(1/2) z_{1-\alpha_t}^2 (1 + o(1)), \qquad t \to \infty. \tag{3.4.30}$$

Hence the following equivalence is valid:

$$\alpha 1' \Leftrightarrow z_{1-\alpha_t} = o(v_t). \tag{3.4.31}$$

Since on the basis of the relation $\lim_{t\to\infty} v_t = \infty$ conditions $\Lambda1$, $\Lambda2$, and $\Lambda3$ are fulfilled, Theorems 2.3.1 and 2.3.3 imply that, in order to prove the relations (3.4.28), it is enough to prove the implications

$$\beta 1 \Rightarrow d1 \Rightarrow \alpha 1'. \tag{3.4.32}$$

Since $z_{1-\alpha_t} \to \infty$ as $t \to \infty$, the relation (3.4.23) implies that $d1 \Rightarrow z_{1-\alpha_t} = o(v_t)$, whence in view of (3.4.31) we get the implication $d1 \Rightarrow \alpha 1'$.

If $\beta 1$ is fulfilled, then in view of equality (3.4.24) we have $v_t - z_{1-\alpha_t} \to \infty$ as $t \to \infty$, and hence by virtue of the asymptotic expansion of $1 - \Phi(x)$ as $x \to \infty$ [**121**] we have

$$\lim_{t\to\infty} (v_t - z_{1-\alpha_t})^{-2} \ln \beta \left(\delta_t^{+,\alpha_t}\right) = -1/2.$$

From this, by virtue of $\beta 1$ we get $z_{1-\alpha_t} = o(v_t)$, and then by virtue of (3.4.31) and the equivalence $d1 \Leftrightarrow \alpha 1'$ proved above we get the implication $\beta 1 \Rightarrow d1$. So, the implication (3.4.32) is proved.

The implications (3.4.29) can be proved in the same way as the implications (3.4.28), and so the proof of (3.4.29) is omitted. □

REMARK 3.4.1. Since condition $\Lambda2$ is fulfilled, it follows from assertion 1) of Theorem 3.4.5 that conditions $\alpha 1'$ and $d1$ in the implications $\Lambda2$ and $\alpha 1' \Rightarrow d1 \Rightarrow \beta 1$ of Theorem 2.3.3 cannot be loosened. Similarly, since condition $\Lambda3$ is also fulfilled, assertion 2) of Theorem 3.4.5 implies that conditions $\alpha 2'$ and $\beta 2$ in $\Lambda3$ and $\alpha 2' \Rightarrow \beta 2 \Rightarrow d2$ also cannot be loosened.

REMARK 3.4.2. Theorem 3.4.5 immediately yields the following assertion: if $\lim_{t\to\infty} v_t = \infty$, then the implications (with $\chi_t = (1/2) v_t^2$)

$$(\alpha 1', \alpha 2') \Leftrightarrow (d1, d2) \Leftrightarrow (\beta 1, \beta 2) \Leftrightarrow z_{1-\alpha_t} = o(v_t) \tag{3.4.33}$$

are valid.

Now let $\lim_{t\to\infty} v_t = v \in [0, \infty)$, that is, the asymptotic distinguishability of type **a** (see (3.4.18)–(3.4.21)) takes place. For $v = 0$ the relationship between the behavior of the level α_t and probability of error of the second kind $\beta(\delta_t^{+,\alpha_t})$ is described by Theorem 2.5.6, and for $v \in (0, \infty)$, by Theorem 2.6.5, namely: for any $\alpha \in [0, 1]$

$$\lim_{t\to\infty} \alpha_t = \alpha \Leftrightarrow \lim_{t\to\infty} \beta \left(\delta_t^{+,\alpha_t}\right) = \Phi(z_{1-\alpha} - v).$$

Now we consider the case of close hypotheses, that is, we assume that $y_t \to \theta$ as $t \to \infty$. In this case any behavior of v_t as $t \to \infty$ is possible; in particular, if the limit $\lim_{t\to\infty} v_t = v$ exists, then v can take any value in $[0, \infty]$.

THEOREM 3.4.6. *Let $y_t \to \theta$ as $t \to \infty$, and let the following conditions be fulfilled:*

1) $s_t(\theta) \in \mathbf{C}_1^0(\Theta)$ *with respect to the variable* θ *for all* $t \in \mathbf{R}_+$ *and* $|\dot{s}(y)|^2 \circ I_t < \infty$ *for* $t \in \mathbf{R}_+$ *for all* y *in a neighborhood of the point* θ, *where* $\dot{s}(y) = \nabla_y s(y)$;
2) *the matrix* $\dot{s}(\theta)\dot{s}(\theta)' \circ I_t$ *is nonsingular for all* $t \in \mathbf{R}_+$ *and* $|\varphi_t(\theta)| \to 0$ *as* $t \to \infty$, *where* $\varphi_t(\theta) = (\dot{s}(\theta)\dot{s}(\theta)' \circ I_t)^{-1/2}$;
3) $\lim_{t\to\infty} \big|\varphi_t(\theta)\big(\int_0^1 \dot{s}(\theta + u\Delta_t)du - \dot{s}(\theta)\big)\big|^2 \circ I_t = 0$.

Then the following representation is valid:

$$\Lambda(y_t, \theta) = u_t'\left(\eta^t + q^t\right) - (1/2)u_t'\left(I_k + p^t\right)u_t, \tag{3.4.34}$$

where

$$u_t = \varphi_t^{-1}(\theta)\Delta_t, \qquad \eta^t = \varphi^t(\theta)\dot{s}(\theta) \cdot w \in \mathcal{M}_{\text{loc}}^c(\mathbf{F}, \mathsf{P}_\theta), \tag{3.4.35}$$

$$\mathcal{L}\left(\eta_t^t | \mathsf{P}_\theta\right) = \mathcal{N}(0, I_k), \tag{3.4.36}$$

$$\mathsf{P}_\theta\text{-}\lim_{t\to\infty}\left(\left|q_t^t\right| + \left|p_t^t\right|\right) = 0. \tag{3.4.37}$$

If, in addition,

4) $\lim_{t\to\infty}\left(\Delta_t' \int_0^1 (\dot{s}(\theta + u\Delta_t) - \dot{s}(\theta))\, du\right)^2 \circ I_t = 0$,

then the representation (3.4.34)–(3.4.37) *is true, in which*

$$\lim_{t\to\infty}\left|p_t^t u_t\right| = 0. \tag{3.4.38}$$

It is enough to apply Theorems 3.4.3 and 3.4.4 and to take into account that, since the function $s(\theta)$ is determinate, the process η^t is a Gaussian continuous martingale with matrix of quadratic characteristics $\langle \eta^t, \eta^t \rangle = \varphi_t(\theta)\dot{s}(\theta)\dot{s}(\theta)' \circ I_t \varphi_t(\theta)$, which implies $\langle \eta^t, \eta^t \rangle_t = I_k$.

Theorem 3.4.6 implies that under conditions 1)–3) in view of Theorem 2.6.5 the behavior of the test δ_t^{+,α_t} is described by (2.6.22), and under conditions 1)–4) in view of Theorem 2.6.6 the behavior is described by (2.6.24).

EXAMPLE 3.4.1. Now let $h = 0$ and $b = 1$, so that $a(\theta) = s(\theta)$. Then we get a scheme of observation "a signal in the white noise", well known in radio engineering [**40, 50**]. In this scheme the function $s(\theta)$ acts as a modulator. Consider the case of frequency modulation, putting $s_t(\theta) = f(\theta t)$, where $\theta \in \Theta = (A, B)$, $0 < A < B < \infty$ and f is a periodic, with period 1, twice continuously differentiable function, $f(t) \neq \text{const}$ (phase and pulse-time modulation can be treated similarly). In view of Lemma 3.5.3 [**40**] for all t large enough and some constant $c > 0$ we have

$$v_t^2 = (s(y_t) - s(\theta))^2 \circ I_t \geq c\Delta_t^2 t^3 \left(1 + \Delta_t^2 t^2\right)^{-1},$$

whence it follows that $\lim_{t\to\infty} v_t = \infty$ if $\Delta_t^2 t^3 \to \infty$ as $t \to \infty$. Furthermore, for $t \to \infty$

$$\varphi_t(\theta) = \left(\int_0^t u^2 \dot{f}^2(\theta u)\, du\right)^{-1/2} = t^{-3/2}\left(\frac{1}{3}\int_0^1 \dot{f}^2(u)\, du\right)^{-1/2}(1 + o(1)), \tag{3.4.39}$$

and hence condition 2) of Theorem 3.4.6 is fulfilled. Putting $\Delta_t \to 0$ as $t \to \infty$, we get the estimate

$$\left|\int_0^1 \dot{s}(\theta + u\Delta_t)\, du - \dot{s}(\theta)\right|^2 \circ I_t \leq \frac{\Delta_t^2 t^5}{15}\int_0^1 \ddot{f}^2(u)\, du(1 + o(1)).$$

From this it follows that condition 3) of Theorem 3.4.6 is fulfilled for $\Delta_t = o(t^{-1})$ and condition 4) for $\Delta_t = o(t^{-5/4})$.

Note that in view of (3.4.39) the contiguous case (type **a**) takes place for $\Delta_t = O(t^{-3/2})$. Therefore conditions 3) and 4) of Theorem 3.4.6 can also be fulfilled in the case of close noncontiguous hypotheses (type **e**). This fact justifies consideration of the case $\overline{\lim}_{t\to\infty} |u_t| = \infty$ in the asymptotic decomposition of $\Lambda(y_t, \theta)$ (condition $\Lambda 8$).

EXAMPLE 3.4.2. In the coefficient $a(\theta) = h + s(\theta)b$ we put $h_t = (1/2)\xi_t$ and $b_t = \xi_t$ for all $t \in \mathbf{R}_+$ and assume that $\mathsf{P}_\theta\{\xi_0 = 0\} = p \in (0,1)$ for all θ. Hence with respect to P_θ the process ξ satisfies the stochastic differential equation

$$d\xi_t = \big((1/2)\xi_t + s_t(\theta)\xi_t\big)\,dt + \xi_t\,dw_t, \tag{3.4.40}$$

whose initial condition ξ_0 satisfies the condition $\mathsf{P}_\theta\{\xi_0 = 0\} = p$. Using the Itô formula, we see that the solution of equality (3.4.40) has the form (P_θ-a.s.)

$$\xi_t = \xi_0 \exp(w_t + s(\theta)\circ I_t), \qquad t \in \mathbf{R}_+. \tag{3.4.41}$$

It is easy to check that the conditions of Theorem 1.2.3 are fulfilled. Obviously,

$$\gamma_t(y,\theta) = \xi_t\xi_t^{\oplus}(s_t(y) - s_t(\theta)), \qquad t \in \mathbf{R}_+. \tag{3.4.42}$$

Taking into account (3.4.41) and (3.4.42), from Theorem 1.2.4 we get (P_θ-a.s.)

$$\Lambda_t(y,\theta) = \xi_t\xi_t^{\oplus}\left(v_t\zeta_t - (1/2)v_t^2\right), \qquad t \in \mathbf{R}_+, \tag{3.4.43}$$

where ζ_t is independent of ξ_0 and

$$v_t^2 = (s(y_t) - s(\theta))^2 \circ I_t, \qquad \mathcal{L}(\zeta_t|\mathsf{P}_\theta) = \mathcal{N}(0,1). \tag{3.4.44}$$

Therefore the following decomposition holds:

$$\mathcal{L}(\Lambda_t(y_t,\theta)|\mathsf{P}_\theta) = p\varepsilon_0 + q\mathcal{N}\left(-v_t^2/2, v_t^2\right), \tag{3.4.45}$$

where $q = 1 - p$ and ε_0 is the Dirac measure concentrated at 0. From here we get

$$\overline{\lim_{t\to\infty}}\, v_t = \infty \Leftrightarrow (H^t)\,\overline{\triangleleft\triangleright}\,(\widetilde{H}^t),\ (H^t)\,\overline{\triangle}\,(\widetilde{H}^t), \tag{3.4.46}$$

$$\overline{\lim_{t\to\infty}}\, v_t < \infty \Leftrightarrow (H^t) \triangleleft\triangleright (\widetilde{H}^t), \tag{3.4.47}$$

$$\lim_{t\to\infty} v_t = 0 \Leftrightarrow (H^t) \cong (\widetilde{H}^t), \tag{3.4.48}$$

that is, there are three types of asymptotic distinguishability of two families of hypotheses, namely:

$$\mathbf{d} \Leftrightarrow \overline{\lim_{t\to\infty}}\, v_t = \infty, \tag{3.4.49}$$

$$\mathbf{a}_1 \Leftrightarrow 0 < \overline{\lim_{t\to\infty}}\, v_t < \infty, \tag{3.4.50}$$

$$\mathbf{a}_0 \Leftrightarrow \lim_{t\to\infty} v_t = 0. \tag{3.4.51}$$

Taking into account (3.4.41) and (3.4.42), from Theorem 1.2.3 we have (P_{y_t}-a.s.)

$$\Lambda_t(y_t,\theta) = \xi_0\xi_0^{\oplus}\left(v_t\widetilde{\zeta}_t + (1/2)v_t^2\right), \tag{3.4.52}$$

where $\widetilde{\zeta}_t$ is independent of ξ_0 and

$$\mathcal{L}(\widetilde{\zeta}_t|\mathsf{P}_{y_t}) = \mathcal{N}(0,1). \tag{3.4.53}$$

The relations (3.4.43), (3.4.44), (3.4.52), and (3.4.53) imply that for every type of asymptotic distinguishability described by relations (3.4.46)–(3.4.51) it is not hard to investigate the behavior of $\beta(\delta_t^{+,\alpha_t})$ in various cases of behavior of the level α_t as $t \to \infty$. For example, if $\lim_{t\to\infty} v_t = \infty$ (type **d**), then from (3.4.43), (3.4.44), (3.4.52), and (3.4.53) we get

$$\lim_{t\to\infty} \alpha_t = 0 \Leftrightarrow \varliminf_{t\to\infty} \beta\left(\delta_t^{+,\alpha_t}\right) \geq p,$$
$$\lim_{t\to\infty} \alpha_t = \alpha \in (0,p) \Leftrightarrow \lim_{t\to\infty} \beta\left(\delta_t^{+,\alpha_t}\right) = p - \alpha,$$
$$\varliminf_{t\to\infty} \alpha_t \geq p \Leftrightarrow \lim_{t\to\infty} \beta\left(\delta_t^{+,\alpha_t}\right) = 0.$$

Homogeneous diffusion processes. Ergodic case. Suppose the coefficients of (1.2.38) are independent of time, that is, $a(t,\xi,\theta) = a(\xi_t,\theta)$, $b(t,\xi) = b(\xi_t)$, and let the initial value be determinate ($\xi_0 = x$, $x \in \mathbf{R}^1$). In addition to the conditions of Theorem 1.2.3 we assume that $b(x) > 0$ for all $x \in \mathbf{R}^1$. We introduce the following notation:

$$\mathcal{J}_-(\theta) = \int_{-\infty}^0 f(x,\theta)\,dx, \qquad \mathcal{J}_+(\theta) = \int_0^\infty f(x,\theta)\,dx,$$
$$D(\theta) = \int_{-\infty}^\infty \left(f(x,\theta)b^2(x)\right)^{-1} dx,$$

where

$$f(x,\theta) = \exp\left\{-2\int_0^x a(z,\theta)b^{-2}(z)\,dz\right\}.$$

Let us take two fixed points $y, \theta \in \Theta$ and consider hypotheses H^t and $\widetilde{H}^t$, with their corresponding distributions P_θ and P_y. Suppose that

$$\mathcal{J}_+(\theta) = \infty, \qquad \mathcal{J}_-(\theta) = \infty, \qquad D(\theta) < \infty. \tag{3.4.54}$$

Then the process ξ is a positively recurrent (ergodic) process with respect to P_θ with a finite invariant measure [**19, 114, 122**]

$$\pi_\theta(A) = (D(\theta))^{-1} \int_A \left(f(x,\theta)b^2(x)\right)^{-1} dx, \qquad A \in \mathcal{B}^1. \tag{3.4.55}$$

In this case by the ergodic theorem ([**19**], Theorem 2, Section 18, Chapter 1; [**122**], Theorem 4.5.1) condition (3.4.3) is fulfilled with

$$\psi_t = t\int \gamma^2(x;y,\theta)\pi_\theta(dx), \qquad c_s = s, \qquad \gamma(x;y,\theta) = (a(x,y) - a(x,\theta))/b(x)$$

if the integral $\int \gamma^2(x;y,\theta)\pi_\theta(dx)$ is finite. Then the law of large numbers (3.4.4) holds.

Consider the conditions on the Hellinger process $h(\varepsilon; \mathsf{P}_y, \mathsf{P}_\theta)$ that ensure the validity of relation (3.2.5) for the Hellinger integral $H_t(\varepsilon; \mathsf{P}_y, \mathsf{P}_\theta)$. To this end we use Theorem 3.2.3. In view of the relation (3.4.5) for $\varepsilon \in (0,1)$ we have

$$h(\varepsilon) = h(\varepsilon; \mathsf{P}_y, \mathsf{P}_\theta) = \frac{\varepsilon(1-\varepsilon)}{2} \gamma^2(y,\theta) \circ I. \tag{3.4.56}$$

Let us introduce the following stopping times:

$$\tau^{(i)} = \Big\{ s \in \mathbf{R}_+; \xi_s^{(i)} = x, \inf_{0<u<s} \xi_u^{(i)} \le x-1 \Big\}, \qquad i = 1, 2, \dots, \tag{3.4.57}$$

where

$$\xi_s^{(i)} = \xi_{\tau^{(i)}+s}^{(i-1)}, \quad i = 1, 2, \dots, \qquad \xi_s^{(0)} = \xi_s.$$

It is known that random variables $\tau^{(1)}, \tau^{(2)}, \dots$ are independent and identically distributed [**19, 122**] and $\varkappa = \mathsf{E}_\theta \, \tau^{(1)} < \infty$. Let ν_t be a positive integer such that

$$\sum_{i=1}^{\nu_t} \tau^{(i)} \le t < \sum_{i=1}^{\nu_t+1} \tau^{(i)}, \qquad t \in \mathbf{R}_+.$$

Then

$$\gamma^2(y,\theta) \circ I_t \ge \sum_{i=1}^{\nu_t} \zeta_i, \qquad \zeta_i = \int_{T_i-1}^{T_i} \gamma^2(\xi_s; y, \theta)\, ds, \tag{3.4.58}$$

where

$$T_i = \sum_{j=1}^{i} \tau^{(j)}, \quad i = 1, 2, \dots, \qquad T_0 = 0.$$

Random variables $\zeta_1, \zeta_2, \dots$ are also independent and identically distributed. Note that the stopping times $\tau^{(i)}$, $i = 1, 2, \dots$, can be introduced by taking any $y \ne x$ in (3.4.57) instead of $x-1$. Then $\tau^{(i)}$ must be the time of first return to the point x of the process $\xi^{(i)}$ after hitting the point y. In this case the random variables $\tau^{(1)}, \tau^{(2)}, \dots$ and $\zeta_1, \zeta_2, \dots$ have the same properties [**114**].

Suppose that the random variable T_1 satisfies the Cramér condition:

$$\mathsf{E}_\theta \exp(\delta T_1) < \infty \quad \text{for some } \delta > 0.$$

Then by the theorem on large deviations for sums of independent identically distributed random variables $\tau^{(1)}, \tau^{(2)}, \dots$ ([**105**], Theorem 10, Chapter 8), for all $\varepsilon > 0$ small enough there exists a constant l such that for all n large enough

$$\mathsf{P}_\theta\Big\{ \sum_{i=1}^{\nu_t} \tau^{(i)} \ge n\left(\varkappa + \sigma\sqrt{\varepsilon}\right) \Big\} = \left(1 - \Phi\left(\sqrt{\varepsilon n}\right)\right) \exp\!\left(\varepsilon^{3/2} n \lambda\left(\sqrt{\varepsilon}\right)\right) \left(1 + l\sqrt{\varepsilon}\right), \tag{3.4.59}$$

where $\sigma^2 = \mathbf{D}_\theta \tau^{(1)}$ and $\lambda(t)$ is the Cramér series for $\tau^{(1)}$ with respect to P_θ.

Let the random variable ζ_1 also satisfy the Cramér condition with respect to P_θ. Then, once again, by the theorem on large deviations for sums of independent identically distributed random variables $\zeta_1, \zeta_2, \dots$ ([**105**], Theorem 10, Chapter 8), for all $\varepsilon > 0$ small enough there exists a constant $\tilde{l}$ such that for all n large enough

$$\mathsf{P}_\theta\Big\{ \sum_{i=1}^{\nu_t} \zeta_i < n\left(\tilde{\varkappa} - \tilde{\sigma}\sqrt{\varepsilon}\right) \Big\} = \left(1 - \Phi\left(\sqrt{\varepsilon n}\right)\right) \exp\!\left(-\varepsilon^{3/2} n \tilde{\lambda}\left(-\sqrt{\varepsilon}\right)\right) \left(1 + \tilde{l}\sqrt{\varepsilon}\right), \tag{3.4.60}$$

where $\widetilde{\varkappa} = \mathsf{E}_\theta\, \zeta_1$, $\widetilde{\sigma}^2 = \operatorname{Var}_\theta \zeta_1$, and $\widetilde{\lambda}(t)$ is the Cramér series for ζ_1 with respect to P_θ. Note that $\widetilde{\varkappa} = \varkappa\gamma$, where $\gamma = \int \gamma^2(x; y, \theta)\, \pi_\theta(dx)$.

We introduce a positive integer n_t such that $n_t(\varkappa + \sigma\sqrt{\varepsilon}) = t + o(t)$ as $t \to \infty$ and put

$$D_{t,\varepsilon} = \{\nu_t \geq n_t\} \cap \left\{ \sum_{i=1}^{n_t} \zeta_i \geq n_t(\widetilde{\varkappa} - \widetilde{\sigma}\sqrt{\varepsilon}\,) \right\}. \tag{3.4.61}$$

Taking into account the equality

$$\{\nu_t < n_t\} = \left\{ \sum_{i=1}^{n} \tau^{(i)} > t \right\}$$

and using the upper bound [**121**]

$$1 - \Phi(x) < \frac{1}{x\sqrt{2\pi}} \exp\left(-\frac{x^2}{2}\right), \qquad x > 0,$$

from the relations (3.4.59)–(3.4.61) we get

$$\mathsf{P}_\theta(D_{t,\varepsilon}^c) \leq \frac{2\,(1 + \bar{l}\sqrt{\varepsilon}\,)}{\sqrt{2\pi\varepsilon n_t}} \exp\Big(-\Big(\frac{\varepsilon}{2} - \varepsilon^{3/2} l(\varepsilon)\Big)\, n_t\Big)\, (1 + o(1)), \tag{3.4.62}$$

where $\bar{l} = |l| \vee |\tilde{l}|$, $l(\varepsilon) = |\lambda(\sqrt{\varepsilon}\,)| \vee |\widetilde{\lambda}(-\sqrt{\varepsilon}\,)|$.

Putting $\psi_t = \gamma t/2$, from the relation (3.4.62) we derive

$$\varlimsup_{t\to\infty} \psi_t^{-1} \ln \mathsf{P}_\theta\left(D_{t,\varepsilon}^c\right) \leq -\frac{\varepsilon - 2\varepsilon^{3/2} l(\varepsilon)}{\gamma\,(\varkappa + \sigma\sqrt{\varepsilon}\,)}.$$

Now if $\varkappa\gamma \leq 1$, then, taking into account that in view of properties of Cramér series $l(\varepsilon) \to l(0) < \infty$ as $\varepsilon \to 0$, we get

$$\varlimsup_{\varepsilon\downarrow 0}\, \varlimsup_{t\to\infty}\, (\varepsilon\psi_t)^{-1} \ln \mathsf{P}_\theta\left(D_{t,\varepsilon}^c\right) \leq -1. \tag{3.4.63}$$

From (3.4.58) we get

$$\inf_{D_{t,\varepsilon}} \gamma^2(y, \theta) \circ I_t \geq n_t\left(\widetilde{\varkappa} - \widetilde{\sigma}\sqrt{\varepsilon}\,\right),$$

whence in view of (3.4.56) for $\psi_t = \gamma t/2$ we obtain

$$\varliminf_{t\to\infty} \psi_t^{-1} \inf_{D_{t,\varepsilon}} h_t(\varepsilon) \geq \frac{\varepsilon(1 - \varepsilon)\,(\widetilde{\varkappa} - \widetilde{\sigma}\sqrt{\varepsilon}\,)}{\gamma\,(\varkappa + \sigma\sqrt{\varepsilon}\,)}.$$

Using the equality $\widetilde{\varkappa} = \varkappa\gamma$, from the above inequality we derive

$$\varliminf_{\varepsilon\downarrow 0}\, \varliminf_{t\to\infty}\, (\varepsilon\psi_t)^{-1} \inf_{D_{t,\varepsilon}} h_t(\varepsilon) \geq 1. \tag{3.4.64}$$

Thus, in view of the relations (3.4.63) and (3.4.64) and Theorem 3.2.3, we have proved the following theorem.

THEOREM 3.4.7. *Let conditions* (3.4.54) *be fulfilled, the random variables* τ_1 *and* ζ_1 *satisfy the Cramér condition, and* $\varkappa\gamma \leq 1$. *Then for* $\psi_t = \gamma t/2$

$$\varlimsup_{\varepsilon\downarrow 0}\, \varlimsup_{t\to\infty}\, (\varepsilon\psi_t)^{-1} \ln H_t(\varepsilon; \mathsf{P}_y, \mathsf{P}_\theta) \leq -1.$$

Now let $y = y_t$ depend on t and $y_t \to \theta$ as $t \to \infty$.

THEOREM 3.4.8. *Let $y_t \to \theta$ as $t \to \infty$, and let the following conditions be fulfilled:*

1) $a(x,\theta) \in \mathbf{C}_1^0(\Theta)$ *with respect to the variable* θ *for every* $x \in \mathbf{R}^1$ *and*

$$|f(y)|^2 \circ I_t < \infty \quad (\mathsf{P}_\theta\text{-}a.s.)$$

for all $t \in \mathbf{R}_+$ *and all* $y \in \Theta$ *in a neighborhood of the point* θ, *where* $f(\theta) = (f(\xi_t,\theta))_{t\in\mathbf{R}_+}$ *and* $f(x,\theta) = \nabla_\theta a(x,\theta)/b(x)$;

2) *elements of the matrix* $f(x,\theta)f(x,\theta)'$ *are integrable with respect to the measure* π_θ, *and the matrix* $I(\theta) = \int f(x,\theta)f(x,\theta)'\pi_\theta(dx)$ *is nonnegative;*

3) *the function*

$$f_t(x) = \left|\int f(x,\theta + u\Delta_t)\,du - f(x,\theta)\right|^2$$

is bounded and $\int f_t(x)\pi_\theta(dx) \to 0$ *as* $t \to \infty$.

Then the representation (3.4.9)–(3.4.12) *is true, in which*

$$\varphi_t(\theta) = (tI(\theta))^{-1/2}, \qquad g_0 = f(\theta).$$

It is enough to apply Theorem 3.4.3 and to take into account that condition 1) implies condition 1) of Theorem 3.4.3, condition 2) implies condition 2) of Theorem 3.4.3 with $\varphi_t(\theta) = (tI(\theta))^{-1/2}$, and condition 3) of Theorem 3.4.3 follows from condition 3) if we apply the following lemma.

LEMMA 3.4.1. *Let condition* (3.4.54) *be fulfilled, and let* $f_t(x)$, $t \in \mathbf{R}_+$, *be the family of Borel functions of* $x \in \mathbf{R}^1$ *such that* $\sup_{t,x}|f_t(x)| < \infty$. *Then*

$$\lim_{t\to\infty}\left(\frac{1}{t}\int_0^1 \mathsf{E}_\theta\, f_t(\xi_s)\,ds - \int f_t(x)\pi_\theta(dx)\right) = 0.$$

PROOF. The lemma can be proved similarly to Theorem 1 ([**19**], Section 18, Part I). □

THEOREM 3.4.9. *Let the conditions of Theorem* 3.4.8 *be fulfilled and, in addition,*

1) $\mathsf{E}_\theta\,\tau_1^2 < \infty$ *and* $\mathsf{E}_\theta(|f(\theta)|^2 \circ I_{\tau_1})^2 < \infty$;

2) *the function* $f_t(x) = \int(\sqrt{t}\,\Delta_t'(f(x,\theta+u\Delta_t) - f(x,\theta)))^2\,du$ *is jointly bounded with respect to the variables and* $\int f_t(x)\,\pi_\theta(dx) \to 0$ *as* $t \to \infty$.

Then the representation (3.4.9)–(3.4.12) *is valid, in which the relation* (3.4.13) *holds.*

PROOF. It is enough to check the conditions of Theorem 3.4.4. In view of Theorem 3.4.8 it suffices to establish the validity of conditions 1) and 2) of Theorem 3.4.4. We have

$$f(\theta)f(\theta)' \circ I_t = \sum_{i=1}^{\nu_t}\widetilde{\zeta}_i + \zeta_t^*,$$

where

$$\widetilde{\zeta}_i = \int_{\tau_i-1}^{\tau_i} f(\xi_s,\theta)f(\xi_s,\theta)'\,ds, \qquad \zeta_t^* = \int_{\tau_{\nu_t}}^{t} f(\xi_s,\theta)f(\xi_s,\theta)'\,ds,$$

$\widetilde{\zeta}_1, \widetilde{\zeta}_2, \ldots$ being independent identically distributed random variables and $\mathsf{E}_\theta\, \widetilde{\zeta}_1 = \varkappa I(\theta)$. From this we have

$$\begin{aligned}\sqrt{t}\left(\frac{1}{t} f(\theta) f(\theta)' \circ I_t - I(\theta)\right)& \\ = \left(\frac{\nu_t}{t}\right)^{1/2} \nu_t^{-1/2} \sum_{i=1}^{\nu_t} (\widetilde{\zeta}_i - \varkappa I(\theta)) + \frac{\nu_t - \varkappa^{-1} t}{\sqrt{t}} \varkappa I(\theta) + \frac{\zeta_t^*}{\sqrt{t}}.&\end{aligned} \tag{3.4.65}$$

Since $\mathsf{E}_\theta\, \tau_1^2 < \infty$, from renewal theory [**10**] we have

$$\mathcal{L}\left(t^{-1/2}(\nu_t - \varkappa^{-1} t) | \mathsf{P}_\theta\right) \xrightarrow{w} \mathcal{N}(0, \sigma^2 \varkappa^{-3}), \qquad t \to \infty. \tag{3.4.66}$$

Since $\mathsf{E}_\theta(|f(\theta)|^2 \circ I_{\tau_1})^2 < \infty$, in view of Theorem 2.6.1 of [**112**] (see also Assertion 3 in [**188**]) elements of the matrix $\nu_t^{-1/2} \sum_{i=1}^{\nu_t} (\widetilde{\zeta}_i - \varkappa I(\theta))$ are asymptotically normal as $t \to \infty$ with zero mean and finite variance with respect to P_θ and

$$\mathsf{P}_\theta\text{-} \lim_{t \to \infty} t^{-1/2} |\zeta_t^*| = 0.$$

Thus condition 1) of Theorem 3.4.4 is proved. Condition 2) of Theorem 3.4.4 follows from condition 2) and Lemma 3.4.1. □

Null recurrent diffusion processes. Under the conditions of the previous subsection we assume that instead of (3.4.54) the following conditions are fulfilled:

$$\mathcal{J}_+(\theta) = \infty, \qquad \mathcal{J}_-(\theta) = \infty, \qquad D(\theta) = \infty. \tag{3.4.67}$$

Then the process ξ is a null recurrent process with respect to P_θ with σ-finite invariant measure [**19, 114, 122**]

$$\pi_\theta(A) = \int_A \left(f(x, \theta) b^2(x)\right)^{-1} dx, \qquad A \in \mathcal{B}^1.$$

We suppose that $\xi_0 = 0$ and introduce the process $\tilde{\xi}_s = h(\xi_s, \theta)$, $s \in \mathbf{R}_+$, where

$$h(x, \theta) = \int_0^x f(y, \theta)\, dy.$$

By the Itô formula the process $\tilde{\xi}$ is a solution of the stochastic differential equation (with respect to P_θ)

$$d\tilde{\xi}_s = \tilde{b}(\tilde{\xi}_s, \theta)\, dw_s, \qquad \tilde{\xi}_0 = 0,$$

where

$$\tilde{b}(x, \theta) = f(c(x, \theta), \theta) b(c(x, \theta))$$

and $c(x, \theta)$ is a function, inverse to the function $h(x, \theta)$ with respect to the variable x for fixed θ.

The process $\tilde{\xi}$ is a null recurrent process. For the process $\tilde{\xi}$ we introduce cycles that start at the point $x = 0$ and last till the time of first return to zero after hitting the point $x = 1$. Let τ_n be the termination time of the nth cycle, $\tau_0 = 0$. Suppose that

$$\mathsf{P}_\theta(\tau_1 \geq x) = c x^{-d}(1 + o(1)), \qquad x \to \infty, \tag{3.4.68}$$

where $c > 0$ and $0 < d < 1$.

Moreover, we assume the integral

$$\bar{\lambda} = \int \lambda^2(x; y, \theta)/(f(x,\theta)b^2(x))\,dx \tag{3.4.69}$$

to be finite. Then in view of Theorem 4.11.1 of [**122**] we get, as $t \to \infty$,

$$\mathsf{P}_\theta\left\{\frac{c\Gamma(1-d)}{2\bar{\lambda}t^d}\int_0^t \lambda^2(\xi_s; y, \theta)\,ds < x\right\} \to 1 - G_d(x^{-1/d})$$

for all $x > 0$, where $\Gamma(\cdot)$ is the gamma-function and $G_d(x)$ is the distribution function of the stable law with exponent d, whose Laplace transform has the form $\exp(-s^d)$ [**122**]. Together with (3.4.7), this implies that as $t \to \infty$ we have

$$\mathcal{L}\left(\psi_t^{-1}\Lambda_t(y,\theta)|\mathsf{P}_\theta\right) \xrightarrow{w} L, \tag{3.4.70}$$

where $\psi_t = (c\Gamma(1-d))^{-1}\bar{\lambda}t^d$ and L is the distribution law on $(-\infty, 0)$ with distribution function $L(x) = G_d((-x)^{-1/d})$, $x < 0$.

We have thus proved the following theorem.

THEOREM 3.4.10. *Let conditions* (3.4.67) *and* (3.4.68) *be fulfilled, and let the integral* (3.4.69) *be finite. Then the weak convergence* (3.4.70) *occurs with* $\psi_t = (c\Gamma(1-d))^{-1}\bar{\lambda}t^d$ *and distribution function* $L(x) = G_d((-x)^{-1/d})$, $x < 0$, *of the limit law* L.

Homogeneous diffusion processes with exact growth at infinity. If one of the conditions $\mathcal{J}_-(\theta) = \infty$ and $\mathcal{J}_+(\theta) = \infty$ is violated, then the process ξ_t extends to infinity as $t \to \infty$. Let, for example, $\mathcal{J}_-(\theta) = \infty$ but $\mathcal{J}_+(\theta) < \infty$. Then, as is well known [**19**], $\mathsf{P}_\theta(\lim_{t\to\infty}\xi_t = \infty) = 1$. Moreover, the following lemma holds, which gives the precise rate of growth of ξ_t as $t \to \infty$.

LEMMA 3.4.2. *Let the following conditions be fulfilled:*

1) $a(x,\theta) = c(\theta)x^q(1 + o(1))$ *as* $x \to \infty$, *where* $q \in (-1, 1)$ *and* $c(\theta) \in (0, \infty)$;
2) *if* $-1 < q < 0$, *then the function* $b(x)$ *is bounded;*
3) *if* $q = 0$, *then the function* $a(x,\theta)$ *is bounded and the function* $b(x)$ *satisfies the inequality* $b^2(x) \le C(1 + |x|^\alpha)$ *for some* $\alpha < 1$, $C > 0$;
4) *if* $0 < q < 1$, *then* $b(x) = o(x^q)$ *and* $b^2(x) = o(x^{1+q})$ *as* $x \to \infty$.

Then the following relation is true:

$$\mathsf{P}_\theta\left\{\lim_{t\to\infty}((1-q)c(\theta)t)^{-1/(1-q)}\xi_t = 1\right\} = 1.$$

PROOF. The proof can be found in [**19**]. □

Based on Lemma 3.4.2, it is easy to establish the law of large numbers (3.4.3) (see Example 3.4.3 below).

Note that for $k > 1$, as $y_t \to \theta$, $t \to \infty$, the representation (3.4.9) has a degenerate form with a singular matrix $I(\theta)$ instead of I_k, with vector η_t^t that has the limit covariance matrix $I(\theta)$ as $t \to \infty$. To be more exact, the following analog of Theorem 3.4.3 holds for $k \ge 1$.

LEMMA 3.4.3. *Let $y_t \to \theta$ as $t \to \infty$, and let the following conditions be fulfilled:*

1) $a(x,\theta) \in \mathbf{C}_1^0(\Theta)$ *with respect to the variable* θ *for all* $x \in \mathbf{R}^1$, *and*

$$|f(y)|^2 \circ I_t < \infty \quad (\mathsf{P}_\theta\text{-}a.s.)$$

for all $t \in \mathbf{R}_+$ *and all* $y \in \Theta$ *in a neighborhood of the point* θ;

2) $f(x,y) = c(y)x^q(1+o(1))$ *as* $x \to \infty$ *for all* $y \in \Theta$ *in a neighborhood of the point* θ;

3) $\int_0^1 \dot c(\theta + u\Delta_t)\,du \to \dot c(\theta)$ *as* $t \to \infty$.

Then the following representation holds:

$$\Lambda(y_t,\theta) = u_t'\left(\eta^t + q^t\right) - (1/2)u_t'\left(I(\theta) + p^t\right)u_t,$$

where

$$u_t = \varphi_t^{-1}(\theta)\Delta_t, \qquad \varphi_t^2(\theta) = (1+q)c(\theta)^{2q/(q-1)}((1-q)t)^{(q+1)/(q-1)},$$
$$I(\theta) = \dot c(\theta)\dot c(\theta)', \qquad \eta^t = \varphi_t(\theta)f(\theta)\cdot w,$$
$$\mathcal{L}\left(\eta_t^t|\mathsf{P}_\theta\right) \xrightarrow{w} \mathcal{N}(0,I(\theta)), \qquad t\to\infty,$$
$$\mathsf{P}_\theta\text{-}\lim_{t\to\infty}\left(|q_t^t| + |p_t^t|\right) = 0.$$

This lemma shows that the behavior of $u_t'\dot c(\theta)$ as $t\to\infty$ is of crucial importance in investigating the asymptotic behavior of δ_t^{+,α_t}.

EXAMPLE 3.4.3. Let $a(x,\theta) = \theta|x|^q$ and $b(x) = 1$, where $0 \le q < 1$. If $\theta < 0$, then $\mathcal{J}_+(\theta) = \infty$ and $\mathcal{J}_-(\theta) < \infty$ and hence $\xi_t \to -\infty$ as $t\to\infty$ (P_θ-a.s.). If $\theta > 0$, then $\mathcal{J}_+(\theta) < \infty$ and $\mathcal{J}_-(\theta) = \infty$ and hence $\xi_t \to \infty$ as $t \to \infty$ (P_θ-a.s.). Finally, if $\theta = 0$, then $\mathcal{J}_+(\theta) = \infty$ and $\mathcal{J}_-(\theta) = \infty$ and hence the process ξ is recurrent with respect to P_θ for $\theta = 0$. Moreover, $D(0) = \infty$ and hence the process ξ is a null recurrent process with respect to P_θ with Lebesgue invariant measure.

Note that the stochastic differential equation (1.2.38) with $a(x,\theta) = \theta|x|^q$ and $b(x) = 1$ has a strong solution for any θ [**35**]. Moreover,

$$\int_0^t |\xi_s|^{2q}\,ds < \infty \quad (\mathsf{P}_\theta\text{-a.s.})$$

for any $\theta \in \mathbf{R}^1$ and hence the measures P_y and P_θ are locally absolutely continuous for any $y,\theta\in\Theta$ and (P_θ-a.s.)

$$\Lambda_t(y,\theta) = (y-\theta)\int_0^t |\xi_s|^q\,dw_s - (1/2)(y-\theta)^2\int_0^t |\xi_s|^{2q}\,ds. \tag{3.4.71}$$

Using Lemma 3.4.2, for an arbitrary $y \in \mathbf{R}^1$ such that $y \ne \theta$ and $\theta > 0$ we get the relation

$$\mathsf{P}_\theta\text{-}\lim_{t\to\infty} t^{-(1+q)/(1-q)}\int_0^t |\xi_s|^{2q}\,ds = (1-q)^{-1}\theta^{2q/(1-q)}(1-q)^{(1+q)/(1-q)}$$

(see the proof in [**54**])). In view of Theorem 3.4.1 the last equality implies the law of large numbers (3.4.4) with

$$c_s = s^{(1+q)/(1-q)}, \qquad \varphi_t = (y-\theta)^2(1-q)\theta^{2q/(1-q)}((1-q)t)^{(1+q)/(1-q)}.$$

The case $\theta < 0$ can be investigated similarly to the case $\theta > 0$.

Now let $y \in \mathbf{R}^1$ be an arbitrary number, $y \neq 0$, and $\theta = 0$. Since

$$t^{-(q+1)} \int_0^t |w_s|^{2q}\, ds = \int_0^1 \left| w_{tu}/t^{1/2} \right|^{2q} du,$$

in view of self-similarity of the Wiener process w we have

$$\mathcal{L}\left(t^{-(q+1)} \int_0^t |\xi_s|^{2q}\, ds \mid \mathsf{P}_0\right) = \mathcal{L}\left(\int_0^1 |w_u|^{2q}\, du \mid \mathsf{P}_0\right). \tag{3.4.72}$$

From this

$$\mathsf{P}_0\text{-}\lim_{t\to\infty} t^{-(q+1)} \int_0^t |\xi_s|^q\, dw_s = 0. \tag{3.4.73}$$

Now, combining (3.4.71)–(3.4.73), we get

$$\mathcal{L}(t^{-q-1}\Lambda_t(y,0)|\mathsf{P}_0) \xrightarrow{w} \mathcal{L}\left(-(1/2)y^2 \int_0^1 |w_u|^{2q}\, du \mid \mathsf{P}_0\right)$$

as $t \to \infty$. Therefore condition $\Lambda4$ (Section 2.4) is fulfilled with

$$\psi_t = t^{q+1}, \qquad L = \mathcal{L}\left(-(1/2)y^2 \int_0^1 |w_u|^{2q}\, du \mid \mathsf{P}_0\right).$$

3.5. Observations of counting processes

General theorems. Let ξ be a counting process with compensators ν and $\widetilde{\nu}$ with respect to the measures P and $\widetilde{\mathsf{P}}$, respectively (see Section 1.2). We assume the compensators ν and $\widetilde{\nu}$ to be continuous a.s. with respect to the measures P and $\widetilde{\mathsf{P}}$, respectively. Moreover, the conditions of Theorem 1.2.5, which ensure the local equivalence of the measures P and $\widetilde{\mathsf{P}}$, are assumed to be fulfilled. By this theorem the process of local density z of the measure $\widetilde{\mathsf{P}}$ with respect to the measure P has the form (P-a.s.)

$$z = \exp\{M - V\}, \tag{3.5.1}$$

where

$$M = \ln\lambda * (\xi - \nu) \in \mathcal{M}^d_{\mathrm{loc}}(\mathbf{F}, \mathsf{P}), \tag{3.5.2}$$
$$M = (\lambda - 1 - \ln\lambda) \circ \nu \in \mathcal{V}_{\mathrm{loc}}(\mathbf{F}, \mathsf{P}). \tag{3.5.3}$$

The following theorem is the law of large numbers for $\Lambda = \ln z$.

THEOREM 3.5.1. *Let $s \in (0,\infty)$ be a fixed number, ψ_t a positive nonrandom function such that $\psi_t \to \infty$ as $t \to \infty$, and let the following conditions be fulfilled:*

1) $\mathsf{P}\text{-}\lim_{t\to\infty} \psi_t^{-1} V_{st} = a_s$, *where a_s is a nonrandom positive function satisfying the condition $a_1 = 1$;*
2) *for any $\varepsilon \in (0,1]$*

$$\mathsf{P}\text{-}\lim_{t\to\infty} \left[\chi\left(\frac{|\ln\lambda|}{\psi_t} \le \varepsilon\right)\left(\frac{\ln\lambda}{\psi_t}\right)^2 \circ \nu_{st} + \chi\left(\frac{|\ln\lambda|}{\psi_t} > \varepsilon\right)\frac{|\ln\lambda|}{\psi_t} \circ \nu_{st}\right] = 0.$$

Then the following relation is valid:

$$\text{P-}\lim_{t\to\infty} \psi_t^{-1}\Lambda_{st} = -a_s. \tag{3.5.4}$$

REMARK 3.5.1. In view of Remark 1.2.6 we have $\ln\lambda \in \mathcal{G}^1_{\mathrm{loc}}(\nu, \mathbf{F}, \mathsf{P})$. From this, taking into account the inequality

$$\chi(|x| \le \varepsilon)|x|^2 + \chi(|x| > \varepsilon)|x| \le |x|, \quad x \in \mathbf{R}^1,\ \varepsilon \in (0,1],$$

we conclude that condition 2) of Theorem 3.5.1 will be fulfilled if

$$\text{P-}\lim_{t\to\infty} \psi_t^{-1}\, |\ln\lambda| \circ \nu_{st} = 0. \tag{3.5.5}$$

If $\ln\lambda \in \mathcal{G}^2_{\mathrm{loc}}(\nu, \mathbf{F}, \mathsf{P})$, then, evidently, for condition 2) of Theorem 3.5.1 to be fulfilled it suffices that

$$\text{P-}\lim_{t\to\infty} \psi_t^{-1} \ln^2\lambda \circ \nu_{st} = 0.$$

The following theorem establishes the law of large numbers (3.5.4) in the case where relation (3.5.5) and condition 2) of Theorem 3.5.5 do not hold in general.

THEOREM 3.5.2. *Let $s \in (0,\infty)$ be a fixed number, ψ_t a nonrandom positive function such that $\psi_t \to \infty$ as $t \to \infty$, and let the following conditions be fulfilled:*

1) *condition* 1) *of Theorem* 3.5.1;
2) $\text{P-}\lim_{t\to\infty} \psi_t^{-1} \ln^{\pm}\lambda \circ \nu_{st} = k_s^{\pm}$, *where k_s^+ and k_s^- are nonnegative finite functions and*

$$\ln^+\lambda = 0 \vee \ln\lambda, \qquad \ln^-\lambda = -(0 \wedge \ln\lambda);$$

3) *if $k_s^+ > 0$, then $\mathsf{E}\sup_{u>0} \Delta\xi_u \ln^+\lambda_u < \infty$, and if $k_s^- > 0$, then*

$$\mathsf{E}\sup_{u>0} \Delta\xi_u \ln^-\lambda_u < \infty.$$

Then the relation (3.5.4) *is valid.*

PROOF. In view of (3.5.1) and condition 1) it is enough to prove that

$$\text{P-}\lim_{t\to\infty} \psi_t^{-1} M_{st} = 0. \tag{3.5.6}$$

By virtue of condition IV) of Theorem 1.2.5 and Remark 1.2.6 we have (P-a.s.)

$$M = A - \tilde{A}, \tag{3.5.7}$$

where $A = \ln\lambda * \xi \in \mathcal{A}_{\mathrm{loc}}(\mathbf{F}, \mathsf{P})$ and $\tilde{A} = \ln\lambda \circ \nu \in \mathcal{A}_{\mathrm{loc}}(\mathbf{F}, \mathsf{P})$. It is evident that

$$A = A^+ - A^-, \qquad \tilde{A} = \tilde{A}^+ - \tilde{A}^-, \tag{3.5.8}$$

where $A^{\pm} = \ln^{\pm}\lambda * \xi$ and $\tilde{A}^{\pm} = \ln^{\pm}\lambda \circ \nu$.

Suppose that $k_s^+ > 0$. Then $\tilde{A}^+_\infty = \infty$ (P-a.s.). Moreover, $A^+ \in \mathcal{A}^+_{\mathrm{loc}}(\mathbf{F}, \mathsf{P})$ and $\tilde{A}^+$ is the compensator of the process A^+ and in view of condition 3), $\mathsf{E}\sup_{t>0} \Delta A^+_t < \infty$. Therefore we can apply Theorem 2.5.12 of [**98**], which gives

$$\lim_{t\to\infty} \left(A^+_{st} - \tilde{A}^+_{st}\right)/\tilde{A}^+_{st} = 0 \quad \text{(P-a.s.)}.$$

From this, taking into account condition 2), we get

$$\text{P-}\lim_{t\to\infty} \psi_t^{-1}\left(A^+_{st} - \tilde{A}^+_{st}\right) = 0. \tag{3.5.9}$$

Now let $k_s^+ = 0$. Since the process $\tilde{A}^+$ is the compensator of the process A^+, the process A^+ is dominated by the process $\tilde{A}^+$. Then in view of the Lenglart inequality (Theorem 1.1.3 with $\xi = A^+$ and $\eta = \tilde{A}^+$) and condition 2) with $k_s^+ = 0$ we get $\mathsf{P}\text{-}\lim_{t\to\infty} \psi_t^{-1} A_{st}^+ = 0$. Hence, taking into consideration condition 2) with $k_s^+ = 0$, we conclude that the relation (3.5.9) is also valid in the case $k_s^+ = 0$.

Similarly,

$$\mathsf{P}\text{-}\lim_{t\to\infty} \psi_t^{-1}\left(A_{st}^- - \tilde{A}_{st}^-\right) = 0. \tag{3.5.10}$$

Combining (3.5.7)–(3.5.10), we get (3.5.6). Thus, (3.5.4) is proved. □

By Theorem 3.1.7 for $\varepsilon \in (0,1)$ the Hellinger process $h(\varepsilon) = h(\varepsilon; \widetilde{\mathsf{P}}, \mathsf{P})$ of order ε for measures $\widetilde{\mathsf{P}}$ and P is of the form

$$h(\varepsilon) = [\varepsilon(\lambda - 1) - (\lambda^\varepsilon - 1)] \circ \nu, \tag{3.5.11}$$

and if $H_t(\varepsilon_0; \widetilde{\mathsf{P}}, \mathsf{P}) < \infty$ for all $t \in \mathbf{R}_+$ for some $\varepsilon_0 < 0$, then for all $\varepsilon \in [\varepsilon_0, 0)$ the Hellinger process is of the form

$$h(\varepsilon) = -[\varepsilon(\lambda - 1) - (\lambda^\varepsilon - 1)] \circ \nu. \tag{3.5.12}$$

In the case of counting processes the statements of Theorems 3.2.2 and 3.2.3 are the same, with the Hellinger process $h(\varepsilon)$ replaced by its expression given by formulas (3.5.11) and (3.5.12), and so they are omitted.

We introduce a family of stochastic processes $Y^t = (Y_s^t)_{s\in\mathbf{R}_+}$, where $Y_s^t = \psi_t^{-1}(\Lambda_{st} + \varphi_{st})$ and the function φ_s is continuous. Obviously, $Y^t \in \mathcal{S}(\mathbf{F}^t, \mathsf{P})$, where $\mathbf{F}^t = (\mathcal{F}_{st})_{s\in\mathbf{R}_+}$ with the triple of predictable characteristics $(B^t, 0, \nu^t)$, and

$$\begin{aligned} B_s^t &= \psi_t^{-1}\left[\varphi_{st} - \chi\left(\left|\psi_t^{-1}\ln\lambda\right| > 1\right)\ln\lambda \circ \nu_{st} - V_{st}\right],\\ \nu^t((0,s], C) &= \chi\left(\psi_t^{-1}\ln\lambda \in C\right) \circ \nu_{st}, \qquad s \in \mathbf{R}_+,\ C \in \mathcal{B}_0. \end{aligned}$$

To state our next theorem, we introduce, on the stochastic basis $(\Omega, \mathcal{F}, \mathbf{F}, \mathsf{P})$, a left quasicontinuous semimartingale $Y \in \mathcal{S}(\mathbf{F}, \mathsf{P})$, $Y_0 = 0$, with independent increments and the triple of predictable characteristics (B^Y, C^Y, ν^Y). Moreover, for any $\delta \in (0,1]$ and $t \in \mathbf{R}_+$ we define a local square-integrable martingale $M^{t,\delta} \in \mathcal{M}_{\mathrm{loc}}^2(\mathbf{F}^t, \mathsf{P})$ of the form

$$M_s^{t,\delta} = \psi_t^{-1}\chi\left(\left|\psi_t^{-1}\ln\lambda\right| \le \delta\right)\ln\lambda * (\xi - \nu)_{st},$$

whose quadratic characteristic is computed by the formula

$$\left\langle M^{t,\delta}\right\rangle_s = \psi_t^{-2}\chi\left(\left|\psi_t^{-1}\ln\lambda\right| \le \delta\right)\ln^2\lambda \circ \nu_{st}.$$

THEOREM 3.5.3. *Let S be a nonempty subset in $\mathbf{R}_+$ and let the following conditions be fulfilled:*

1) *for any bounded function $f = f(x)$, continuous on $R_0 \setminus (\{-1\} \cup \{1\})$ and such that $f(x) = \chi(|x| > \varepsilon) f(x)$ for some $\varepsilon > 0$,*

$$\mathsf{P}\text{-}\lim_{t\to\infty} f\left(\psi_t^{-1}\ln\lambda\right) \circ \nu_{st} = f * \nu_s^Y, \qquad s \in S;$$

2) $\mathsf{P}\text{-}\lim_{t\to\infty} B_s^t = B_s^Y$, $s \in S$;

3) *for any $\varepsilon > 0$ and $s \in S$*

$$\lim_{\delta\to 0}\overline{\lim_{t\to\infty}} \mathsf{P}\left\{\left|\left\langle M^{t,\delta}\right\rangle_s - C_s^Y\right| > \varepsilon\right\} = 0.$$

Then $Y^t \xrightarrow{w_f(S)} Y$ as $t \to \infty$.

PROOF. The proof follows from Theorems 3.2.4 and 3.2.5. □

Theorem 3.5.3 holds also in the case of the triangular array scheme (see Remark 3.2.2).

Now we use a parametric approach to counting processes (see Section 1.2). Suppose on a space with filtration $(\Omega, \mathcal{G}, (\mathcal{G}_t))$ a parametric family of probability measures $(\mathsf{P}_\theta, \theta \in \Theta)$ is given, where Θ is an open subset in $\mathbf{R}^k$, $k \geq 1$. For every $\theta \in \Theta$ let a counting process ξ on the stochastic basis $(\Omega, \mathcal{F}, \mathbf{F}, \mathsf{P}_\theta)$ have a compensator $\nu(\theta)$, which is assumed to be (P_θ-a.s.) continuous. Suppose that the conditions of Theorem 1.2.5 are fulfilled with $\mathsf{P} = \mathsf{P}_\theta$ and $\widetilde{\mathsf{P}} = \mathsf{P}_y$ for any $y, \theta \in \Theta$, which ensure local equivalence of the measures P_θ and P_y. From Theorem 1.2.5 we get the following form for the process of local density $z(y,\theta)$ of the measure P_y with respect to the measure P_θ:

$$z(y,\theta) = \exp\{M(y,\theta) - V(y,\theta)\}, \tag{3.5.13}$$

where

$$M(y,\theta) = \ln \lambda(y,\theta) * (\xi - \nu(\theta)) \in \mathcal{M}_{\mathrm{loc}}(\mathbf{F}, \mathsf{P}_\theta), \tag{3.5.14}$$

$$V(y,\theta) = (\lambda(y,\theta) - 1 - \ln \lambda(y,\theta)) \circ \nu(\theta) \in \mathcal{V}_{\mathrm{loc}}(\mathbf{F}, \mathsf{P}_\theta). \tag{3.5.15}$$

Let $\theta \in \Theta$ be independent of t and let $y_t \in \Theta$ depend on t so that $y_t \to \theta$ as $t \to \infty$. In the notation of Section 3.3, from Theorem 3.5.2 we get the following result.

THEOREM 3.5.4. *Let $y_t \to \theta$ as $t \to \infty$ and let the following conditions be fulfilled:*

1) *the function $\ln \lambda(y,\theta)$ belongs to $\mathbf{C}_1^0(\Theta)$ as a function of the variable $y \in \Theta$, all the other variables being fixed, and for all y in a neighborhood of the point θ we have $l(y,\theta), f(y,\theta) \in \mathcal{G}^2_{\mathrm{loc}}(\nu(\theta), \mathbf{F}, \mathsf{P}_\theta)$;*

2) *there exists a positive definite symmetric matrix $\varphi_t(\theta)$ such that $|\varphi_t(\theta)| \to 0$ as $t \to \infty$ and*

$$\mathsf{P}_\theta\text{-}\lim_{t\to\infty} \varphi_t(\theta) l_0 l_0' \circ \nu(\theta)_t \varphi_t(\theta) = I_k;$$

3) *for any $\varepsilon \in (0,1]$*

$$\mathsf{P}_\theta\text{-}\lim_{t\to\infty} \chi\big(|\varphi_t(\theta) l_0| > \varepsilon\big) |\varphi_t(\theta) l_0|^2 \circ \nu(\theta)_t = 0;$$

4) $\mathsf{P}_\theta\text{-}\lim_{t\to\infty} \int_0^1 (|\bar{l}_v^t|^2 + |\bar{f}_v^t|^2)\, dv \circ \nu(\theta)_t = 0.$

Then the following representation takes place:

$$\Lambda(y_t,\theta) = u_t' \left(\eta^t + q^t\right) - (1/2) u_t' \left(I_k + p^t\right) u_t, \tag{3.5.16}$$

where

$$u_t = \varphi_t^{-1}(\theta)\Delta_t, \qquad \eta^t = \varphi_t(\theta) l_0 * (\xi - \nu(\theta)) \in \mathcal{M}^2_{\mathrm{loc}}(\mathbf{F}, \mathsf{P}_\theta), \tag{3.5.17}$$

$$\mathcal{L}\left(\eta_t^t | \mathsf{P}_\theta\right) \xrightarrow{w} \mathcal{N}(0, I_k), \qquad t \to \infty, \tag{3.5.18}$$

$$\mathsf{P}_\theta\text{-}\lim_{t\to\infty} \left(|q_t^t| + |p_t^t|\right) = 0. \tag{3.5.19}$$

Theorem 3.3.5 implies the following theorem.

THEOREM 3.5.5. *Let the conditions of Theorem 3.5.4 be fulfilled and in addition,*

1) $\mathsf{P}_\theta\text{-}\lim_{t\to\infty}[\varphi_t(\theta)l_0 l_0' \circ \nu(\theta)_t \varphi_t(\theta) - I_k]u_t = 0$;
2) $\mathsf{P}_\theta\text{-}\lim_{t\to\infty}\int_0^1[(\Delta_t'\tilde{f}_v^t)^2 + (\Delta_t'\tilde{l}_v^t)^2]\,dv \circ \nu(\theta)_t = 0$, *where* $\tilde{f}_v^t = f_v^t - f_0$, $\tilde{l}_v^t = l_v^t - l_0$.

Then the representation (3.5.16)–(3.5.19) *is valid, in which*

$$\mathsf{P}_\theta\text{-}\lim_{t\to\infty}|p_t'u_t| = 0. \tag{3.5.20}$$

Counting processes with determinate compensators. Let the compensators (ν and $\widetilde{\nu}$ in the nonparametric case and $\nu(\theta)$ and $\theta \in \Theta$ in the parametric case) of a counting process ξ be determinate and continuous in time. Then the process ξ is a process with independent increments that has the Poisson distribution

$$\mathsf{P}_\theta\{\xi_t - \xi_s = n\} = \frac{(\nu_t(\theta) - \nu_s(\theta))^n}{n!}e^{-(\nu_t(\theta)-\nu_s(\theta))}, \qquad n = 0, 1, 2, \dots,$$

where $0 \le s < t < \infty$.

In this case the process ξ is called *a Poisson process*. If $\nu_t(\theta) = c(\theta)t$, $0 < c(\theta) < \infty$, the process ξ is called *homogeneous*, otherwise, *inhomogeneous*.

In this case the conditions of Theorems 3.5.1–3.5.5 get a trivial form, so we confine ourselves to investigating properties of the Hellinger process $h(\varepsilon) = h(\varepsilon; \widetilde{\mathsf{P}}, \mathsf{P})$ defined by (3.5.11) and (3.5.12) and to considering some examples.

Let ν and $\widetilde{\nu}$ be the compensators of the process ξ with respect to the measures P and $\widetilde{\mathsf{P}}$, respectively. Consider the behavior of the Hellinger process $h_t(\varepsilon)$ in relation to the behavior of ν_t and λ_t as $t \to \infty$. The following theorem provides sufficient conditions for the validity of the relations

$$\lim_{\varepsilon\downarrow 0}\varepsilon^{-1}\lim_{t\to\infty}\psi_t^{-1}h_t(\varepsilon) = 1, \tag{3.5.21}$$

$$\lim_{\varepsilon\uparrow 0}\varepsilon^{-1}\lim_{t\to\infty}\psi_t^{-1}h_t(\varepsilon) = -1, \tag{3.5.22}$$

where ψ_t is a function such that $\psi_t \to \infty$ as $t \to \infty$ (its form is given in the theorem).

THEOREM 3.5.6. *Let the limit*

$$\lim_{t\to\infty}\lambda_t = \lambda \tag{3.5.23}$$

exist and let the following conditions be fulfilled:

1) *if* $\lambda = \infty$, *then* $\widetilde{\nu}_t \to \infty$ *as* $t \to \infty$;
2) *if* $\lambda \in (0,1) \cup (1,\infty)$, *then* $\nu_t \to \infty$ *as* $t \to \infty$;
3) *if* $\lambda = 1$, *then* $\chi(|\ln\lambda \le 1|)\ln^2\lambda \circ \nu_t \to \infty$ *as* $t \to \infty$.

Then the relations (3.5.21) *and* (3.5.22) *are valid, in which*

$$\psi_t = \begin{cases} \widetilde{\nu}_t, & \text{if } \lambda = \infty, \\ (\lambda - 1 - \ln\lambda)\nu_t, & \text{if } \lambda \in (0,1)\cup(1,\infty), \\ \tfrac{1}{2}\chi(|\ln\lambda| \le 1)\ln^2\lambda \circ \nu_t, & \text{if } \lambda = 1. \end{cases} \tag{3.5.24}$$

If $\lambda = 0$ *and* $\nu_t \to \infty$, *then for any* $\varepsilon \in [0, 1)$

$$\lim_{t\to\infty}\nu_t^{-1}h_t(\varepsilon) = \begin{cases} 0, & \text{if } \varepsilon = 0, \\ 1-\varepsilon, & \text{if } \varepsilon \in (0,1), \end{cases} \tag{3.5.25}$$

and for any $\varepsilon \in [\varepsilon_0, 0)$

$$h_t(\varepsilon) = \lambda^\varepsilon \circ \nu_t(1 + o(1)) \tag{3.5.26}$$

as $t \to \infty$.

PROOF. In view of (3.5.11) and (3.5.12) we have

$$h_t(\varepsilon) = \operatorname{sgn}(\varepsilon)[\varepsilon(\lambda - 1) - (\lambda^\varepsilon - 1)] \circ \nu_t, \qquad t \in \mathbf{R}_+, \tag{3.5.27}$$

where $\varepsilon \in [\varepsilon_0, 1)$ and $\varepsilon_0 < 0$ is a constant in Theorem 3.1.7.

If $\lambda = \infty$ and $\widetilde{\nu}_t \to \infty$, then from (3.5.27) we have

$$h_t(\varepsilon) = \varepsilon \operatorname{sgn}(\varepsilon)\lambda \circ \nu_t(1 + o(1)) = \varepsilon \operatorname{sgn}(\varepsilon)\widetilde{\nu}_t(1 + o(1))$$

as $t \to \infty$. From this we get the required relations (3.5.21) and (3.5.22) for $\psi_t = \widetilde{\nu}_t$.

Now let $\lambda \in (0, 1) \cup (1, \infty)$ and $\nu_t \to \infty$ as $t \to \infty$. Then for $t \to \infty$ from (3.5.27) we obtain

$$h_t(\varepsilon) = \operatorname{sgn}(\varepsilon)[\varepsilon(\lambda - 1) - (\lambda^\varepsilon - 1)]\nu_t + o(\nu_t).$$

Taking into consideration the equality

$$\frac{\partial}{\partial \varepsilon}[\varepsilon(\lambda - 1) - (\lambda^\varepsilon - 1)] = \lambda - 1 - \lambda^\varepsilon \ln \lambda,$$

we get the relations (3.5.21) and (3.5.22) with $\psi_t = (\lambda - 1 - \ln \lambda)\nu_t$.

Finally, let $\lambda = 1$. We take an arbitrary fixed $\delta > 0$ such that $1 - \delta > 1/e$. Since $\lambda = 1$, there exists $s_0 = s_0(\delta)$ such that $|\lambda_s - 1| < \delta$ for all $s > s_0$. Let us represent the Hellinger process $h_t(\varepsilon)$ for $t > s_0$ in the following manner:

$$h_t(\varepsilon) = h_{s_0}(\varepsilon) + \operatorname{sgn}(\varepsilon) \int_{s_0}^t [\varepsilon(\lambda_s - 1) - (\lambda_s^\varepsilon - 1)]\, d\nu_s. \tag{3.5.28}$$

Now we expand the integral on the right-hand side of equality (3.5.28) for fixed $t > s_0$ into the Taylor series as a function of ε in a neighborhood of the point $\varepsilon = 0$. We obtain

$$\begin{aligned} &\int_{s_0}^t [\varepsilon(\lambda_s - 1) - (\lambda_s^\varepsilon - 1)]\, d\nu_s \\ &\quad = \varepsilon \int_{s_0}^t (\lambda_s - 1 - \ln \lambda_s)\, d\nu_s - \frac{\varepsilon^2}{2} \int_{s_0}^t \ln^2 \lambda_s\, d\nu_s - \frac{\varepsilon^3}{6} \int_{s_0}^t \lambda_s^{\tilde{\varepsilon}} \ln^3 \lambda_s\, d\nu_s, \end{aligned} \tag{3.5.29}$$

where $\tilde{\varepsilon}$ is a point between 0 and ε.

Note that all integrals on the right of the equality (3.5.29) are finite by virtue of condition IV) of Theorem 1.2.5 and Remark 1.2.6, since $|\ln \lambda_s| < 1$ for all $s > s_0$, where, for the third integral, $\lambda_s^{\tilde{\varepsilon}} < (1 + \delta)^\varepsilon$ for $\varepsilon > 0$ and $\lambda_s^{\tilde{\varepsilon}} < (1 - \delta)^\varepsilon$ for $\varepsilon < 0$ for all $s > s_0$. Since $\lambda_s \to 1$ as $s \to \infty$, for $t \to \infty$ we have

$$\int_{s_0}^t (\lambda_s - 1 - \ln \lambda_s)\, d\nu_s = \frac{1}{2} \int_{s_0}^t (\lambda_s - 1)^2\, d\nu_s(1 + o(1)), \tag{3.5.30}$$

$$\int_{s_0}^t \ln^2 \lambda_s\, d\nu_s = \int_{s_0}^t (\lambda_s - 1)^2\, d\nu_s(1 + o(1)), \tag{3.5.31}$$

$$\int_{s_0}^t \lambda_s^{\tilde{\varepsilon}} \ln^3 \lambda_s\, d\nu_s = o\left(\int_{s_0}^t \ln^2 \lambda_s\, d\nu_s\right). \tag{3.5.32}$$

Combining (3.5.28)–(3.5.32), we obtain the required relations (3.5.21) and (3.5.22) with $\psi_t = \chi(|\ln\lambda| \le 1)\ln^2\lambda \circ \nu_t$.

Thus the relations (3.5.21) and (3.5.22) are proved in the case $\lambda > 0$ with ψ_t defined by formula (3.5.24).

Let $\lambda = 0$ and $\nu_t \to \infty$ as $t \to \infty$. Then for $\varepsilon > 0$ from (3.5.27) we get $h_t(\varepsilon) = (1-\varepsilon)\nu_t + o(\nu_t)$ as $t \to \infty$, whence by virtue of the equality $h_t(0) = 0$ the relation (3.5.25) follows. If $\varepsilon < 0$, then the relation (3.5.26) follows from (3.5.27) in an obvious way. □

REMARK 3.5.2. Under the conditions of Theorem 3.5.6 in view of Theorem 3.1.6 for $\varepsilon \in [\varepsilon_0, 1)$ and $\lambda > 0$ we get

$$\lim_{t\to\infty} \psi_t^{-1} \ln H_t(\varepsilon; \widetilde{\mathsf{P}}, \mathsf{P}) = \Psi(\varepsilon), \tag{3.5.33}$$

where ψ_t is given by formula (3.5.24) and the function $\Psi(\varepsilon)$ is defined as follows:

$$\Psi(\varepsilon) = \begin{cases} -\varepsilon, & \text{if } \lambda = \infty, \\ \frac{\lambda^\varepsilon - 1 - \varepsilon(\lambda-1)}{\lambda - 1 - \ln\lambda}, & \text{if } \lambda \in (0,1)\cup(1,\infty), \\ -\varepsilon(1-\varepsilon), & \text{if } \lambda = 1. \end{cases}$$

Hence for $\lambda \in (0,\infty)$ condition Λ^* (Section 2.3) is fulfilled with $\varepsilon_- = \varepsilon_0$, $\varepsilon_+ = 1$, and $\widetilde{\chi}_t = \psi_t$. It means that for $\lambda \in (0,\infty)$ the theorem on large deviations (Lemma 2.3.4) holds with $\gamma_0 = -1$, $\gamma_- = \Psi'(\varepsilon_0)$, $\gamma_+ = \Psi'(1)$, and $I(\gamma) = \gamma\varepsilon_\gamma - \Psi(\varepsilon_\gamma)$, where ε_γ is the unique solution of the equation $\Psi'(\varepsilon) = \gamma$. Observe that for $\lambda = \infty$ condition Λ^* is not fulfilled since the function $\Psi(\varepsilon)$ is not strictly convex. Theorem 3.5.6 implies also that for $\lambda = 0$ condition Λ^* is not fulfilled since

$$\lim_{t\to\infty} \nu_t^{-1} \ln H_t(\varepsilon; \widetilde{\mathsf{P}}, \mathsf{P}) = \begin{cases} \varepsilon - 1, & \text{if } \varepsilon \in (0,1), \\ 0, & \text{if } \varepsilon = 0, \\ \infty, & \text{if } \varepsilon \in [\varepsilon_0, 0). \end{cases}$$

Furthermore, for $\lambda > 0$ the relation (3.5.33) implies the relation

$$\lim_{\varepsilon\to 0} \varepsilon^{-1} \lim_{t\to\infty} \psi_t^{-1} \ln H_t(\varepsilon; \widetilde{\mathsf{P}}, \mathsf{P}) = -1, \tag{3.5.34}$$

that is, conditions $\Lambda 2$ and $\Lambda 3$ (Section 2.3) are fulfilled with $\chi_t = \psi_t$ for all $\lambda \in (0,\infty]$, though for $\lambda = \infty$ the condition Λ^* is not fulfilled.

EXAMPLE 3.5.1. Let $\nu_s = cs^a$ and $\lambda_s = ds^b$ for all $s \in \mathbf{R}_+$, where a, b, c, d are some positive numbers. Then, evidently,

$$\widetilde{\nu}_t = \lambda \circ \nu_t = acd \int_0^t s^{a+b-1}\, ds = \frac{acd}{a+b} t^{a+b}, \qquad t \in \mathbf{R}_+.$$

Therefore $\lambda_t \to \infty$ and $\widetilde{\nu}_t \to \infty$ as $t \to \infty$. By Theorem 3.5.6 this means that the relations (3.5.21) and (3.5.22) are fulfilled for the Hellinger process $h(\varepsilon)$, and in view of Remark 3.5.2 the relation (3.5.34) is valid for the Hellinger integral $H_t(\varepsilon; \widetilde{\mathsf{P}}, \mathsf{P})$ with $\psi_t = \widetilde{\nu}_t$.

For $t \to \infty$ we have $V_{st} = \frac{acd}{a+b}(st)^{a+b}(1 + o(1))$, and therefore condition 1) of Theorem 3.5.6 is fulfilled for $\psi_t = \widetilde{\nu}_t$ and $a_s = s^{a+b}$. Then for fixed $s \in (0,\infty)$ we have, as $t \to \infty$,

$$\ln^+ \lambda \circ \nu_{st} = \int_{d^{-1/b}}^{st} \ln \lambda_u \, d\nu_u = bc(st)^a(\ln(st))(1+o(1)), \tag{3.5.35}$$

$$\ln^- \lambda \circ \nu_{st} = -\int_0^{d^{-1/b}} \ln \lambda_u \, d\nu_u = \frac{bc}{a} d^{-a/b}. \tag{3.5.36}$$

From this it follows that condition 2) of Theorem 3.5.2 is fulfilled with $k_s^{\pm} = 0$. Since $k_s^{\pm} = 0$, condition 3) of Theorem 3.5.2 is trivial. Thus the assertion of Theorem 3.5.2 is valid, in view of which for any $s \in (0, \infty)$

$$\mathsf{P}\text{-}\lim_{t\to\infty} \psi_t^{-1} \Lambda_{st} = -s^{a+b}, \tag{3.5.37}$$

where $\psi_t = acd(a+b)^{-1}t^{a+b}$.

Note that the relations (3.5.35) and (3.5.36) imply the validity of (3.5.5) with $\psi_t = \widetilde{\nu}_t$, whence by Remark 3.5.1 condition 2) of Theorem 3.5.1 also holds.

Now we consider the family of processes $Y^t = (Y_s^t)_{s\in\mathbf{R}_+}$, $t \in \mathbf{R}_+$, where

$$Y_s^t = \psi_t^{-1}(\Lambda_{st} + \varphi_{st}), \qquad s \in \mathbf{R}_+, \tag{3.5.38}$$

with $\varphi_{st} = V_{st}$ and $\psi_t^2 = cb^2 t^a \ln^2 t$. Observe that the normalizing factor ψ_t was chosen on the basis of the fact that $Y^t \in \overline{\mathcal{M}}^2(\mathbf{F}^t, \mathsf{P})$ with quadratic characteristic

$$\langle Y^t \rangle_s = \ln^2 \lambda \circ \nu_{st} = cb^2(st)^a(\ln t)^2(1+o(1)), \qquad t \to \infty.$$

Now we apply Theorem 3.5.3. Since

$$B_s^t = -\psi_t^{-1} \chi\left(|\psi_t^{-1} \ln \lambda| > 1\right) \ln \lambda \circ \nu_{st},$$

for fixed $s \in (0, \infty)$ there exists $t_0 = t_0(s)$ such that for all $t > t_0$ we have

$$B_s^t = -\psi_t^{-1} \chi(\ln \lambda > -\psi_t) \ln \lambda \circ \nu_{st}.$$

From this it follows that $B_s^t \to 0$ as $t \to \infty$. This means that condition 2) of Theorem 3.5.3 is fulfilled with $B_s^Y = 0$. Reasoning similarly, we establish that condition 1) of Theorem 3.5.3 is fulfilled with $\nu^Y = 0$. Furthermore, for fixed $s \in (0, \infty)$ and $\delta \in (0, 1)$ and $t \to \infty$ we get

$$\chi\left(|\psi_t^{-1} \ln \lambda| \le \delta\right) \left(\psi_t^{-1} \ln \lambda\right)^2 \circ \nu_{st} = \left(\psi_t^{-1} \ln \lambda\right)^2 \circ \nu_{st} = s^\alpha + o(1),$$

that is, condition 3) of Theorem 3.5.3 is fulfilled with $C_s^Y = s^\alpha$. Therefore Theorem 3.5.3 implies that for $t \to \infty$

$$Y^t \xrightarrow{w_f(S)} Y, \tag{3.5.39}$$

where $Y = (Y_s)$ is a continuous Gaussian martingale with $\langle Y \rangle_s = s^\alpha$ and $Y_0 = 0$.

Thus under the conditions of this example the law of large numbers (3.5.37), the central limit theorem (3.5.39), and relation (3.5.34) for the Hellinger integral $H_t(\varepsilon; \widetilde{\mathsf{P}}, \mathsf{P})$ are fulfilled for Λ_t.

EXAMPLE 3.5.2. Let $\nu_s = a\ln(1+bs)$ and $\lambda_s = \exp(c+ds)$ for all $s \in \mathbf{R}_+$, where a, b, c, d are some positive numbers. As in Example 3.5.1, $\widetilde{\nu}_s \to \infty$ and $\lambda_s \to \infty$, and hence in view of Theorem 3.5.6 and Remark 3.5.2 for $\psi_t = \widetilde{\nu}_t$ the relation (3.5.34) is fulfilled for the Hellinger integral $H_t(\varepsilon; \widetilde{\mathsf{P}}, \mathsf{P})$. In this case the law of large numbers is also easily established for Λ_t with $\psi_t = \widetilde{\nu}_t$ (Theorem 3.5.1 with $s=1$). Now we find the form of the limit process Y for processes $Y^t = (Y^t_s)$ given by relation (3.5.38), in which $\varphi_{st} = V_{st}$ and $\psi_t^2 = a(dt)^2/2$. To that end we apply Theorem 3.5.3. Obviously, as $t \to \infty$,

$$B^t_s = -\psi_t^{-1}\chi(\ln\lambda > \psi_t)\ln\lambda \circ \nu_{st} \to -\sqrt{2\alpha}\left(s - \sqrt{a/2}\right)^+$$

and condition 2) of Theorem 3.5.3 holds with $B^Y_s = -\sqrt{2\alpha}\,(s-\sqrt{a/2})^+$. For any $s \in \mathbf{R}_+$ and $C \in \mathcal{B}_0$ we have, as $t \to \infty$,

$$\begin{aligned}\chi_C\left(\frac{\ln\lambda}{\psi_t}\right)\circ\nu_{st} &= \int_0^s \chi_C\left(\frac{c+dut}{dt\sqrt{a/2}}\right)\frac{abt\,du}{1+btu} \to \int_0^s \chi_C\left(\frac{u}{\sqrt{a/2}}\right)\frac{a\,du}{u}\\ &= \int_0^s\int_{R_0}\chi_C(x)\nu^Y(du,dx),\end{aligned}$$

where

$$\nu^Y(du,dx) = \frac{a}{u}\,du\,\varepsilon_{\left\{u/\sqrt{a/2}\right\}}(dx) \tag{3.5.40}$$

and $\varepsilon_{\{z\}}(dx)$ is the Dirac measure at the point z. Hence condition 1) of Theorem 3.5.3 is fulfilled for the compensator ν^Y of the jump measure of the limit process Y. For any fixed $\varepsilon \in (0,1]$ and $s \in \mathbf{R}_+$ as $t \to \infty$ we have

$$\left(\frac{\ln\lambda}{\psi_t}\right)^2\chi\left(\left|\frac{\ln\lambda}{\psi_t}\right| \le \varepsilon\right)\circ\nu_{st} = \left(\frac{\ln\lambda}{\psi_t}\right)^2\circ\nu_{(s\wedge\delta_t)t} = (s\wedge\delta)^2 + o(1),$$

where $\delta = \varepsilon\sqrt{a/2}$ and $\delta_t = \varepsilon\sqrt{a/2} - c/(td)$. From this it follows that condition 3) of Theorem 3.5.3 is fulfilled with $C^Y_s = 0$, and hence $Y^c = 0$.

We have thus proved that for $t \to \infty$ the convergence (3.5.39) occurs, where Y is a semimartingale with canonical representation

$$Y_s = -\sqrt{2a}\left(1-\sqrt{a/2}\right)^+ + \int_0^s\int_{|x|\le 1} xd\left(\mu - \nu^Y\right) + \int_0^s\int_{|x|>1} x\,d\mu.$$

Here μ is the jump measure of the process Y with compensator ν^Y of the form (3.5.40). After simple computations for the characteristic function of the process Y we get

$$\mathsf{E}\exp(iuY_s) = \exp\left\{-iu\sqrt{2a}\,s + a\int_0^s\left(e^{iu\sqrt{2/a}\,x} - 1\right)\frac{dx}{x}\right\}, \qquad u \in \mathbf{R}^1.$$

In deriving the last equality we used of the following well-known result ([**198**], Theorem 4.4.1).

THEOREM 3.5.7$'$. *Let Y be a semimartingale with independent increments, $Y_0 = 0$, with time-continuous triple (B^Y, C^Y, ν^Y). Then for $s < t$ (P-a.s.) we have*

$$\mathsf{E}\{\exp(iu(Y_t - Y_s))/\mathcal{F}_s\} = \exp\{G_t(u) - G_s(u)\},$$

where $G_t(u)$ is the cumulant of the process Y defined by

$$G_t(u) = iuB_t^Y - \frac{u^2}{2}C_t^Y + \int_0^t\int_{R_0} \left(e^{iux} - 1 - iux\,\chi(|x| \le 1)\right) d\nu^Y.$$

It is seen from Theorem 3.5.7$'$ that weak convergence of Y^t to Y, as $t \to \infty$, can be proved directly using the form of the characteristic function for Y_s^t, which is given by this theorem.

EXAMPLE 3.5.3. Let $\nu_s = \ln(1+s^b)$ and $\lambda_s = \exp(-s^a)$, where a and b are some positive numbers. Consider a family of stochastic processes $Y^t = (Y_s^t)_{s\in\mathbf{R}_+}$, $t \in \mathbf{R}_+$, where $Y_s^t = t^{-a}\Lambda_{st}$. Writing the canonical representation for a semimartingale and applying Theorem 3.5.7$'$, we get

(3.5.41)
$$\begin{aligned}\mathsf{E}\exp(iut^{-a}\Lambda_{st}) = \exp\Big\{&-iut^{-a}(\lambda - s - \chi(|\ln\lambda| \le t^a)\ln\lambda) \circ \nu_{st}\\ &+ \left(\exp\left(iu\frac{\ln\lambda}{t^a}\right) - 1 - iu\frac{\ln\lambda}{t^a}\chi(|\ln\lambda| \le t^a)\right)\circ\nu_{st}\Big\}.\end{aligned}$$

For fixed $s \in (0, \infty)$ and $t \to \infty$ we have

$$\begin{aligned}&t^{-a}(\lambda - s - \chi(|\ln\lambda| \le t^a)\ln\lambda)\circ\nu_{st}\\ &\quad= t^{-a}\int_0^s (\lambda_{vt} - 1 - \chi(|\ln\lambda_{vt}| \le t^a)\ln\lambda_{vt})\,d\nu_{vt}\\ &\quad= \int_0^s \left(\frac{\exp(-v^a t^a) - 1}{t^a} + v^a\chi(v \le 1)\right)\frac{tb(vt)^{b-1}}{1 + (vt)^b}\,dv \qquad (3.5.42)\\ &\quad\to b\int_0^s v^{a-1}\chi(v \le 1)\,dv = \frac{b}{a}(s \wedge 1)^a,\end{aligned}$$

and

$$\begin{aligned}&\left(\exp\left(iu\frac{\ln\lambda}{t^a}\right) - 1 - iu\frac{\ln\lambda}{t^a}\chi(|\ln\lambda| \le t^a)\right)\circ\nu_{st}\\ &\quad= \int_0^s (\exp(-iuv^a) - 1 + iuv^a\chi(v \le 1))tb(vt)^{b-1}\left(1 + (vt)^b\right)^{-1}dv \qquad (3.5.43)\\ &\quad\to b\int_0^s (\exp(-iuv^a) - 1 + iuv^a\chi(v \le 1))\,\frac{dv}{v}.\end{aligned}$$

For $t \to \infty$ from the relations (3.5.41)–(3.5.43) we obtain

(3.5.44)
$$\begin{aligned}&\mathsf{E}\exp(iut^{-a}\Lambda_{st}) \to \mathsf{E}\exp(iuY_s)\\ &\quad= \exp\left\{-iu\frac{b}{a}(s \wedge 1)^a + b\int_0^s (\exp(-iuv^a) - 1 + iuv^a\chi(v \le 1))\,\frac{dv}{v}\right\}.\end{aligned}$$

From this it is seen that the limit process Y is a semimartingale with canonical representation

$$Y_s = B_s^Y + \int_0^s \int_{|x|\le 1} x\, d\left(\mu - \nu^Y\right) + \int_0^s \int_{|x|>1} x\, d\mu,$$

where μ is the jump measure of Y with compensator ν^Y, and

$$B_s^Y = -\frac{b}{a}(s\wedge 1)^a, \qquad \nu^Y(dv, dx) = \frac{b}{v}\, dv \varepsilon_{\{-v^a\}}(dx).$$

The relation (3.5.44) implies that the distribution of the random variable $t^{-a}\Lambda_t$ weakly converges, as $t\to\infty$, to the distribution with characteristic function

$$g(u) = \exp\left\{-\frac{b}{a}\int_{-1}^{0} (e^{iux}-1)\,\frac{dx}{x}\right\}.$$

From [**23, 108**] we see that $g(u)$ is the characteristic function of a one-vertex infinitely divisible distribution that has a density and is concentrated on the interval $(-\infty, 0)$.

EXAMPLE 3.5.4. Let $\nu_s = 1-(1+\ln(1+s))^{-1}$, $\lambda_s = (1+s)^{-a}$ and $a>0$. Evidently, $\widetilde{\nu}_\infty < \nu_\infty = 1$. Using Theorem 3.5.7$'$, for $t\to\infty$ we get

$$\begin{aligned}\mathsf{E}\exp(iu\Lambda_t) &= \exp\Bigl\{-iu\bigl(\lambda - 1 - \chi(|\ln\lambda|\le 1)\ln\lambda\bigr)\circ\nu_t \\ &\qquad\qquad + \bigl(e^{iu\ln\lambda} - 1 - \chi(|\ln\lambda|\le 1)\ln\lambda\bigr)\circ\nu_t\Bigr\} \\ &\to \exp\left\{iu(1-\widetilde{\nu}_\infty) + \int_0^\infty \left(e^{-iua\ln(1+s)} - 1\right) d\nu_s\right\} \\ &= \exp\left\{iu(1-\widetilde{\nu}_\infty) + a\int_0^\infty (e^{iux}-1)\,\frac{dx}{(a-x)^2}\right\}.\end{aligned}$$

Therefore for $t\to\infty$ the random variable Λ_t weakly converges to a random variable Λ that takes values in the interval $(-\infty, 1-\widetilde{\nu}_\infty)$, where $1-\widetilde{\nu}_\infty > 0$ and, as is easy to see, $\mathsf{E}\,\Lambda = -\infty$.

EXAMPLE 3.5.5. Let the compensators ν and $\widetilde{\nu}$ be of the form

$$\nu_t = \int_0^t e^{a+bs}\, ds, \qquad \widetilde{\nu}_t = \int_0^t e^{c+ds}\, ds, \qquad t\in\mathbf{R}_+,$$

where b is a positive number, and a, c, and d are arbitrary numbers such that $a\ne c$ if $b=d$. It is not hard to show that for $s=1$ condition 1) of Theorem 3.5.1 is fulfilled with ψ_t of the form

$$\psi_t = \begin{cases} \frac{1}{d}\, e^{c+dt}, & \text{if } d>b, \\ \frac{1}{b}(\lambda - 1 - \ln\lambda)e^{a+bt}, & \text{if } d=b, \\ \left(1-\frac{d}{b}\right) te^{a+bt}, & \text{if } d<b, \end{cases} \tag{3.5.45}$$

where $\lambda = e^{c-a}$.

For $t\to\infty$ we have

$$\ln^2\lambda\circ\nu_t = \begin{cases} b^{-1}(d-b)^2 t^2 e^{a+bt}(1+o(1)), & \text{if } d\ne b, \\ b^{-1}(c-a)^2 e^{a+bt}(1+o(1)), & \text{if } d=b. \end{cases}$$

From this it follows that for $s = 1$ condition 2) of Theorem 3.5.1 is also fulfilled. Hence for $s = 1$ the law of large numbers is fulfilled with normalizing factor ψ_t of the form (3.5.45). Note that if the level of the test δ_t^{+,α_t} satisfies conditions $\alpha 1$ and $\alpha 2$, then in view of Theorem 2.3.1 for the probability of error of the second kind $\beta(\delta_t^{+,\alpha_t})$ we have $\lim_{t\to\infty} \psi_t^{-1} \ln \beta(\delta_t^{+,\alpha_t}) = -1$. Now, taking (3.5.45) into consideration, we get

$$\lim_{t\to\infty} \frac{1}{t} \ln\ln(\beta(\delta_t^{+,\alpha_t}))^{-1} = b \vee d.$$

These relations imply that $\beta(\delta_t^{+,\alpha_t})$ decreases according to the double exponential law.

Now we consider the parametric case assuming the compensators $\nu(\theta)$ to be determinate and time-continuous for all $\theta \in \Theta$. With regard to Theorems 3.5.4 and 3.5.5, we only make the following remark concerning condition 2) of Theorem 3.5.4. Obviously, for this condition to be fulfilled, it is sufficient that the matrix $I_t(\theta) = l_0 l_0' \circ \nu(\theta)_t$ be positive definite with finite elements for all $t \in \mathbf{R}_+$, $|I_t^{-1}(\theta)| \to 0$ as $t \to \infty$ and $\varphi_t(\theta) = I_t^{-1/2}(\theta)$.

EXAMPLE 3.5.6. Let $\nu(\theta)$ be a compensator of the form

$$\nu_t(\theta) = \int_0^t e^{\theta' + \theta'' s}\, ds, \qquad \theta = (\theta', \theta'')' \in \mathbf{R}^1 \times (0, \infty).$$

In this case the following relations are valid:

$$\lambda_t(y,\theta) = \exp\{(y' - \theta') + t(y'' - \theta'')\}, \qquad l_t(y,\theta) = \begin{pmatrix} 1 \\ t \end{pmatrix} \lambda_t(y,\theta),$$

where $y = (y', y'')'$.

It is not hard to see that condition 1) of Theorem 3.5.4 is fulfilled. The matrix $I_t(\theta) = l_0 l_0' \circ \nu(\theta)_t$ is of the form

$$I_t = \frac{e^{\theta' + \theta'' t}}{\theta''} \begin{pmatrix} 1 - e^{-\theta'' t} & t - \frac{1}{\theta''}\left(t - e^{-\theta'' t}\right) \\ t - \frac{1}{\theta''}\left(1 - e^{-\theta'' t}\right) & t^2 - \frac{2t}{\theta''} + \frac{2}{(\theta'')^2}\left(1 - e^{-\theta'' t}\right) \end{pmatrix},$$

and hence the matrix $I_t(\theta)$ is nonsingular. Solving the equation

$$\varphi_t(\theta) I_t(\theta) \varphi_t(\theta) = I_k$$

for $t \to \infty$, we obtain

$$\varphi_t(\theta) = (\theta'')^{3/2} e^{-(\theta' + \theta'' t)/2} \begin{pmatrix} t + o(t) & -1 + o(1) \\ -1 + o(1) & \left(1 + \frac{1}{\theta''} + o(1)\right) \frac{1}{t} \end{pmatrix}, \tag{3.5.46}$$

and hence condition 2) of Theorem 3.5.4 is fulfilled with matrix $\varphi_t(\theta)$ of the form (3.5.46). For a given $\varepsilon > 0$ for all $s \in [0, t]$ and for all t large enough the following inequality holds:

$$\begin{aligned} |\varphi_t(\theta) l_s(\theta,\theta)|^2 = Ce^{-\theta'' t}\Big\{ &\big[t(1 + o(1)) - s(1 + o(1))\big]^2 \\ &+ \left[\left(1 + \frac{1}{\theta''}\right)\frac{s}{t}(1 + o(1)) - 1 + o(1)\right]^2 \Big\} \\ \le \widetilde{C} t^2 e^{-\theta'' t} \le \varepsilon, \end{aligned}$$

where the constants C and $\widetilde{C}$ are independent of s and t. This implies that condition 3) of Theorem 3.5.4 is fulfilled. Next, for $u \in [0,1]$, $s \in [0,t]$, and t large enough,

$$\begin{aligned}
&\left|\varphi_t(\theta)\big(l_s(\theta+u\Delta_t,\theta)-l_s(\theta,\theta)\big)\right|^2 \\
&\quad = (\theta'')^3 e^{-\theta-\theta'' t} \\
&\qquad \times \left\{\big[t(1+o(1))-s(1+o(1))\big]^2+\left[\frac{C's}{t}(1+o(1))-1+o(1)\right]^2\right\} \\
&\qquad \times \left(e^{u\Delta_t'+us\Delta_t''}-1\right)^2 \\
&\quad \le \overline{C}t^2 e^{-\theta'' t}\left(e^{|\Delta_t'|+t|\Delta_t''|}-1\right)^2,
\end{aligned}$$

where $C' = 1+1/\theta''$, $\Delta_t = (\Delta_t', \Delta_t'')'$, and the constant $\overline{C}$ is independent of u, s, and t. Therefore, for t large enough,

$$\sup_{0\le u\le 1} |\bar{l}_u^t|^2 \circ \nu(\theta)_t \le C'' t^2 \left(\exp\{|\Delta_t'|+t|\Delta_t''|\}-1\right)^2,$$

where the constant C'' is independent of t. Taking into account that $f_s(y,\theta) = l_s(\theta,\theta)$ for all s, y, θ, we see that condition 4) of Theorem 3.5.4 is fulfilled if

$$|\Delta_t'| + t|\Delta_t''| = o(t^{-1}), \qquad t\to\infty.$$

Thus under the restriction $|\Delta_t'| + t|\Delta_t''| = o\left(t^{-1}\right)$ all conditions of Theorem 3.5.4 are fulfilled, and hence the representation (3.5.16)–(3.5.19) is valid in which

$$\eta_t^t = \varphi_t(\theta)\left(\xi_t - \nu_t(\theta), \int_0^t s\,d(\xi_s - \nu_s(\theta))\right).$$

If $\varphi_t(\theta) = I_t^{-1/2}(\theta)$, then condition 1) of Theorem 3.5.5 is trivial. Taking into account the obvious estimate $((\tilde{l}_v^t)'\Delta_t)^2 \le (|\Delta_t'| + t|\Delta_t''|)^4$, valid for all $v \in [0,1]$, and equality $\tilde{f}_v^t = 0$, we see that condition 2) of Theorem 3.5.5 holds if

$$|\Delta_t'| + t|\Delta_t''| = o(\exp\{-\theta'' t/4\}).$$

Thus under this condition the representation (3.5.16)–(3.5.19) holds, in which the relation (3.5.20) is valid.

Renewal processes. Let $T_n = \sum_{k=1}^n \tau_k$, $n = 1, 2, \ldots$, be jump times of a counting process ξ, where τ_k, $k = 1, 2, \ldots$, are independent random variables with continuous distribution functions $F_k(t)$ and $\widetilde{F}_k(t)$ with respect to the measures P and $\widetilde{\mathsf{P}}$, respectively, so that $F_k(t) = \mathsf{P}\{\tau_k < t\}$, $\widetilde{F}_k(t) = \widetilde{\mathsf{P}}\{\tau_k < t\}$. In renewal theory the process ξ is called a *renewal process* [**22, 47, 48**].

It is well known [**30, 45**] that P- and $\widetilde{\mathsf{P}}$-compensators of the process ξ have the form

$$\nu_t = \sum_{k=1}^{\infty} \int_{t\wedge T_{k-1}}^{t\wedge T_k} \frac{d\Phi_k(s)}{1-\Phi_k(s-)}, \qquad \widetilde{\nu}_t = \sum_{k=1}^{\infty} \int_{t\wedge T_{k-1}}^{t\wedge T_k} \frac{d\widetilde{\Phi}_k(s)}{1-\widetilde{\Phi}_k(s-)},$$

where $T_0 = 0$ and $\Phi_k(s)$ and $\widetilde{\Phi}_k(s)$ are conditional distribution functions of the form

$$\begin{aligned}
\Phi_k(s) &= \mathsf{P}\{T_k \le s / T_1, T_2, \ldots, T_{k-1}\}, \\
\widetilde{\Phi}_k(s) &= \widetilde{\mathsf{P}}\{T_k \le s / T_1, T_2, \ldots, T_{k-1}\}.
\end{aligned}$$

Since random variables $\tau_1, \tau_2, \dots$ are independent, we have

$$\Phi_k(s) = F_k(s - T_{k-1}+), \qquad \widetilde{\Phi}_k(s) = \widetilde{F}_k(s - T_{k-1}+).$$

Since the functions $F_k(s)$ and $\widetilde{F}_k(s)$ are continuous, we get the following representations:

$$\nu_t = \sum_{k=1}^{\xi_{t-}} \pi_k(\tau_k) + \pi_{\xi_{t-}+1}(t - T_{\xi_{t-}}), \tag{3.5.47}$$

$$\widetilde{\nu}_t = \sum_{k=1}^{\xi_{t-}} \widetilde{\pi}_k(\tau_k) + \widetilde{\pi}_{\xi_{t-}+1}(t - T_{\xi_{t-}}), \tag{3.5.48}$$

where we put $\sum_{k=1}^{0} = 0$, $\xi_{0-} = 0$, and

$$\pi_k(t) = \ln(1 - F_k(t))^{-1}, \qquad \widetilde{\pi}_k(t) = \ln(1 - \widetilde{F}_k(t))^{-1}. \tag{3.5.49}$$

We assume that for each $k = 1, 2, \dots$ the probability measure on $\mathbf{R}_+$ generated by the distribution function $\widetilde{F}_k(t)$ is absolutely continuous with respect to the probability measure generated by the distribution function $F_k(t)$. Under this assumption we use the notation $\widetilde{F}_k \ll F_k$ and $\rho_k(t) = d\widetilde{F}_k/dF_k(t)$, $t \in \mathbf{R}_+$. Then in view of (3.5.49), $\widetilde{\pi}_k \ll \pi_k$ and the corresponding density has the form

$$l_k(t) = \frac{d\widetilde{\pi}_k}{d\pi_k}(t) = \frac{1 - F_k(t)}{1 - \widetilde{F}_k(t)} \rho_k(t), \qquad t \in \mathbf{R}_+. \tag{3.5.50}$$

Under this assumption condition 1) of Theorem 1.2.5 holds and in view of (3.5.47), (3.5.48), and (3.5.50) we have

$$\lambda_t = \frac{d\widetilde{\nu}}{d\nu}(t) = l_{\xi_{t-}+1}(t - T_{\xi_{t-}}), \qquad t \in \mathbf{R}_+. \tag{3.5.51}$$

By (3.5.47) and (5.5.51) we have

$$\begin{aligned} (1 - \lambda^{1/2})^2 \circ \nu_t &\le \sum_{k=1}^{\xi_{t-}+1} \int_{T_{k-1}}^{T_k} \left(1 - l_k^{1/2}(s - T_{k-1})\right)^2 d\pi_k(s - T_{k-1}) \\ &= \sum_{k=1}^{\xi_{t-}+1} \int_0^{\tau_k} \left(1 - l_k^{1/2}(s)\right)^2 d\pi_k(s). \end{aligned} \tag{3.5.52}$$

Since $F_k(0) = \widetilde{F}_k(0) = 0$ for all k, this implies that $\xi_{t-} < \infty$ (P-a.s.) and ($\widetilde{\mathsf{P}}$-a.s.) for all $t \in \mathbf{R}_+$. Since $\tau_k < \infty$ (P-a.s.) and ($\widetilde{\mathsf{P}}$-a.s.) for all k, it follows that if

$$\int_0^t \left(1 - l_k^{1/2}(s)\right)^2 d\pi_k(s) < \infty$$

for all $t \in \mathbf{R}_+$ and all $k = 1, 2, \dots$, then in view of (3.5.52) we have $(1 - \lambda^{1/2})^2 \circ \nu_t < \infty$ (P-a.s.) and ($\widetilde{\mathsf{P}}$-a.s.) for all $t \in \mathbf{R}_+$. This means that condition III) of Theorem 1.2.5 is fulfilled.

Thus the following theorem holds.

THEOREM 3.5.7. *Let the distribution functions $F_k(t)$ and $\widetilde{F}_k(t)$ satisfy the following conditions*:

1) $\widetilde{F}_k \ll F_k$ *with density* $\rho_k(t) = d\widetilde{F}_k/dF_k(t)$ *for all* $k = 1, 2, \ldots$;
2) *for each* $k = 1, 2, \ldots$ *and all* $t \in \mathbf{R}_+$

$$\int_0^t \left(1 - l_k^{1/2}(s)\right)^2 d\pi_k(s) < \infty.$$

Then $\widetilde{\mathsf{P}} \overset{\text{loc}}{\ll} \mathsf{P}$ *and the process of local density* z *of the measure* $\widetilde{\mathsf{P}}$ *with respect to the measure* P *has the form* (1.2.39), *in which* ν *and* λ *are given by* (3.5.47) *and* (3.5.51), *respectively. If, in addition, conditions* 1) *and* 2) *hold, in which* F_k *and* $\widetilde{F}_k$ *are interchanged, and the condition*

3) *for all* $k = 1, 2, \ldots$ *and all* $t \in \mathbf{R}_+$

$$\int_0^t |\ln l_k(s)|\, d\pi_k(s) < \infty$$

holds, then the process of local density z *has the representation* (1.2.42)–(1.2.44).

PROOF. The local absolute continuity $\widetilde{\mathsf{P}} \overset{\text{loc}}{\ll} \mathsf{P}$ has already been proved, whereas the form of the local density z follows from Theorem 1.2.5 since conditions 1) and 2) imply conditions I)–III) of Theorem 1.2.5. To prove the theorem under the additional condition 3), it is enough to note that in view of conditions 1) and 3),

$$|\ln \lambda| \circ \nu_t \leq \sum_{k=1}^{\xi_{t-}+1} \int_0^{\tau_k} |\ln l_k(s)|\, d\pi_k(s) < \infty$$

(P-a.s. and $\widetilde{\mathsf{P}}$-a.s.) for all $t \in \mathbf{R}_+$ and to apply Theorem 1.2.5. □

Below we consider the behavior of z_t as $t \to \infty$, but only in the case where $\tau_1, \tau_2, \ldots$ are identically distributed random variables. Then

$$\begin{aligned} &F(t) = F_k(t), \qquad &&\widetilde{F}(t) = \widetilde{F}_k(t), \qquad &&\pi(t) = \pi_k(t), \\ &\widetilde{\pi}(t) = \widetilde{\pi}_k(t), \qquad &&\rho(t) = \rho_k(t), \qquad &&l(t) = l_k(t). \end{aligned}$$

Renewal processes. Laws of large numbers and large deviations. First we establish the law of large numbers for $\Lambda_t = \ln z_t$, in the form given by Theorem 3.5.1. We need the strong law of large numbers for ξ_t as $t \to \infty$ [**10**].

LEMMA 3.5.1. *If* $0 < a = \mathsf{E}\tau_1 < \infty$, *then*

$$\mathsf{P}\left\{\lim_{t\to\infty} t^{-1}\xi_t = a^{-1}\right\} = 1.$$

Moreover, we need the limit theorem on convergence of a sequence of random variables indexed with random variables ([**188**], Theorem 1).

LEMMA 3.5.2. *Let* $(Y_n)_{n\geq 1}$ *be a sequence of random variables such that*

$$\mathsf{P}\left\{\lim_{t\to\infty} Y_n = Y\right\} = 1,$$

where Y *is a random variable.*

If N_t, $t \in \mathbf{R}_+$, are positive random variables such that $\mathsf{P}\{\lim_{t\to\infty} N_t = \infty\} = 1$, *then* $\mathsf{P}\{\lim_{t\to\infty} Y_{N_t} = Y\} = 1$.

We introduce the random variables

$$\delta_k = \int_0^{\tau_k} (l(s) - 1 - \ln l(s))\, d\pi(s), \qquad k = 1, 2, \dots. \tag{3.5.53}$$

It is not hard to see that

$$\mathsf{E}\,\delta_k = \int_0^{\infty} \big(l(s) - 1 - \ln l(s)\big)\, dF(s) = \mathsf{E}\big(l(\tau_k) - 1 - \ln l(\tau_k)\big), \qquad k = 1, 2, \dots.$$

THEOREM 3.5.8. *Let random variables $\tau_1, \tau_2, \dots$ be identically distributed, let the conditions of Theorem 3.5.7 be fulfilled, and*

1) $0 < a = \mathsf{E}\,\tau_1 < \infty$;
2) $b = \mathsf{E}\,\delta_1 < \infty$, $\mathsf{E}\,|\ln l(\tau_1)| < \infty$.

Then for any $s \in (0, \infty)$ the following relation is true:

$$\lim_{t\to\infty} \psi_t^{-1} \Lambda_{st} = -s \quad (\mathsf{P}\text{-}a.s.), \tag{3.5.54}$$

where $\psi_t = a^{-1} bt$.

PROOF. Theorem 3.5.7 implies that

$$\psi_t^{-1} \Lambda_{st} = \psi_t^{-1} M_{st} - \psi_t^{-1} V_{st}, \tag{3.5.55}$$

where

$$M_{st} = \ln \lambda * (\xi - \nu)_{st}, \qquad V_{st} = (\lambda - 1 - \ln \lambda) \circ \nu_{st}.$$

First we prove the relation

$$\lim_{t\to\infty} \psi_t^{-1} V_{st} = s \quad (\mathsf{P}\text{-a.s.}). \tag{3.5.56}$$

Taking into account (3.5.47), (3.5.50), and (3.5.51), we get the estimates

$$\sum_{k=1}^{\xi_{st-}} \delta_k \le V_{st} < \sum_{k=1}^{\xi_{st-}+1} \delta_k, \tag{3.5.57}$$

where δ_k are random variables of the form (3.5.53).

Consider the representation

$$\psi_t^{-1} \sum_{k=1}^{\xi_{st-}} \delta_k = \frac{\xi_{st-}}{\psi_t} \xi_{st-}^{-1} \sum_{k=1}^{\xi_{st-}} \delta_k. \tag{3.5.58}$$

By condition 2) we have $b = \mathsf{E}\,\delta_k < \infty$. Then by the strong law of large numbers

$$\lim_{n\to\infty} \frac{1}{n} \sum_{k=1}^{n} \delta_k = b \quad (\mathsf{P}\text{-a.s.}).$$

In view of condition 1) and Lemma 3.5.1 we have $\xi_t \to \infty$ (P-a.s.) as $t \to \infty$. Therefore by Lemma 3.5.2

$$\lim_{t\to\infty} \xi_{st-}^{-1} \sum_{k=1}^{\xi_{st-}} \delta_k = b \quad (\mathsf{P}\text{-a.s.}). \tag{3.5.59}$$

By Lemma 3.5.1,

$$\lim_{t\to\infty} \psi_t^{-1}\xi_{st-} = s/b \quad (\mathsf{P}\text{-a.s.}). \tag{3.5.60}$$

Now we combine (3.5.58)–(3.5.60) to obtain

$$\lim_{t\to\infty} \psi_t^{-1} \sum_{k=1}^{\xi_{st-}} \delta_k = s \quad (\mathsf{P}\text{-a.s.}). \tag{3.5.61}$$

Using the same argument, one can prove that

$$\lim_{t\to\infty} \psi_t^{-1} \sum_{k=1}^{\xi_{st-}+1} \delta_k = s \quad (\mathsf{P}\text{-a.s.}). \tag{3.5.62}$$

Now from inequalities (3.5.57) and relations (3.5.61) and (3.5.62) we get the desired relations (3.5.56).

The relations (3.5.55) and (3.5.56) imply that for proving (3.5.54) it suffices to show that

$$\lim_{t\to\infty} \psi_t^{-1} M_{st} = 0 \quad (\mathsf{P}\text{-a.s.}). \tag{3.5.63}$$

By virtue of condition 3) of Theorem 3.5.7 we have

$$M = A - \tilde{A}, \tag{3.5.64}$$

where $A = \ln\lambda * \xi$ and $\tilde{A} = \ln\lambda \circ \nu$.

Consider the representations

$$A = A_+ - A_-, \qquad \tilde{A} = \tilde{A}_+ - \tilde{A}_-, \tag{3.5.65}$$

where $A_\pm = \ln^\pm \lambda * \xi$ and $\tilde{A}_\pm = \ln^\pm \lambda \circ \nu$.

Obviously,

$$\psi_t^{-1} A_{st,\pm} = \frac{\xi_{st}}{\psi_t}\, \xi_{st}^{-1} \sum_{k=1}^{\xi_{st}} \ln^\pm l(\tau_k).$$

Using Lemmas 3.5.1 and 3.5.2, we get

$$\lim_{t\to\infty} \psi_t^{-1} A_{st,\pm} = \frac{c_\pm}{b} s \quad (\mathsf{P}\text{-a.s.}),$$

where $c_\pm = \mathsf{E} \ln^\pm l(\tau_1)$. In view of the first equality in (3.5.65), this implies the relation

$$\lim_{t\to\infty} \psi_t^{-1} A_{st} = \frac{c_+ - c_-}{b} s \quad (\mathsf{P}\text{-a.s.}). \tag{3.5.66}$$

The following representation is true:

$$\tilde{A}_{st,\pm} = \sum_{k=1}^{\xi_{st-}} \int_0^{\tau_k} \ln^\pm l(u)\, d\pi(u) + \int_0^{t-T_{\xi_{st-}}} \ln^\pm l(u)\, d\pi(u),$$

where, as is easy to see, $\mathsf{E}\int_0^{\tau_1} \ln^\pm l(s)\, d\pi(s) = c_\pm$. Using upper and lower bounds for $\tilde{A}_{st,\pm}$, which can be obtained in a similar way as the estimates (3.5.57), and applying once again Lemmas 3.5.1 and 3.5.2, we get

$$\lim_{t\to\infty} \psi_t^{-1} \tilde{A}_{st,\pm} = \frac{c_\pm}{b} s \quad (\mathsf{P}\text{-a.s.}).$$

In view of (3.5.65) this implies that

$$\lim_{t\to\infty} \psi_t^{-1} \tilde{A}_{st} = \frac{c_+ - c_-}{b} s \quad (\mathsf{P}\text{-a.s.}). \tag{3.5.67}$$

Combining (3.5.64), (3.5.66) and (3.5.67), we get the required relation (3.5.63). □

Consider Theorem 3.2.3 using the form of the Hellinger process $h(\varepsilon)$ in the case of a counting process. Now we employ the elementary inequality for some $\varepsilon_0 \in (0,1)$,

$$\varepsilon(x-1) - (x_\varepsilon - 1) \geq \varepsilon(x - 1 - \ln x) - \frac{\varepsilon^2}{2}(1 \vee x)^{\varepsilon_0} \ln^2 x$$

with $x > 0$ and $\varepsilon \in (0, \varepsilon_0)$, formula (3.5.11), and equalities (3.5.47), (3.5.50), and (3.5.51) to obtain the estimate

$$h_t(\varepsilon) \geq \varepsilon \sum_{k=1}^{\xi_{t-}} \left(\delta_k - \frac{\varepsilon}{2}\zeta_k\right), \tag{3.5.68}$$

where

$$\zeta_k = \int_0^{\tau_k} (1 \vee l(s))^{\varepsilon_0} \ln^2 l(s)\, d\pi(s), \qquad k = 1, 2, \dots,$$

and δ_k is defined by formula (3.5.53).

Suppose that τ_1 satisfies the Cramér condition with respect to the measure P; by the theorem on large deviations for sums of independent random variables $\tau_1, \tau_2, \dots$ ([**105**], Chapter 8, Theorem 10), for each sufficiently small $\varepsilon > 0$ there exists a constant l such that for all n large enough

$$\mathsf{P}\left\{\sum_{k=1}^{n} \tau_k \geq n\left(a + \sigma\sqrt{\varepsilon}\right)\right\} = \left(1 - \Phi\left(\sqrt{\varepsilon n}\right)\right) e^{\varepsilon^{3/2} n \lambda(\sqrt{\varepsilon})} \left(1 + l\sqrt{\varepsilon}\right), \tag{3.5.69}$$

where $\sigma^2 = \operatorname{Var} \tau_1$ and $\lambda(t)$ is the Cramér series for τ_1 with respect to P.

Let δ_1 and ζ_1 also satisfy the Cramér condition with respect to the measure P. Then once again, by the theorem on large deviations ([**105**], Chapter 8, Theorem 10), for each sufficiently small $\varepsilon > 0$ there exist constants l_1 and l_2 such that for all sufficiently large n

(3.5.70)
$$\mathsf{P}\left\{\sum_{k=1}^{n} \delta_k < n\left(b - \tilde{\sigma}\sqrt{\varepsilon}\right)\right\} = \left(1 - \Phi\left(\sqrt{\varepsilon n}\right)\right) e^{-\varepsilon^{3/2} n \tilde{\lambda}(-\sqrt{\varepsilon})} \left(1 + l_1\sqrt{\varepsilon}\right),$$

(3.5.71)
$$\mathsf{P}\left\{\sum_{k=1}^{n} \zeta_k > n\left(c + \bar{\sigma}\sqrt{\varepsilon}\right)\right\} = \left(1 - \Phi\left(\sqrt{\varepsilon n}\right)\right) e^{\varepsilon^{3/2} n \bar{\lambda}(\sqrt{\varepsilon})} \left(1 + l_2\sqrt{\varepsilon}\right),$$

where $b = \mathsf{E}\,\delta_1$, $c = \mathsf{E}\,\zeta_1$, $\tilde{\sigma}^2 = \operatorname{Var}\delta_1$, $\bar{\sigma}^2 = \operatorname{Var}\zeta_1$, and $\tilde{\lambda}(t)$ and $\bar{\lambda}(t)$ are the Cramér series for δ_1 and ζ_1, respectively, with respect to the measure P.

We introduce a positive integer n_t such that $n_t(a + \sigma\sqrt{\varepsilon}) = t + o(t)$ as $t \to \infty$ and put

$$D_{t,\varepsilon} = \{\xi_{t-} \geq n_t\} \cap \left\{\sum_{k=1}^{n_t} \delta_k \geq n_t\left(b - \tilde{\sigma}\sqrt{\varepsilon}\right)\right\} \cap \left\{\sum_{k=1}^{n_t} \zeta_k \leq n_t\left(c + \bar{\sigma}\sqrt{\varepsilon}\right)\right\}.$$

Taking into account the equality

$$\{\xi_{t-} < n\} = \Big\{\sum_{k=1}^{n} \tau_k \geq t\Big\},$$

for $t \to \infty$ from the relations (3.5.69)–(3.5.71) we get the estimate

$$\mathsf{P}(D^c_{t,\varepsilon}) \leq \frac{3\,(1+\bar{l}\sqrt{\varepsilon})}{\sqrt{2\pi n_t \varepsilon}} \exp\Big\{-\Big(\frac{\varepsilon}{2} - \varepsilon^{3/2}\bar{l}(\varepsilon)\Big)\, n_t\Big\}\,(1+o(1)),$$

where

$$\bar{l} = |l| \vee |l_1| \vee |l_2|, \qquad \bar{l}(\varepsilon) = \left|\lambda\left(\sqrt{\varepsilon}\right)\right| \vee \left|\tilde{\lambda}\left(-\sqrt{\varepsilon}\right)\right| \vee \left|\lambda\left(\sqrt{\varepsilon}\right)\right|.$$

This implies that as $t \to \infty$, we have

$$\varlimsup_{t\to\infty} \psi_t^{-1} \ln \mathsf{P}\left(D^c_{t,\varepsilon}\right) \leq -\frac{\varepsilon - 2\varepsilon^{3/2}\bar{l}(\varepsilon)}{2a^{-1}b\,(a+\sigma\sqrt{\varepsilon})}.$$

Putting $2b \leq 1$ and taking into consideration that $\bar{l}(\varepsilon) \to \bar{l}(0) < \infty$ as $\varepsilon \to 0$, we get the relation (3.2.8) for $\psi'_t = \psi_t = a^{-1}bt$.

By inequality (3.5.68) we have

$$\inf_{D_{t,\varepsilon}} h_t(\varepsilon) \geq \varepsilon n_t \left(b - \widetilde{\sigma}\sqrt{\varepsilon} - \frac{\varepsilon}{2}\left(c + \bar{\sigma}\sqrt{\varepsilon}\right)\right),$$

whence for $\psi_t = a^{-1}bt$ we get

$$\varliminf_{t\to\infty} \psi_t^{-1} \inf_{D_{t,\varepsilon}} h_t(\varepsilon) \geq \frac{\varepsilon\,(b + \widetilde{\sigma}\sqrt{\varepsilon} - (1/2)\varepsilon\,(c+\bar{\sigma}\sqrt{\varepsilon}\,))}{a^{-1}b\,(a+\sigma\sqrt{\varepsilon}\,)}.$$

This implies the relation (3.2.9) for $\psi'_t = \psi_t$.

Thus the following theorem has been proved.

THEOREM 3.5.9. *Let the random variables τ_1, δ_1, and ζ_1 satisfy the Cramér condition with respect to the measure* P *and $b \leq \frac{1}{2}$. Then for $\psi_t = a^{-1}bt$ the following relation holds*:

$$\varlimsup_{\varepsilon\downarrow 0}\,\varlimsup_{t\to\infty} (\varepsilon\psi_t)^{-1} \ln H(\varepsilon; \widetilde{\mathsf{P}}, \mathsf{P}) \leq -1.$$

Renewal processes. Theorems on weak convergence. We impose the following condition on a distribution function $F(t)$:

$$1 - F(t) = t^{-\alpha} h(t), \qquad t \to \infty, \tag{3.5.72}$$

where $0 < \alpha < 2$ and $h(t)$ is a slowly varying function in the Karamata sense, that is, for any positive constant c

$$\lim_{t\to\infty} h(ct)/h(t) = 1. \tag{3.5.73}$$

Denote by $G_\alpha(x)$ the distribution function of a stable law with parametr α whose characteristic function is of the form

$$g_\alpha(z) = \exp\left\{-|z|^\alpha \left(\cos\frac{\pi\alpha}{2} - i\,\sin\frac{\pi\alpha}{2}\,\mathrm{sgn}(z)\right)\Gamma(1-\alpha)\right\},$$

where $\Gamma(\cdot)$ is the Euler gamma-function.

It is well known [**36**] that in the case $0 < \alpha < 1$ the equality $G_\alpha(x) = 0$ holds for all $x \le 0$, whereas in the case $\alpha \ge 1$ the inequality $G_\alpha(x) > 0$ holds for all $x \in \mathbf{R}^1$.

In what follows we shall need the following well-known result ([**145**], Theorem 7).

LEMMA 3.5.3. *Let a distribution function $F(t)$ satisfy conditions* (3.5.72) *and* (3.5.73). *Then*

1) *if $0 < \alpha < 1$, then for any $x > 0$*

$$\lim_{t\to\infty} \mathsf{P}\{\xi_t \ge x/c_t\} = G_\alpha(x^{-1/\alpha}), \tag{3.5.74}$$

where

$$c_t = 1 - F(t); \tag{3.5.75}$$

2) *if $1 < \alpha < 2$, then for any $x \in \mathbf{R}^1$*

$$\lim_{t\to\infty} \mathsf{P}\left\{\xi_t \ge \frac{t}{a} - \frac{b_t}{a^{1+1/\alpha}}x\right\} = G_\alpha(x), \tag{3.5.76}$$

where $a = \mathsf{E}\,\tau_1$ and b_t is a constant depending on t and such that as $t \to \infty$

$$1 - F(b_t) = \frac{1}{t}(1 + o(1)). \tag{3.5.77}$$

The following theorem describes conditions for weak convergence of $\psi_t^{-1}\Lambda_{st}$ as $t \to \infty$ for $s \in (0,\infty)$ with normalizing factor ψ_t.

THEOREM 3.5.10. *Let the condition of Theorem* 3.5.7, *condition* 2) *of Theorem* 3.5.8, *and conditions* (3.5.72) *and* (3.5.73) *with $0 < \alpha < 1$ be fulfilled. Then for any $s \in (0,\infty)$ and $t \to \infty$ the following weak convergence takes place with respect to the measure* P:

$$\psi_t \Lambda_{st} \xrightarrow{w} -s^\alpha \gamma, \tag{3.5.78}$$

where $\psi_t = bt^\alpha/h(t)$, $b = \mathsf{E}\,\delta_1$, $h(t)$ is the slowly varying function from condition (3.5.72), *and γ is a positive random variable with distribution*

$$\mathsf{P}\{\gamma > x\} = G_\alpha(x^{-1/\alpha}), \qquad x > 0. \tag{3.5.79}$$

PROOF. We follow the proof of Theorem 3.5.8, indicating the corresponding changes. First the representation (3.5.55) is valid, where $\psi_t = bt^\alpha/h(t)$ and the estimates (3.5.57) hold. In view of condition 2) of Theorem 3.5.8 we have $b = \mathsf{E}\,\delta_k < \infty$. By condition (3.5.74), which follows from Lemma 3.5.3 and conditions (3.5.72) and (3.5.73), we get $\xi_t \to \infty$ (P-a.s.) as $t \to \infty$. Then in view of Lemma 3.5.2 we obtain the relation (3.5.59). By (3.5.74) from (3.5.58) and (3.5.59) we conclude that for any $s \in (0,\infty)$ and $t \to \infty$ we have

$$\psi_t^{-1} \sum_{k=1}^{\xi_{st-}} \delta_k \xrightarrow{w} s^\alpha \gamma \tag{3.5.80}$$

with respect to the measure P, where γ is a positive random variable with distribution (3.5.79). Similarly, for any $s \in (0, \infty)$ and $t \to \infty$

$$\psi_t^{-1} \sum_{k=1}^{\xi_{st-}+1} \delta_k \xrightarrow{w} s^\alpha \gamma \tag{3.5.81}$$

with respect to the measure P. Combining (3.5.57), (3.5.80), and (3.5.81), we get that for any $s \in (0, \infty)$ and $t \to \infty$

$$\psi_t^{-1} V_{st} \xrightarrow{w} s^\alpha \gamma \tag{3.5.82}$$

with respect to the measure P.

Following the argument in the proof of Theorem 3.5.8 used to prove relation (3.5.63), on the basis of Lemma 3.5.2 we obtain that for any $s \in (0, \infty)$

$$\lim_{t\to\infty} \xi_{st}^{-1} M_{st} = 0 \quad (\mathsf{P}\text{-a.s.}). \tag{3.5.83}$$

Furthermore, taking into account the equality

$$\psi_t^{-1} M_{st} = \frac{s^\alpha}{b} \frac{h(t)}{h(st)} \frac{h(st)\xi_{st}}{(st)^\alpha} \frac{M_{st}}{\xi_{st}}$$

and relations (3.5.24), (3.5.74), and (3.5.83) we see that for any $s \in (0, \infty)$

$$\mathsf{P}\text{-}\lim_{t\to\infty} \psi_t^{-1} M_{st} = 0. \tag{3.5.84}$$

Combining (3.5.55), (3.5.82), and (3.5.84), we now arrive at the desired relation (3.5.78). □

The following theorem is an analog of Theorem 3.5.10 in the case where for $1 < \alpha < 2$ conditions (3.5.72) and (3.5.73) are fulfilled.

THEOREM 3.5.11. *Let the conditions of Theorem* 3.5.7, *conditions* (3.5.72) *and* (3.5.73) *for* $1 < \alpha < 2$, *as well as the additional conditions*

$$\mathsf{E}\,\delta_1^2 < \infty, \qquad \mathsf{E} \ln^2 l(\tau_1) < \infty \tag{3.5.85}$$

hold. Then the following weak convergence with respect to the measure P *takes place:*

$$\psi_t^{-1}(\Lambda_t + \varphi_t) \xrightarrow{w} b\gamma_\alpha, \qquad t \to \infty, \tag{3.5.86}$$

where

$$\varphi_t = a^{-1} bt, \qquad a = \mathsf{E}\,\tau_1, \qquad b = \mathsf{E}\,\delta_1, \qquad \psi_t = a^{-1-1/\alpha} b_t,$$

b_t *is a constant satisfying* (3.5.77), *and* γ_α *is a random variable that has a stable distribution with exponent* α *and distribution function* $G_\alpha(x)$.

PROOF. By Theorem 3.5.7 we have

$$\psi_t^{-1}(\Lambda_t + \varphi_t) = \psi_t^{-1} M_t - \psi_t^{-1}(V_t - \varphi_t). \tag{3.5.87}$$

Evidently, the following representation is true:

$$\psi_t^{-1}(V_t - \varphi_t) = \psi_t^{-1}\left(\xi_{t-} - a_t^{-1} t\right) \xi_{t-}^{-1} V_t + a^{-1} t \xi_{t-}^{-1} \psi_t^{-1}\left(V_t - b\xi_{t-}\right). \tag{3.5.88}$$

Since in view of condition (3.5.85) $b = \mathsf{E}\,\delta_1 < \infty$, applying Lemma 3.5.2 we see that $\xi_{t-}^{-1} V_t \to b$ (P-a.s.) as $t \to \infty$. Now, taking into account the relation (3.5.76) from Lemma 3.5.3, as $t \to \infty$ we have

$$\psi_t^{-1}\left(\xi_{t-} - a^{-1}t\right)\xi_{t-}^{-1} V_t \xrightarrow{w} -b\gamma_\alpha \tag{3.5.89}$$

with respect to the measure P.

Using (3.5.57) we get the estimates

$$\psi_t^{-1}\sum_{k=1}^{\xi_{t-}}(\delta_k - b) \le \psi_t^{-1}(V_t - b\xi_{t-}) \le \psi_t^{-1}\sum_{k=1}^{\xi_{t-}+1}(\delta_k - b) + \psi_t^{-1} b. \tag{3.5.90}$$

Since $\mathsf{E}(\delta_k - b) = 0$ in view of condition (3.5.85) and Wald's identity [**127**], we have

$$\mathsf{E}\left(\psi_t^{-1}\sum_{k=1}^{\xi_{t-}}(\delta_k - b)\right)^2 = \psi_t^{-2}\,\mathsf{E}\,\xi_{t-}\;\mathsf{E}(\delta_1 - b)^2. \tag{3.5.91}$$

But by the renewal theorem, $t^{-1}\,\mathsf{E}\,\xi_{t-} \to a^{-1}$ as $t \to \infty$ [**10**], and hence

$$\psi_t^{-2}\,\mathsf{E}\,\xi_{t-} \to 0, \qquad t \to \infty.$$

Here we used the fact that in view of (3.5.77) $\psi_t = t^{1/\alpha}\tilde{h}(t)$, where $\tilde{h}(t)$ is a slowly varying function in the Karamata sense. Thus (3.5.91) implies the equality

$$\mathsf{P}\text{-}\lim_{t\to\infty} \psi_t^{-1}\sum_{k=1}^{\xi_{t-}}(\delta_k - b) = 0. \tag{3.5.92}$$

Now, combining (3.5.90) and (3.5.92), we get

$$\mathsf{P}\text{-}\lim_{t\to\infty} \psi_t^{-1}(V_t - b\xi_{t-}) = 0. \tag{3.5.93}$$

Finally, the relations (3.5.88), (3.5.89), and (3.5.93) imply that as $t \to \infty$,

$$\psi_t^{-1}(V_t - \varphi_t) \xrightarrow{w} -b\gamma_\alpha. \tag{3.5.94}$$

Obviously, by condition (3.5.85) $M \in \overline{\mathcal{M}}^2(\mathbf{F}, \mathsf{P})$ and

$$\langle M\rangle_t = \ln^2 l \circ \nu_t \le \sum_{k=1}^{\xi_{t-}+1}\int_0^{\tau_k} \ln^2 l(s)\, d\pi(s). \tag{3.5.95}$$

Using condition (3.5.85) and Lemmas 3.5.1 and 3.5.2, we have (P-a.s.)

$$\lim_{t\to\infty}\frac{1}{t}\sum_{k=1}^{\xi_{t-}+1}\int_0^{\tau_k} \ln^2 l(s)\, d\pi(s) = \frac{d}{a}. \tag{3.5.96}$$

The relations (3.5.95) and (3.5.96) imply

$$\mathsf{P}\text{-}\lim_{t\to\infty} \psi_t^{-1} M_t = 0. \tag{3.5.97}$$

Combining (3.5.87), (3.5.94), and (3.5.97), we get the desired relation (3.5.86). □

Renewal processes. The parametric case. Let $\tau_1, \tau_2, \dots$ be independent identically distributed random variables with the continuous distribution function $F(t;\theta) = \mathsf{P}_\theta\{\tau_1 < t\}$, $\theta \in \Theta \subset \mathbf{R}^k$. Putting $F(t;\theta) = F(t)$ and $F(t;y) = \widetilde{F}(t)$ for some $\theta, y \in \Theta$, we assume that the conditions of Theorem 3.5.7 are fulfilled, which ensure the local equivalence of the measures P_y and P_θ and give the form of the local density $z(y,\theta)$ of the measure P_y with respect to the measure P_θ (see also (3.5.13)–(3.5.15)). We introduce the notation

$$\rho(t;y,\theta) = \frac{dF(\,\cdot\,;y)}{dF(\,\cdot\,;\theta)}(t), \qquad l(t;y,\theta) = \frac{1-F(t;\theta)}{1-F(t;y)}\rho(t;y,\theta),$$
$$\pi(t;\theta) = \ln[1-F(t;\theta)]^{-1}.$$

Then the relations (3.5.47) and (3.5.48) imply

$$\lambda_t(y,\theta) = \frac{d\nu(y)}{d\nu(\theta)}(t) = l(t - T_{\xi_{t-}}; y, \theta).$$

Now we establish the validity of Theorem 3.5.4 in this case. Assume that $y = y_t$ depends on t, $y_t \to \theta$ as $t \to \infty$, and $\Delta_t = y_t - \theta$. We also assume that $\ln l(s;y,\theta) \in \mathbf{C}_1^0(\Theta)$ with respect to the variable $y \in \Theta$ for fixed s and θ. Now let $\nabla_y l(s;y,\theta) = L(s;y,\theta)$ and $\nabla_y \ln l(s;y,\theta) = \varkappa(s;y,\theta)$. Moreover, suppose that for all y in a neighborhood of the point θ and for all $t \in \mathbf{R}_+$ the integrals

$$\int_0^t |L(s;y,\theta)|^2\, d\pi(s;\theta), \qquad \int_0^t |\varkappa(s;y,\theta)|^2\, d\pi(s;\theta)$$

are finite. Then condition 1) of Theorem 3.5.4 is fulfilled.

The matrix $I_t(\theta) = l_0 l_0' \circ \nu(\theta)_t$ in condition 2) of Theorem 3.5.4 can be represented in the form

$$I_t(\theta) = \sum_{k=1}^{\xi_{t-}} I^{(k)}(\theta) + \tilde{I}_t(\theta), \tag{3.5.98}$$

where

$$I^{(k)}(\theta) = \int_0^{\tau_k} L(s;\theta,\theta)L(s;\theta,\theta)'\, d\pi(s;\theta), \qquad k = 1,2,\dots,$$
$$\tilde{I}_t(\theta) = \int_0^{t - T_{\xi_{t-}}} L(s;\theta,\theta)L(s;\theta,\theta)'\, d\pi(s;\theta).$$

Assuming that $\mathsf{E}_\theta |L(\tau_1;\theta,\theta)|^2 < \infty$ and using Lemmas 3.5.1 and 3.5.2, from (3.5.98) we get that $\lim_{t\to\infty} t^{-1} I_t(\theta) = a^{-1} I(\theta)$ (P_θ-a.s.), where $I(\theta) = \mathsf{E}_\theta I^{(1)}(\theta)$.

Therefore, in order that condition 2) of Theorem 3.5.4 be fulfilled, it is enough to require that the matrix $I(\theta)$ be nonsingular and to put $\varphi_t(\theta) = (a^{-1} t I(\theta))^{-1/2}$.

We have the following estimate:

$$\begin{aligned}
&\chi\big(|\varphi_t(\theta) l_0| > \varepsilon\big) |\varphi_t(\theta) l_0|^2 \circ \nu(\theta)_t \\
&\quad \le \sum_{k=1}^{\xi_{t-}+1} \int_0^{\tau_k} \chi(|\varphi_t(\theta) L(s;\theta,\theta)| > \varepsilon) |\varphi_t(\theta) L(s;\theta,\theta)|^2\, d\pi(s;\theta) \\
&\quad \le \frac{a}{t} |I^{-1}(\theta)| \sum_{k=1}^{\xi_{t-}+1} \int_0^{\tau_k} \chi\left(\left|I^{-1/2}(\theta) L(s;\theta,\theta)\right| > \varepsilon\sqrt{\frac{t}{a}}\right) |L(s;\theta,\theta)|^2\, d\pi(s;\theta).
\end{aligned}$$

Now we apply Wald's identity to obtain the estimate

$$\begin{aligned}
&\mathsf{E}_\theta\, \chi\big(|\varphi_t(\theta)l_0| > \varepsilon\big)|\varphi_t(\theta)l_0|^2 \circ \nu(\theta)_t \\
&\qquad \leq a|I^{-1}(\theta)|\frac{\mathsf{E}_\theta(\xi_{t-}+1)}{t}\, \mathsf{E}_\theta\, \chi\left(\left|I^{-1/2}(\theta)L(\tau_1;\theta,\theta)\right| > \varepsilon\sqrt{\frac{t}{a}}\right)|L(\tau_1;\theta,\theta)|^2.
\end{aligned}$$

Taking into account the renewal theorem and the condition $\mathsf{E}_\theta\,|L(\tau_1;\theta,\theta)|^2 < \infty$, we conclude that condition 3) of Theorem 3.5.4 holds.

Using the same estimates, Wald's identity, and the renewal theorem, we get condition 4) of Theorem 3.5.4 provided that

$$\lim_{t\to\infty} \mathsf{E}_\theta \int_0^1 |L(\tau_1;\theta+s\Delta_t,\theta) - L(\tau_1;\theta,\theta)|^2\,ds = 0,$$

$$\lim_{t\to\infty} \mathsf{E}_\theta \int_0^1 |\varkappa(\tau_1;\theta+s\Delta_t,\theta) - \varkappa(\tau_1;\theta,\theta)|^2\,ds = 0.$$

Thus we have established the conditions that ensure the validity of the assertion of Theorem 3.5.4. Similar arguments allow one to obtain the conditions that ensure the validity of Theorem 3.5.5.

CHAPTER 4

Asymptotic Estimation of Parameters

4.1. Formulation of the problem

Let $(X^t, \mathfrak{B}^t, (P^t_\theta, \theta \in \Theta))$, $t \in \mathbf{R}_+$, be the family of parametric statistical experiments generated by observations ξ^t of an arbitrary nature (see Section 1.2), and let $\Theta \subset \mathbf{R}^k$ be an open convex parameter set, $k \geq 1$.

We consider the problem of estimating the parameter θ from observation ξ^t of the distribution, about which we know only that it belongs to the family $(P^t_\theta, \theta \in \Theta)$.

DEFINITION 4.1.1. Any statistic $\theta_t = (\theta_t(x))_{x \in X^t}$ that is a measurable mapping of the space $(X^t, \mathfrak{B}^t)$ into the space $(\mathbf{R}^k, \mathcal{B}^k)$ is called a *statistical estimate* (or, briefly, an *estimate*) of the parameter θ.

In what follows the random variable $\theta_t(\xi^t)$ is denoted θ_t and called an estimate.

In this chapter we consider only the well-known Bayes and maximum likelihood estimates. The measures P^t_y and P^t_θ are assumed to be mutually absolutely continuous for all $\theta, y \in \Theta$ and $t \in \mathbf{R}_+$.

DEFINITION 4.1.2. By a *maximum likelihood estimate* of the parameter θ we mean a statistic $\hat{\theta}_t$ defined by the formula

$$\Lambda_t(x; \hat{\theta}_t(x), y) = \sup_{\theta \in \overline{\Theta}} \Lambda_t(x; \theta, y), \qquad x \in X^t, \tag{4.1.1}$$

where y is a fixed point in Θ, $\overline{\Theta}$ is the closure of the set Θ, and $\Lambda_t(x; \theta, y) = \ln dP^t_\theta / dP^t_y(x)$, $x \in X^t$.

If the equation (4.1.1) has no solution, then we put $\hat{\theta}_t(x) = \theta'$, where θ' is any fixed point in Θ. If (4.1.1) has several solutions, then, unless otherwise specified, we take $\hat{\theta}_t(x)$ to be any of them.

Note that in what follows the maximum likelihood method is used only when the set Θ is bounded and the function $\Lambda_t(x; \theta, y)$ is a continuous function of θ for fixed $x \in X^t$, $y \in \Theta$, and $t \in \mathbf{R}_+$, that is, when the equation (4.1.1) has a solution.

DEFINITION 4.1.3. A *Bayes estimate* with respect to a loss function $l_t = (l_t(y))_{y \in \mathbf{R}^k}$ and an a priori density $q = (q(y))_{y \in \Theta}$ is the statistic $\tilde{\theta}_t$ defined by the equation

$$\int_\Theta l_t\left(\tilde{\theta}_t(x) - u\right) p_t(u/x)\, du = \inf_{\theta \in \Theta} \int_\Theta l_t(\theta - u) p_t(u/x)\, du, \qquad x \in X^t, \tag{4.1.2}$$

where

$$p_t(u/x) = \left(\int_\Theta \rho_t(x; \theta, y) q(\theta)\, d\theta\right)^{-1} \rho_t(x; u, y) q(u)$$

is the a posteriori density, y is any fixed point in Θ, and $\rho_t(x; y, \theta)$ is the density of the measure P_y^t with respect to the measure P_θ^t.

If the equation (4.1.2) has several solutions, then $\tilde{\theta}_t(x)$ can be taken to be any of them.

Note that the maximum likelihood estimate $\hat{\theta}_t$ and the Bayes estimate $\tilde{\theta}_t$ are independent of the point $y \in \Theta$.

DEFINITION 4.1.4. A family of estimates $(\theta_t^*, t \in \mathbf{R}_+)$ is said to be (l_t)-*asymptotically efficient* in $K \subset \Theta$ (*asymptotically efficient with respect to the family of loss functions* $(l_t, t \in \mathbf{R}_+)$) if for any nonempty open set $U \subset K$ we have

$$\lim_{t\to\infty} \left(\inf_{\theta_t} \sup_{u\in U} \mathsf{E}_u\, l_t(\theta_t - u) - \sup_{u\in U} \mathsf{E}_u\, l_t(\theta_t^* - u) \right) = 0. \tag{4.1.3}$$

Note that if the estimate θ_t^* satisfies (4.1.3), then it is called an *asymptotically minimax estimate* in any nonempty open set $U \subset K$.

We introduce some classes of loss functions to be used below.

Denote by $\mathbf{W}$ the class of functions $L(u, v)$ defined on $\Theta \times \Theta$ and satisfying the conditions:

1) $L(u, v) = l(u - v)$;
2) the function $l(u)$ is defined on $\mathbf{R}^k$, is nonnegative, continuous at the point $u = 0$, $l(0) = 0$, and does not vanish identically;
3) the function $l(u)$ is even, that is, $l(-u) = l(u)$;
4) the set $\{u; l(u) < c\}$ is convex for all $c > 0$.

A function $l(u), u \in \mathbf{R}^k$ satisfying the above conditions 2)–4) is also called a *loss function*, and the set of such loss functions is also denoted by $\mathbf{W}$.

By $\mathbf{W}'$ we denote the class of functions in $\mathbf{W}$ that satisfy conditions 1)–4′), where

4′) the set $\{u; l(u) < c\}$ is convex for all $c > 0$ and bounded for all sufficiently small $c > 0$.

By $\mathbf{W}_p$ (respectively, $\mathbf{W}'_p$) we denote the class of those functions in $\mathbf{W}$ (respectively, $\mathbf{W}'$) that have a polynomial majorant.

By $\widetilde{\mathbf{W}}'_p$ we denote the class of functions $\mathbf{W}'_p$, for which there exists $\gamma > 0$ such that for some $A_0 > 0$ and all $A \geq A_0$ we have

$$\inf_{|u|\geq A} l(u) \geq \sup_{|u|\geq A^\gamma} l(u).$$

Note that the class $\mathbf{W}'_p$ consists of all loss functions of the form $|u - v|^r$, where $r > 0$ and, in particular, it contains the quadratic loss function $|u - v|^2$.

In what follows we use the notation

$$Z_{t,\theta}(u) = \rho_t(\xi^t; \theta + \varphi_t(\theta)u, \theta) = \frac{dP^t_{\theta+\varphi_t(\theta)u}}{dP^t_\theta}(\xi^t), \tag{4.1.4}$$

where $\varphi_t(\theta)$ is a $(k \times k)$-matrix that will be chosen below, and $u \in \mathbf{R}^k$ is such that $\theta + \varphi_t(\theta)u \in \Theta$.

DEFINITION 4.1.5. A family $(P_\theta^t, \theta \in \Theta)$ is said to be *locally asymptotically normal* (LAN) at a point $\theta \in \Theta$ as $t \to \infty$ if for some nonsingular $(k \times k)$-matrix $\varphi_t(\theta)$

and any $u \in \Theta_{t,\theta} = \varphi_t^{-1}(\theta)(\Theta - \theta)$ the following representation is valid:

$$Z_{t,\theta}(u) = \exp\left\{u'\eta_{t,\theta} - \frac{1}{2}|u|^2 + r_t(u,\theta)\right\}, \tag{4.1.5}$$

where

$$\mathcal{L}\left(\eta_{t,\theta}|P_\theta^t\right) \xrightarrow{w} \mathcal{N}(0, I_k), \qquad t \to \infty, \tag{4.1.6}$$

$$P_\theta^t\text{-}\lim_{t\to\infty} r_t(u,\theta) = 0. \tag{4.1.7}$$

The family $(P_\theta^t, \theta \in \Theta)$ for which the representation (4.1.5)–(4.1.7) holds, is sometimes said to satisfy the *local asymptotic normality condition* (LANC) at θ.

The measure density $\rho_t(x; y, \theta)$ is called a *likelihood ratio*, a likelihood ratio of the form (4.1.4) is called a *normalizing likelihood ratio*, whereas the matrix $\varphi_t(\theta)$ is called a *normalizing matrix*.

In this chapter the *uniform variant of the LANC* is of importance. Let us introduce the proper definition.

DEFINITION 4.1.6. A family $(P_\theta^t, \theta \in \Theta)$ is said to be *uniformly locally asymptotically normal* in $K \subset \Theta$ if for some nonsingular matrix $\varphi_t(\theta)$ for any sequences (θ_n), (t_n), and (u_n) such that $\theta_n \in K$ and $\theta_n + \varphi_{t_n}(\theta_n)u_n \in K$ for all n and $t_n \to \infty$, and $u_n \to u \in \mathbf{R}^k$ as $n \to \infty$, the following representation is valid:

$$Z_{t_n,\theta_n}(u_n) = \exp\left\{u_n'\eta_{t_n,\theta_n} - \frac{1}{2}|u_n|^2 + r_{t_n}(u_n,\theta_n)\right\}, \tag{4.1.8}$$

where

$$\mathcal{L}\left(\eta_{t_n,\theta_n}|P_{\theta_n}^{t_n}\right) \xrightarrow{w} \mathcal{N}(0, I_k), \qquad n \to \infty, \tag{4.1.9}$$

$$P_{\theta_n}^{t_n}\text{-}\lim_{t\to\infty} r_{t_n}(u_n,\theta_n) = 0. \tag{4.1.10}$$

In this case we say that the family $(P_\theta^t, \theta \in \Theta)$ *satisfies the LANC uniformly in* $K \subset \Theta$.

Now we formulate the conditions under which the properties of statistical estimates are studied.

K1. The family $(P_\theta^t, \theta \in \Theta)$ satisfies the LANC uniformly in any compact set $K \subset \Theta$ with normalizing matrix $\varphi_t(\theta)$.

K2. For any compact set $K \subset \Theta$,

$$\lim_{t\to\infty} \sup_{\theta \in K} |\varphi_t(\theta)| = 0, \tag{4.1.11}$$

and uniformly with respect to $\theta, y \in K$,

$$\lim_{t\to\infty} \varphi_t^{-1}(y)\varphi_t(\theta) = B(y,\theta), \tag{4.1.12}$$

where the matrix $B(y,\theta)$ is continuous with respect to the variable y.

K3. For any compact set $K \subset \Theta$ there exist constants $\beta > 0$, $m > 0$, $B = B(K) > 0$, and $a = a(K) \geq 0$ such that for any $R > 0$ and $t \geq t_0 = t_0(K)$

$$|u - v|^{-\beta}\, \mathsf{E}_\theta \left|Z_{t,\theta}^{1/m}(u) - Z_{t,\theta}^{1/m}(v)\right|^m \leq B(1 + R^a)$$

for all $\theta \in K$ and $u, v \in \Theta_{t,\theta}$ such that $|u| \leq R$ and $|v| \leq R$.

K4. For any compact set $K \subset \Theta$ and any $N > 0$ there are constants $t_0 = t_0(N, K)$ and $C = C(N, K)$ such that

$$|u|^N \mathsf{E}_\theta Z_{t,\theta}^{1/2}(u) \le C < \infty$$

for all $\theta \in K$, $u \in \Theta_{t,\theta}$, and $t > t_0$.

Conditions **K1**, (4.1.11), **K3**, and **K4** completely coincide with the conditions in [**40**], and condition (4.1.12) determines the class $\Phi(K)$ of normalizing matrices $\varphi_t(\theta)$ defined in [**40**].

The following two theorems describe asymptotic properties of statistical estimates $\hat{\theta}_t$ and $\tilde{\theta}_t$ in the scheme of general statistical experiments.

THEOREM 4.1.1. *Let conditions* **K1–K4** *be fulfilled with $\beta > k$ in condition* **K3**. *Then for any compact set $K \subset \Theta$ the maximum likelihood estimate $\hat{\theta}_t$ possesses the properties*:

1) *as $t \to \infty$,*

$$\mathcal{L}(\varphi_t^{-1}(\theta)(\hat{\theta}_t - \theta)|P_\theta^t) \xrightarrow{w} \mathcal{N}(0, I_k)$$

uniformly with respect to $\theta \in K$;

2) *for any function $l(x)\colon \mathbf{R}^k \to \mathbf{R}^1$ that has a polynomial majorant,*

$$\lim_{t\to\infty} \mathsf{E}_\theta\, l(\varphi_t^{-1}(\theta)(\hat{\theta}_t - \theta)) = \mathsf{E}\, l(\eta)$$

uniformly with respect to $\theta \in K$, where η is a k-dimensional random variable with $\mathcal{L}(\eta|\mathsf{P}) = \mathcal{N}(0, I_k)$;

3) *the family of estimates $(\hat{\theta}_t, \theta \in \mathbf{R}_+)$ is asymptotically efficient in K with respect to the family of loss functions $l_t(x) = l(\varphi_t^{-1}(\theta_0)x)$, $t \in \mathbf{R}_+$, where $l \in \mathbf{W}_p$ and θ_0 is any fixed point in Θ.*

PROOF. The proof can be found in [**40**] (Theorems 3.1.1 and 3.1.3).

THEOREM 4.1.2. *Let conditions* **K1–K4** *be fulfilled, and let $\tilde{\theta}_t$ be a Bayes estimate with respect to a continuous positive a priori density $q(\theta)$ and loss function $l_t(x) = l(\varphi_t^{-1}(\theta_0)x)$, $x \in \mathbf{R}^k$, $t \in \mathbf{R}_+$, where $l \in \mathbf{W}_p$ and θ_0 is an arbitrary point in Θ. Then the estimate $\tilde{\theta}_t$ possesses the properties* 1)–3) *in Theorem* 4.1.1 (*with $\tilde{\theta}_t$ instead of $\hat{\theta}_t$*).

PROOF. See the proof in [**40**] (Theorem 3.2.1).

Theorems 4.1.1 and 4.1.2 imply that for the estimates $\hat{\theta}_t$ and $\tilde{\theta}_t$ to possess the properties 1)–3) formulated in Theorem 4.1.1 it is sufficient that the family of probability measures $(P_\theta^t, \theta \in \Theta)$ satisfy conditions **K1–K4**. This chapter is dedicated, in fact, to the establishment of conditions **K1–K4** for statistical experiments generated by observations of semimartingales.

It should be observed that the fulfillment of conditions **K1–K4** enables one to construct asymptotically optimal Bayes tests for distinguishing between complex parametric hypotheses [**8, 44**].

4.2. Properties of the normalized likelihood ratio for semimartingales

Consider a parametric family of statistical experiments $(\mathbf{D}^t, \mathfrak{B}^t, (P_\theta^t, \theta \in \Theta))$, $t \in \mathbf{R}_+$, generated by observations ξ^t of a semimartingale ξ. On the stochastic basis $(\Omega, \mathcal{F}, \mathbf{F}, \mathsf{P}_\theta)$ the semimartingale ξ has a triplet of predictable characteristics

$(\alpha(\theta), \langle m(\theta)\rangle, \nu(\theta))$. With a fixed $\theta, y \in \Theta$ we associate two probability measures P_θ and P_y. Putting $\mathsf{P} = \mathsf{P}_\theta$ and $\widetilde{\mathsf{P}} = \mathsf{P}_y$, we assume that the conditions of Theorems 1.2.1 and 1.2.2 hold, which ensure the local equivalence of measures P_θ and P_y and define the form of the local density process $z(y,\theta)$ of the measure P_y with respect to the measure P_θ (see Sections 1.2 and 3.3). Using the representation (3.3.1)–(3.3.3) of the process $z(y,\theta)$, we get the expression for the normalized likelihood ratio $Z_{t,\theta}(u) = z_t(\theta + \varphi_t(\theta)u, \theta)$ which is used in this section.

Local asymptotic normality. Theorem 3.3.1 implies the following theorem.

THEOREM 4.2.1. *Let θ be a fixed point in Θ, and let all conditions of Theorem* 3.3.1 *be fulfilled with $y_t = \theta + \varphi_t(\theta)u$, where $\varphi_t(\theta)$ is the matrix in condition* 2) *of Theorem* 3.3.1, *u is independent of t, and $u \in \varphi_t^{-1}(\theta)(\Theta - \theta)$. Then the family $(P_\theta^t, \theta \in \Theta)$ possesses the LAN property at the point $\theta \in \Theta$ as $t \to \infty$ with normalizing matrix $\varphi_t(\theta)$.*

In the following theorem more severe but simpler constraints for the LANC to hold are formulated. This theorem follows from Theorem 3.3.2.

THEOREM 4.2.2. *Let θ be a fixed point in Θ and let the conditions of Theorem* 3.3.2 *be fulfilled with $y_t = \theta + \varphi_t(\theta)u$, where $\varphi_t(\theta)$ is the matrix in condition* 2) *of Theorem* 3.3.2, *u is independent of t, and $u \in \varphi_t^{-1}(\theta)(\Theta - \theta)$. Then the family $(P_\theta^t, \theta \in \Theta)$ possesses the LAN property at the point $\theta \in \Theta$ as $t \to \infty$ with normalizing matrix $\varphi_t(\theta)$.*

The following theorem is a uniform version of Theorem 4.2.2.

THEOREM 4.2.3. *For any compact set $K \subset \Theta$ let the conditions of Theorem* 3.3.2 *be fulfilled for all $\theta \in K$ with $y_t = \theta + \varphi_t(\theta)u$, where $\varphi_t(\theta)$ is the matrix in condition* 2) *of Theorem* 3.3.2, *conditions* 2)–4) *hold uniformly with respect to $\theta \in K$ and in addition, condition* 4) *holds uniformly with respect to $|u| \le c$ for any $c \in (0,\infty)$. Moreover, let the following additional condition be fulfilled:*

$$\lim_{N\to\infty} \sup_{\theta\in K,\, t\in\mathbf{R}_+} \mathsf{P}_\theta\left\{\operatorname{Tr}\varphi_t(\theta)\left[g_0 g_0' \circ \langle m(\theta)\rangle_t + l_0 l_l' * \nu(\theta)_t\right]\varphi_t(\theta) > N\right\} = 0.$$

Then the family $(P_\theta^t, \theta \in \Theta)$ possesses the LAN property uniformly in K as $t \to \infty$ for any compact set $K \subset \Theta$ with normalizing matrix $\varphi_t(\theta)$, that is, condition **K1** *is fulfilled.*

PROOF. It suffices to repeat the arguments of the proof of Theorem 3.3.2 using Theorem 1.3.5 instead of Theorem 1.3.3. □

Condition K3 with $\beta = m = 2$. In what follows we need the following lemma, which is of independent interest.

LEMMA 4.2.1. *Let the conditions of Theorem* 1.2.1 *be fulfilled, and let $z = (z_t)$ be the local density of the measure P_y with respect to the measure P_θ. Moreover, let $g = (g_t)$ be a nonnegative $\mathbf{F}$-predictable process and $k = (k_{t,x})$ a nonnegative $\mathcal{P}(\mathbf{F}) \times \mathcal{B}_0$-measurable function such that*

$$A = g \circ \langle m(\theta)\rangle + k * \nu(\theta) \in \mathcal{V}_{\mathrm{loc}}(\mathbf{F}, \mathsf{P}_\theta).$$

Then the following equality holds:

$$\mathsf{E}_\theta\, z_- \circ A_t = \mathsf{E}_y\, A_t, \qquad t \in \mathbf{R}_+. \tag{4.2.1}$$

PROOF. We introduce the stopping times

$$\tau_n = \inf\Big\{t\colon z_-^2\gamma^2(y,\theta)\circ\langle m(\theta)\rangle_t + z_-^2\chi\big(z_-|\lambda(y,\theta)-1|\le 1\big)(\lambda(y,\theta)-1)^2 * \nu(\theta)_t + z_-\chi\big(z_-|\lambda(y,\theta)-1|>1\big)|\lambda(y,\theta)-1| * \nu(\theta)_t \ge n\Big\}, \qquad n=1,2,\dots.$$

First, let $A_t \le c < \infty$ (P_θ-a.s.) for all $t\in\mathbf{R}_+$. Since $z\in\overline{\mathcal{M}}(\mathbf{F},\mathsf{P}_\theta)$, we have

$$\begin{aligned}\mathsf{E}_y A_t &= \lim_{n\to\infty}\mathsf{E}_y A_{t\wedge\tau_n} = \lim_{n\to\infty}\mathsf{E}_\theta z_t A_{t\wedge\tau_n}\\ &= \lim_{n\to\infty}\mathsf{E}_\theta A_{t\wedge\tau_n}\mathsf{E}_\theta\{z_t/\mathcal{F}_{t\wedge\tau_n}\}\\ &= \lim_{n\to\infty}\mathsf{E}_\theta A_{t\wedge\tau_n} z_{t\wedge\tau_n}.\end{aligned}\tag{4.2.2}$$

By the Itô formula (Theorem 1.2.1), for all $t\in\mathbf{R}_+$ we have

$$z_tA_t = A\cdot z_t + z_-\circ A_t.\tag{4.2.3}$$

Since $z_t = 1 + z_-\cdot M_t$ for all $t\in\mathbf{R}_+$, where

$$M = \gamma(y,\theta)\cdot m(\theta) + (\lambda(y,\theta)-1)*(\mu-\nu(\theta))\in\mathcal{M}_{\mathrm{loc}}(\mathbf{F},\mathsf{P}_\theta)$$

(see the proof of Theorem 1.2.1), from the definition of the stopping time τ_n and boundness of A_t we get

$$\mathsf{E}_\theta A\cdot z_{t\wedge\tau_n} = \mathsf{E}_\theta Az_-\cdot M_{t\wedge\tau_n} = 0, \qquad n=1,2,\dots.$$

From this and (4.2.3) it follows that

$$\mathsf{E}_\theta z_{t\wedge\tau_n}A_{t\wedge\tau_n} = \mathsf{E}_\theta z_-\circ A_{t\wedge\tau_n}, \qquad n=1,2,\dots.\tag{4.2.4}$$

The relations (4.2.2) and (4.2.4) imply the equation (4.2.1) in the case $A_t\le c<\infty$ (P_θ-a.s.). The general case is obtained by passing to the limit. □

THEOREM 4.2.4. *Let the set Θ be convex, and for every compact set $K\subset\Theta$ the following conditions hold:*

1) *$\gamma(y,\theta)\in\mathbf{C}_1^0(\Theta)$ and $\ln\lambda(y,\theta)\in\mathbf{C}_1^0(\Theta)$ for all $\theta\in K$ with respect to the variable y, all the other variables being fixed;*

2) *for some constants $a=a(K)\ge 0$ and $B=B(K)>0$ and all $N,t\in\mathbf{R}_+$ (here $n(y,\theta)=l(y,\theta)/\lambda^{1/2}(y,\theta)$)*

$$\sup_{\theta\in K}\sup_{\substack{u,v\in\Theta_{t,\theta}\\|u|,|v|\le N}}\mathsf{E}_{\theta+\varphi_t(\theta)u}\Big\{\big|\varphi_t(\theta)g(\theta+\varphi_t(\theta)v,\theta+\varphi_t(\theta)u)\big|^2\circ\langle m(\theta)\rangle_t + \big|\varphi_t(\theta)n(\theta+\varphi_t(\theta)v,\theta+\varphi_t(\theta)u)\big|^2 * \nu(\theta+\varphi_t(\theta)u)_t\Big\} \le B(1+N^a).$$

Then the family $(P_\theta^t,\theta\in\Theta)$ satisfies condition **K3** *for $m=\beta=2$.*

PROOF. Obviously,

$$\mathsf{E}_\theta \left| Z_{t,\theta}^{1/2}(u) - Z_{t,\theta}^{1/2}(v) \right|^2 = 2(1 - \mathsf{E}_y \zeta_t), \tag{4.2.5}$$

where

$$\zeta = (\zeta_s) = (z_s^{1/2}(x, y)), \qquad y = \theta + \varphi_t(\theta)u, \qquad x = \theta + \varphi_t(\theta)v.$$

Here $z(x, y)$ is the local density of the measure P_x with respect to the measure P_y in the stochastic basis $(\Omega, \mathcal{F}, \mathbf{F}, \mathsf{P}_y)$. Using the Itô formula and making some simple calculations, we get (P_y-a.s.)

$$\begin{aligned} \zeta = 1 &+ (1/2)\zeta_- \gamma(x, y) \cdot m(y) - (1/8)\zeta_- \gamma^2(x, y) \circ \langle m(y) \rangle \\ &+ \zeta_-(\lambda^{1/2}(x, y) - 1) * (\mu - \nu(y)) \\ &- (1/2)\zeta_-(\lambda^{1/2}(x, y) - 1)^2 * \nu(y), \end{aligned} \tag{4.2.6}$$

where in view of condition 2)

$$r = (1/2)\zeta_- \gamma(x, y) \cdot m(y) + \zeta_-(\lambda^{1/2}(x, y) - 1) * (\mu - \nu(y)) \in \overline{\mathcal{M}}^2(\mathbf{F}, \mathsf{P}_y).$$

Actually, using first Lemma 4.2.1 and then condition 2), for any $t \in \mathbf{R}_+$ we get

$$\begin{aligned} \mathsf{E}_y \langle r \rangle_t &= \mathsf{E}_y \left[(1/4)\zeta_-^2 \gamma^2(x, y) \circ \langle m(y) \rangle_t + \zeta_-^2 (\lambda^{1/2}(x, y) - 1)^2 * \nu(y)_t \right] \\ &= \mathsf{E}_x \left[(1/4)\gamma^2(x, y) \circ \langle m(y) \rangle_t + (\lambda^{1/2}(x, y) - 1)^2 * \nu(y)_t \right] \\ &= \mathsf{E}_x \left[(1/4)\gamma^2(y, x) \circ \langle m(x) \rangle_t + (\lambda^{1/2}(y, x) - 1)^2 * \nu(x)_t \right] \\ &= \mathsf{E}_x \left[(1/4)\gamma^2(y, x) \circ \langle m(\theta) \rangle_t + (\lambda^{1/2}(y, x) - 1)^2 * \nu(x)_t \right] < \infty, \end{aligned}$$

whence it follows that $r \in \overline{\mathcal{M}}^2(\mathbf{F}, \mathsf{P}_y)$. Here we take into account the equality $\zeta^2 = z(x, y)$ and evident equalities

$$\mathsf{E}_x(\lambda^{1/2}(x, y) - 1)^2 * \nu(y)_t = \mathsf{E}_x(\lambda^{1/2}(y, x) - 1)^2 * \nu(x)_t, \tag{4.2.7}$$

$$\mathsf{E}_x \gamma^2(x, y) \circ \langle m(y) \rangle_t = \mathsf{E}_x \gamma^2(y, x) \circ \langle m(x) \rangle_t = \mathsf{E}_x \gamma^2(y, x) \circ \langle m(\theta) \rangle_t. \tag{4.2.8}$$

Thus equality (4.2.6) implies

$$\mathsf{E}_y \zeta_t = 1 - (1/8)\,\mathsf{E}_y \zeta_- \gamma^2(x, y) \circ \langle m(y) \rangle_t - (1/2)\,\mathsf{E}_y \zeta_-(\lambda^{1/2}(x, y) - 1)^2 * \nu(y)_t. \tag{4.2.9}$$

Using the Cauchy–Schwarz–Bunyakovskii inequality and Lemma 4.2.1 and taking into account the equality $\zeta^2 = z(x, y)$, we get

$$\mathsf{E}_y \zeta_- \gamma^2(x, y) \circ \langle m(y) \rangle_t \leq (\mathsf{E}_x \gamma^2(x, y) \circ \langle m(y) \rangle_t \, \mathsf{E}_y \gamma^2(x, y) \circ \langle m(y) \rangle_t)^{1/2} \tag{4.2.10}$$

and

$$\begin{aligned} &\mathsf{E}_y \zeta_-(\lambda^{1/2}(x, y) - 1)^2 * \nu(y)_t \\ &\qquad \leq (\mathsf{E}_x(\lambda^{1/2}(x, y) - 1)^2 * \nu(y)_t \, \mathsf{E}_y(\lambda^{1/2}(x, y) - 1)^2 * \nu(y)_t)^{1/2}. \end{aligned} \tag{4.2.11}$$

Now we use condition 1), relation (3.3.4), the inequality $|a'\varphi_t(\theta)b| \le |a| \cdot |\varphi_t(\theta)b|$ for any vectors $a \in \mathbf{R}^k$ and $b \in \mathbf{R}^k$ and the Fubini theorem to obtain

(4.2.12)

$$\mathsf{E}_y\,\gamma^2(x,y)\circ\langle m(y)\rangle_t = \mathsf{E}_y\left[(\varphi_t(\theta)(v-u))'\int_0^1 g(x_{s,y},y)\,ds\right]^2 \circ\langle m(y)\rangle_t$$
$$\le |v-u|^2\int_0^1 \mathsf{E}_y\big|\varphi_t(\theta)g(x_{s,y},y)\big|^2\circ\langle m(y)\rangle_t\,ds,$$

(4.2.13)

$$\mathsf{E}_y(\lambda^{1/2}(x,y)-1)^2 * \nu(y)_t = \frac{1}{4}\mathsf{E}_y\left[(\varphi_t(\theta)(v-u))'\int_0^1 n(x_{s,y},y)\,ds\right]^2 * \nu(y)_t$$
$$\le \frac{1}{4}|v-u|^2\int_0^1 \mathsf{E}_y\big|\varphi_t(\theta)n(x_{s,y},y)\big|^2 * \nu(y)_t\,ds,$$

where $x_{s,y} = y + s(x-y) = \theta + \varphi_t(\theta)[sv + (1-s)u]$. Combining equalities (4.2.5), (4.2.9), estimates (4.2.10)–(4.2.13), equalities (4.2.7), (4.2.8), condition 2), and the convexity of the set Θ, we obtain the desired inequality in condition **K3**. □

REMARK 4.2.1. Theorem 3.1.7 implies that the process

$$h(x,y) = (1/8)\gamma^2(x,y)\circ\langle m(y)\rangle + (1/2)(\lambda^{1/2}(x,y)-1)^2 * \nu(y) \tag{4.2.14}$$

is a Hellinger process of order $\frac{1}{2}$ for the measures P_x and P_y. Therefore the equality (4.2.6) takes the form

$$\zeta = 1 - \zeta_-\circ h(x,y) + (1/2)\zeta_-\gamma(x,y)\cdot m(y) + \zeta_-(\lambda^{1/2}(x,y)-1) * (\mu - \nu(y)).$$

From this it follows that the conditions of Theorem 4.2.4 can be formulated in terms of the Hellinger process $h(x,y)$ as follows:

1) for all $\theta \in K$ we have $h(y,\theta) \in \mathbf{C}_1^0(\Theta)$ with respect to the variable y, all the other variables being fixed;
2) for some constants $a = a(K) \ge 0$ and $B = B(K) > 0$ and all $t, N \in \mathbf{R}_+$

$$\sup_{\theta\in K}\ \sup_{\substack{u,v\in\Theta_{t,\theta}\\ |u|,|v|\le N}} \mathsf{E}_{\theta+\varphi_t(\theta)u}\Big|\varphi_t(\theta)\tilde{h}(\theta+\varphi_t(\theta)v,\theta+\varphi_t(\theta)u)\Big|^2 \le B(1+N^a),$$

where $\tilde{h}(y,\theta) = \nabla_y h(y,\theta)$ is the gradient with respect to the variable y.

Estimates for the Hellinger integral. Let us establish the validity of condition **K4**, which, as is easily seen, consists in finding an appropriate estimate for the Hellinger integral $H_t(\frac{1}{2};\mathsf{P}_y,\mathsf{P}_\theta)$, where $y = \theta + \varphi_t(\theta)u$.

THEOREM 4.2.5. *Let the set Θ be bounded, and for any compact set $K \subset \Theta$ the following condition hold:*

1) *for some positive constants b_1, b_2, b_3 and a function χ_t such that $\chi_t \to \infty$ as $t \to \infty$ there exist sets $D_t(y,\theta) \in \mathcal{F}_t$, $y, \theta \in \Theta$, $t \in \mathbf{R}_+$, such that*

$$\inf\{|u|^{-2}\bar{h}_t(\theta + \varphi_t(\theta)u, \theta); \theta \in K, u \in \Theta_{t,\theta}, |u| \le \chi_t\} \ge c_1 > 0, \tag{4.2.15}$$

$$\inf\left\{\left|\varphi_t^{-1}(\theta)\right|^{-b_1} \bar{h}_t(\theta + \varphi_t(\theta)u, \theta); \theta \in K, u \in \Theta_{t,\theta}, |u| > \chi_t\right\} \ge c_2 > 0, \tag{4.2.16}$$

$$\inf\{|u|^{-b_2} \ln \mathsf{P}_\theta^{-1}(D_t^c(\theta + \varphi_t(\theta)u, \theta)); \theta \in K, u \in \Theta_{t,\theta}, |u| \le \chi_t\} \ge c_3 > 0, \tag{4.2.17}$$

$$\inf\{|u|^{-b_3} \ln \mathsf{P}_\theta^{-1}(D_t^c(\theta + \varphi_t(\theta)u, \theta)); \theta \in K, u \in \Theta_{t,\theta}, |u| > \chi_t\} \ge c_4 > 0, \tag{4.2.18}$$

where $D_t^c(y,\theta) = \Omega \setminus D_t(y,\theta)$, the constants c_i depend on K,

$$\bar{h}_t(y,\theta) = \inf\{h_t(y,\theta) : \omega \in D_t(y,\theta)\}, \tag{4.2.19}$$

and $h(x,y)$ is the Hellinger process of order $\frac{1}{2}$ defined by (4.2.14).

Then the family $(P_\theta^t, \theta \in \Theta)$ satisfies condition **K4**.

PROOF. Putting $y = \theta + \varphi_t(\theta)u$ and applying Theorem 3.1.3 with $\varepsilon = \frac{1}{2}$ for measures $\mathsf{P} = \mathsf{P}_\theta$ and $\widetilde{\mathsf{P}} = \mathsf{P}_y$, we get the following multiplicative decomposition:

$$Z_{t,\theta}^{1/2}(u) = N_{t,\theta}(u)e^{-h_t(y,\theta)}, \tag{4.2.20}$$

where $\mathsf{E}_\theta N_{t,\theta}(u) \le 1$ and $h(y,\theta)$ is the Hellinger process of order $\frac{1}{2}$ for the measures P_y and P_θ.

Taking into account the equality $\mathsf{E}_\theta Z_{t,\theta}(u) = 1$ and using the Cauchy–Schwarz–Bunyakovskii inequality, we get

$$\begin{aligned} \mathsf{E}_\theta Z_{t,\theta}^{1/2}(u) &= \mathsf{E}_\theta \chi(D_t(y,\theta))Z_{t,\theta}^{1/2}(u) + \mathsf{E}_\theta \chi(D_t^c(y,\theta))Z_{t,\theta}^{1/2}(u) \\ &\le \mathsf{E}_\theta \chi(D_t(y,\theta))Z_{t,\theta}^{1/2}(u) + \left(\mathsf{P}_\theta(D_t^c(y,\theta))\right)^{1/2}. \end{aligned} \tag{4.2.21}$$

Since $\mathsf{E}_\theta N_{t,\theta}(u) \le 1$, by (4.2.20) and (4.2.19) we have

$$\begin{aligned} \mathsf{E}_\theta \chi(D_t(y,\theta))Z_{t,\theta}^{1/2}(u) &= \mathsf{E}_\theta \chi(D_t(y,\theta))N_{t,\theta}(u)e^{-h_t(y,\theta)} \\ &\le e^{-\bar{h}_t(y,\theta)} \mathsf{E}_\theta \chi(D_t(y,\theta))N_{t,\theta}(u) \\ &\le e^{-\bar{h}_t(y,\theta)}. \end{aligned} \tag{4.2.22}$$

Combining (4.2.21) and (4.2.22), we get the estimate

$$\mathsf{E}_\theta Z_{t,\theta}^{1/2}(u) \le e^{-\bar{h}_t(y,\theta)} + \left(\mathsf{P}_\theta(D_t^c(y,\theta))\right)^{1/2}. \tag{4.2.23}$$

The boundedness of the set Θ implies that $|u| \le c|\varphi_t^{-1}(\theta)|$, where c is a constant. Hence in view of condition (4.2.16) for $\theta \in K$, $u \in \Theta_{t,\theta}$, $|u| > \chi_t$ we get

$$\bar{h}_t(y,\theta) \ge c_2 \left|\varphi_t^{-1}(\theta)\right|^{b_1} \ge c_2 c^{-1}|u|^{b_1}.$$

From this and condition (4.2.15) we obtain

$$\exp\left(-\bar{h}_t(y,\theta)\right) \le \exp\left(-C(|u|^2 \wedge |u|^{b_1})\right), \tag{4.2.24}$$

where C is a positive constant.

Taking into consideration the elementary inequality

$$\ln(x+y) \le \ln 2 + (\ln x) \vee (\ln y), \qquad x>0,\ y>0,$$

from conditions (4.2.17) and (4.2.18) and estimates (4.2.23) and (4.2.24) we easily derive condition **K4**. □

Condition K3 with arbitrary β and m. Theorem 4.1.1 implies that the assertion of Theorem 4.2.4 is not sufficient for validity of properties 1)–3) for maximum likelihood estimates $\hat\theta_t$ for the multidimensional parameter θ ($k>1$). This being the case, we shall prove a theorem that provides sufficient conditions for the validity of **K3** with $\beta > k$ and makes it possible to apply Theorem 4.1.1 in the case $k>1$. To prove this theorem, we need some auxiliary general results, which are formulated in the notation of Section 1.1. The first two lemmas provide estimates for moments of stochastic integrals.

LEMMA 4.2.2. *If $\zeta = (\zeta^1, \zeta^2, \dots, \zeta^k)' \in \mathcal{M}^c_{\mathrm{loc}}(\mathbf{F}, \mathsf{P})$, then for any $p>0$ there exists a constant C_p, dependent only on p and such that for all $t \in \mathbf{R}_+$*

$$\mathsf{E}\sup_{s\le t} |\zeta_s|^p \le C_p\, \mathsf{E}\big(\mathrm{Tr}\,\langle \zeta, \zeta\rangle_t\big)^{p/2}. \tag{4.2.25}$$

In particular, if $f = (f^1, f^2, \dots, f^k)' \in \mathcal{L}^2_{\mathrm{loc}}(m, \mathbf{F}, \mathsf{P})$, where $m \in \mathcal{M}^c_{\mathrm{loc}}(\mathbf{F}, \mathsf{P})$, then

$$\mathsf{E}\sup_{s\le t} |f\cdot m_s|^p \le C_p\, \mathsf{E}\left(|f|^2 \circ \langle m\rangle_t\right)^{p/2}. \tag{4.2.26}$$

PROOF. For $\zeta \in \mathcal{M}^c(\mathbf{F}, \mathsf{P})$ and $k=1$ the inequality (4.2.25) was proved in [**20**]. For $\zeta \in \mathcal{M}^c_{\mathrm{loc}}(\mathbf{F}, \mathsf{P})$ and $k=1$, (4.2.25) can be obtained using localization, from where the general case is readily available. The inequality (4.2.26) follows immediately from (4.2.25). □

LEMMA 4.2.3. *Let μ be an $\mathbf{F}$-optional integer-valued random measure with compensator ν, continuous in t, and let $f = (f^1, f^2, \dots, f^k) \in \mathcal{G}^2_{\mathrm{loc}}(\nu, \mathbf{F}, \mathsf{P})$. Then for $p \ge 2$ and any $t \in \mathbf{R}_+$*

$$\mathsf{E}\sup_{s\le t} |f*(\mu-\nu)_s|^p \le c_p\, \mathsf{E}\left(|f|^2 * \nu_t\right)^{p/2} + C_p\, \mathsf{E}\,|f|^p * \nu_t, \tag{4.2.27}$$

where the constants c_p and C_p depend only on p.

PROOF. If $\nu(dt, dx) = dt\Pi(dx)$, the estimate (4.2.27) was established in [**104**]. The general case can be proved by repeating the arguments in [**104**]. □

The following two lemmas deal with the change of the order of stochastic and ordinary integration, similarly to the Fubini theorem.

LEMMA 4.2.4. *Let $(U, \mathfrak{A}, \pi)$ be a probability space, $m \in \mathcal{M}^2_{\mathrm{loc}}(\mathbf{F}, \mathsf{P})$, and $f = (f(\omega, t, u))$ a real function defined on $\Omega \times \mathbf{R}_+ \times U$ and satisfying the conditions:*

1) *the function f is $\mathcal{P}(\mathbf{F}) \times \mathfrak{A}$-measurable and $f(u) \in \mathcal{L}^2_{\mathrm{loc}}(m, \mathbf{F}, \mathsf{P})$ for every $u \in U$, where $f(u) = (f(\omega, t, u))_{(\omega,t)\in\Omega\times\mathbf{R}_+}$;*
2) *$\int_U f^2(u)\,\pi(du) \circ \langle m\rangle_t < \infty$ (P-a.s.) for every $t \in \mathbf{R}_+$.*

Then for any $t \in \mathbf{R}_+$ *the integrals*

$$\left(\int_U f(u)\,\pi(du)\right)\cdot m_t, \qquad \int_U (f(u)\cdot m_t)\,\pi(du)$$

are well defined and the following equality holds (P-*a.s.*):

$$\left(\int_U f(u)\,\pi(du)\right)\cdot m_t = \int_U (f(u)\cdot m_t)\,\pi(du). \tag{4.2.28}$$

PROOF. Assuming that

$$\mathsf{E}\int_U f^2(u)\,\pi(du)\circ\langle m\rangle_t < \infty, \qquad t\in\mathbf{R}_+, \tag{4.2.29}$$

equality (4.2.28) can be established for $m \in \mathcal{M}^2(\mathbf{F},\mathsf{P})$ and continuous characteristic $\langle m\rangle$ exactly as in [**161**]. It should be observed that the Doléans theorem [**140**] (which is necessary to prove (4.2.28)) on the existence of a version of the stochastic integral $f(u)\cdot m_t$ measurable in u, remains true for $m\in\mathcal{M}^2_{\rm loc}(\mathbf{F},\mathsf{P})$ with an arbitrary characteristic ([**192**], Assertion 5). If condition (4.2.29) does not hold, then (4.2.28) can be established by repeating the arguments in the proof of Lemma 2 in [**33**]. □

LEMMA 4.2.5. *Let* $(U,\mathfrak{A},\pi)$ *be a probability space,* μ *an* $\mathbf{F}$*-optional integer-valued random measure with compensator* ν*, continuous in* t*, and* $f=(f(\omega,t,x,u))$ *a real function defined on* $(\Omega\times\mathbf{R}_+\times R_0^m\times U)$ *and satisfying the conditions*:

1) *the function* f *is* $\widetilde{\mathcal{P}}(\mathbf{F})\times\mathfrak{A}$*-measurable and*

$$f(u) = (f(\omega,t,x,u))_{(\omega,t,x)\in\Omega\times\mathbf{R}_+\times R_0^m} \in \mathcal{G}^2_{\rm loc}(\nu,\mathbf{F},\mathsf{P})$$

for every $u\in U$;

2) $\int_U f^2(u)\,\pi(du)*\nu_t<\infty$ (P-*a.s.*) *for every* $t\in\mathbf{R}_+$.

Then for every $t\in\mathbf{R}_+$ *the integrals*

$$\left(\int_U f(u)\,\pi(du)\right)*(\mu-\nu)_t, \qquad \int_U \big(f(u)*(\mu-\nu)_t\big)\,\pi(du)$$

are well defined and the following equality holds (P-*a.s.*):

$$\left(\int_U f(u)\,\pi(du)\right)*(\mu-\nu)_t = \int_U \big(f(u)*(\mu-\nu)_t\big)\,\pi(du).$$

PROOF. The lemma can be proved exactly as Lemma 4.2.4. □

THEOREM 4.2.6. *For any compact set* $K\subset\Theta$ *let the following conditions hold*:

1) *condition* 1) *of Theorem* 3.3.2 *for all* $\theta\in K$;

2) *the functions* $\gamma(y,\theta)$ *and* $\lambda(y,\theta)$ *are* $\mathcal{P}(\mathbf{F})\times\mathcal{B}(\Theta)$*- and* $\widetilde{\mathcal{P}}(\mathbf{F})\times\mathcal{B}(\Theta)$*-measurable with respect to the variables* (ω,t,y) *and* (ω,t,x,y)*, respectively, for all* $\theta\in K$;

3) *for any* $\theta \in K$ *and* $t \in \mathbf{R}_+$ *there exist numbers* $r \geq l > k$, $r \geq 2$ *and a function* $H(y)\colon \mathbf{R}^k \to \mathbf{R}_+$, *bounded on every compact set, dependent on* θ *and* t *and such that for all* $y, y+h \in \Theta$

$$\begin{aligned}\mathsf{E}_\theta\Big\{\big(|g(y,\theta)|^2 \circ \langle m(\theta)\rangle_t\big)^{r/2} + \big(|\gamma(y,\theta)g(y,\theta)| \circ \langle m(\theta)\rangle_t\big)^r \\ + \big(|l(y,\theta)| * \nu(\theta)_t\big)^r + \big(|f(y,\theta)| * \nu(\theta)_t\big)^r \\ + \big(|f(y,\theta)|^2 * \nu(\theta)_t\big)^{r/2} + |f(y,\theta)|^r * \nu(\theta)_t\Big\} \\ \leq H(y),\end{aligned} \tag{4.2.30}$$

and

(4.2.31)

$$\begin{aligned}&\mathsf{E}_\theta\Big\{\Big(\big|g(y+h,\theta) - g(y,\theta)\big|^2 \circ \langle m(\theta)\rangle_t\Big)^{r/2} + \big(\big|l(y+h,\theta) - l(y,\theta)\big| * \nu(\theta)_t\big)^r \\ &\quad + \big(\big|\gamma(y+h,\theta)g(y+h,\theta) - \gamma(y,\theta)g(y,\theta)\big| \circ \langle m(\theta)\rangle_t\big)^r \\ &\quad + \big(\big|f(y+h,\theta) - f(y,\theta)\big| * \nu(\theta)_t\big)^r \\ &\quad + \Big(\big|f(y+h,\theta) - f(y,\theta)\big|^2 * \nu(\theta)_t\Big)^{r/2} + |f(y+h,\theta) - f(y,\theta)|^r * \nu(\theta)_t\Big\} \\ &\quad \leq H(y)|h|^l;\end{aligned}$$

4) *for some* $a = a(K) \geq 0$, $B = B(K) > 0$, *and* $n \geq 2$

$$\begin{aligned}\sup_{\theta\in K}\ \sup_{u\in\Theta_{t,\theta},|u|\leq N} \mathsf{E}_{\theta+\varphi_t(\theta)u}\Big[\Big(\big|\varphi_t(\theta)g\big(\theta+\varphi_t(\theta)u,\theta\big)\big|^2 \circ \langle m(\theta)\rangle_t\Big)^{n/2} \\ + \Big(\big|\varphi_t(\theta)n\big(\theta+\varphi_t(\theta)u,\theta\big)\big|^2 * \nu(\theta)_t\Big)^{n/2} \\ + \big|\varphi_t(\theta)f\big(\theta+\varphi_t(\theta)u,\theta\big)\big|^n * \nu(\theta+\varphi_t(\theta)u)_t\Big] \\ \leq B(1+N^a).\end{aligned}$$

Then the family $(P_\theta^t, \theta \in \Theta)$ *satisfies condition* **K3** *for* $m = \beta = n$.

PROOF. Obviously, the following equality is valid (P_θ-a.s.):

$$\rho_t^{1/n}(y,\theta) - \rho_t^{1/n}(z,\theta) = (y-z)'\int_0^1 \nabla_y \rho_t^{1/n}(y_{s,z},\theta)\,ds, \tag{4.2.32}$$

where $y_{s,z} = z + s(y-z)$ if $\rho_t^{1/n}(y,\theta) \in \mathbf{C}_1^0(\Theta)$ (P_θ-a.s.) with respect to y.

Under this condition we have

(4.2.33)

$$\begin{aligned}\nabla_y \rho_t^{1/n}(y,\theta) = \frac{1}{n}\rho_t^{1/n}(y,\theta)\Big\{\nabla_y\big(\gamma(y,\theta)\cdot m(\theta)_t\big) - \frac{1}{2}\nabla_y\big(\gamma^2(y,\theta)\circ\langle m(\theta)\rangle_t\big) \\ + \nabla_y\big(\ln\lambda(y,\theta) * (\mu - \nu(\theta))_t\big) \\ - \nabla_y\big[\big(\lambda(y,\theta) - 1 - \ln\lambda(y,\theta)\big) * \nu(\theta)_t\big]\Big\}.\end{aligned}$$

This implies that for $\nabla_y \rho_t^{1/n}(y,\theta)$ (P_θ-a.s.) to exist it is sufficient that the derivatives in braces on the right-hand side of (4.2.33) exist (P_θ-a.s.). Let us show that these derivatives do exist and (P_θ-a.s.)

$$
(4.2.34)\qquad \begin{aligned} \nabla_y\big(\gamma(y,\theta)\cdot m(\theta)_t\big) &= g(y,\theta)\cdot m(\theta)_t,\\ \nabla_y\big(\gamma^2(y,\theta)\circ\langle m(\theta)\rangle_t\big) &= 2\gamma(y,\theta)g(y,\theta)\circ\langle m(\theta)\rangle_t,\end{aligned}
$$

$$
(4.2.35)\qquad \nabla_y\big(\ln\lambda(y,\theta)*(\mu-\nu(\theta))_t\big) = f(y,\theta)*(\mu-\nu(\theta))_t,
$$

$$
(4.2.36)\qquad \nabla_y\left[\big(\lambda(y,\theta)-\ln\lambda(y,\theta)\big)*\nu(\theta)_t\right] = \big(l(y,\theta)-f(y,\theta)\big)*\nu(\theta)_t.
$$

Note that the function $\mathfrak{z}(y) = g(y,\theta)\cdot m(\theta)_t$ is continuous in $y\in\Theta$ (P_θ-a.s.). Indeed, in view of Lemma 4.2.2 and condition 3) for $y, y+h\in\Theta$ we have

$$
\mathsf{E}_\theta\,|\mathfrak{z}(y)|^r \le C_r\,\mathsf{E}_\theta\big(|g(y,\theta)|^2\circ\langle m(\theta)\rangle_t\big)^{r/2} \le C_r H(y),
$$

$$
\mathsf{E}_\theta\,|\mathfrak{z}(y+h)-\mathfrak{z}(y)|^r \le C_r\,\mathsf{E}_\theta\big(|g(y+h,\theta)-g(y,\theta)|^2\circ\langle m(\theta)\rangle_t\big)^{r/2} \le C_r H(y)|h|^l,
$$

where the constant C_r depends only on r. From this and Theorem II.1.19 in [**40**] the continuity (P_θ-a.s.) of $\mathfrak{z}(y)$ follows. Due to the last estimate the random function $\mathfrak{z}(y)$ is stochastically continuous and by Theorem 4.3.1 in [**18**] there exists a measurable separable modification of the function $\mathfrak{z}(y)$, which justifies the applicability of Theorem II.1.19. Using now Lemma 4.2.4, for every $y', y''\in\Theta$ such that $y' = (y_1, y_2,\dots,y_{i-1},a,y_{i+1},\dots,y_k)$, $y'' = (y_1,y_2,\dots,y_{i-1},b,y_{i+1},\dots,y_k)$, and $a<b$ we have

$$
\int_a^b (g^i(y,\theta)\cdot m(\theta)_t)\,dy_i = \left(\int_a^b g^i(y,\theta)\,dy_i\right)\cdot m(\theta)_t = \mathfrak{z}(y'')-\mathfrak{z}(y'),
$$

where $g^i(y,\theta)$ is the ith coordinate of the vector $g(y,\theta)$ (the condition

$$
\int_a^b |g^i(y,\theta)|^2\circ\langle m(\theta)\rangle_t\,dy_i < \infty \quad (\mathsf{P}_\theta\text{-a.s.})
$$

follows from (4.2.30)). Since $\mathfrak{z}(y)$ is continuous (P_θ-a.s.), the derivative $\nabla_y(\gamma(y,\theta)\cdot m(\theta)_t)$ exists (P_θ-a.s.) and the first equality in (4.2.34) is valid. The existence (P_θ-a.s.) of the derivative $\nabla_y(\ln\lambda(y,\theta)*(\mu-\nu(\theta))_t)$ and equality (4.2.35) can be proved similarly using Lemmas 4.2.3 and 4.2.5 instead of Lemmas 4.2.2 and 4.2.4, respectively. Furthermore, the second equality in (4.2.34), equality (4.2.36), and the existence (P_θ-a.s.) of the derivatives $\nabla_y(\gamma^2(y,\theta)\circ\langle m(\theta)\rangle_t)$ and $\nabla_y[(\lambda(y,\theta)-1-\ln\lambda(y,\theta))*\nu(\theta)_t]$ can also be proved in the same way; the proofs do not require Lemmas 4.2.2–4.2.5.

We have thus proved the existence (P_θ-a.s.) of the derivatives in braces in (4.2.33) and hence of the derivative $\nabla_y\rho_t^{1/n}(y,\theta)$. Now, in order to show that $\rho_t^{1/n}(y,\theta)\in\mathbf{C}_1^0(\Theta)$ (P_θ-a.s.) with respect to the variable y, it suffices to demonstrate that the functions

$$
\begin{aligned}
\gamma(s) &= \gamma(y+sh,\theta)\cdot m(\theta)_t,\\
\widetilde{\gamma}(s) &= \gamma^2(y+sh,\theta)\circ\langle m(\theta)\rangle_t,\\
\overline{\gamma}(s) &= \ln\lambda(y+sh,\theta)*(\mu-\nu(\theta))_t,\\
\overline{\overline{\gamma}}(s) &= \big(\lambda(y+sh,\theta)-1-\ln\lambda(y+sh,\theta)\big)*\nu(\theta)_t
\end{aligned}
$$

are absolutely continuous (P_θ-a.s.) with respect to $s \in [0,1]$ for any $y, y+h \in \Theta$. In view of the equality (4.2.34) we have

$$\dot{\gamma}(s) = \frac{d\gamma(s)}{ds} = h'\nabla_y\big(\gamma(y+sh,\theta)\cdot m(\theta)_t\big) = h'g(y+sh,\theta)\cdot m(\theta)_t = h'\mathfrak{z}(y+sh),$$

whence in view of continuity (P_θ-a.s.) of the function $\mathfrak{z}(y)$ we get the continuity (P_θ-a.s.) of the function $\dot{\gamma}(s)$. This means that the function $\gamma(s)$ is absolutely continuous (P_θ-a.s.) [**102**]. The absolute continuity (P_θ-a.s.) of the functions $\widetilde{\gamma}(s), \overline{\gamma}(s)$, and $\overline{\overline{\gamma}}(s)$ is demonstrated similarly. Therefore we have established that $\rho_t^{1/n}(y,\theta) \in \mathbf{C}_1^0(\Theta)$ (P_θ-a.s.) with respect to the variable y. From this we can conclude that the equality (4.2.32) is true.

Putting $y = \theta + \varphi_t(\theta)u$ and $z = \theta + \varphi_t(\theta)v$, from (4.2.32) we get

(4.2.37)
$$\begin{aligned}
\mathsf{E}_\theta\left|Z_{t,\theta}^{1/n}(u) - Z_{t,\theta}^{1/n}(v)\right|^n &= \mathsf{E}_\theta\left|(u-v)'\int_0^1 \varphi_t(\theta)\nabla_y\rho_t^{1/n}(y_{s,z},\theta)\,ds\right|^n \\
&\le |u-v|^n\int_0^1 \mathsf{E}_\theta\left|\varphi_t(\theta)\nabla_y\rho_t^{1/n}(y_{s,z},\theta)\right|^n ds \\
&= n^{-n}|u-v|^n\int_0^1 \mathsf{E}_{y_{s,z}}\Big|\varphi_t(\theta)\Big\{g(y_{s,z},\theta)\cdot m(\theta)_t - \gamma(y_{s,z},\theta)g(y_{s,z},\theta)\circ\langle m(\theta)\rangle_t \\
&\qquad + f(y_{s,z},\theta)*(\mu-\nu(\theta))_t \\
&\qquad - (l(y_{s,z},\theta) - f(y_{s,z},\theta))*\nu(\theta)_t\Big\}\Big|^n ds.
\end{aligned}$$

Using the transformation of predictable characteristics of semimartingales in the case of locally absolutely continuous change of the measure [**46, 155**], we see that (P_y-a.s.)

$$\begin{aligned}
g(y,\theta)\cdot m(\theta)_t - g(y,\theta)\gamma(y,\theta)\circ\langle m(\theta)\rangle_t &= g(y,\theta)\cdot m(y)_t, \\
f(y,\theta)*(\mu-\nu(\theta))_t - (l(y,\theta)-f(y,\theta))*\nu(\theta)_t &= f(y,\theta)*(\mu-\nu(y))_t.
\end{aligned}$$

Putting $y = \theta + \varphi_t(\theta)u$ and $z = \theta + \varphi_t(\theta)v$, from this and (4.2.37) we get

$$\begin{aligned}
&\mathsf{E}_\theta\left|Z_{t,\theta}^{1/n}(u) - Z_{t,\theta}^{1/n}(v)\right|^n \\
&\le \frac{|u-v|^n}{n^n}\int_0^1 \mathsf{E}_{y_{s,z}}\Big|\varphi_t(\theta)\Big\{g(y_{s,z},\theta)\cdot m(y_{s,z})_t \\
&\qquad + f(y_{s,z},\theta)*(\mu-\nu(y_{s,z}))_t\Big\}\Big|^n ds.
\end{aligned} \tag{4.2.38}$$

In view of Lemmas 4.2.2 and 4.2.3, the estimate (4.2.38) and condition 4) imply the validity of condition **K3** for $m = \beta = n$. □

Statistical applications. The properties of $Z_{t,\theta}(u)$ established in Theorems 4.2.1–4.2.6 allow us to use Theorems 4.1.1 and 4.1.2 on asymptotic properties of maximum likelihood estimates $\hat{\theta}_t$ and Bayes estimates $\tilde{\theta}_t$.

THEOREM 4.2.7. *Let the conditions of Theorems* 4.2.3, 4.2.4, *and* 4.2.6 *be fulfilled. Moreover, suppose $n > k$ in condition* 4) *of Theorem* 4.2.6 *and the matrix*

$\varphi_t(\theta)$ in Theorem 4.2.3 satisfies condition **K2**. *Then the assertions* 1)–3) *of Theorem* 4.1.1 *are true for the maximum likelihood estimate $\hat{\theta}_t$.*

Theorem 4.2.7 follows from Theorems 4.1.1, 4.2.3, 4.2.4, and 4.2.6.

THEOREM 4.2.8. *Let the conditions of Theorems* 4.2.3–4.2.5 *be fulfilled, the matrix $\varphi_t(\theta)$ in Theorem* 4.2.3 *satisfying condition* **K2**. *Let $\tilde{\theta}_t$ be a Bayes estimate with respect to a continuous positive aligned density $q(\theta)$ and a loss function $l_t(x) = l(\varphi_t^{-1}(\theta_0)x)$, $x \in \mathbf{R}^k$, where $l \in \widetilde{\mathbf{W}}'_p$ and θ_0 is an arbitrary fixed point in Θ. Then the assertions* 1)–3) *of Theorem* 4.1.1 *are valid for the estimate $\tilde{\theta}_t$ (with $\hat{\theta}_t$ replaced by $\tilde{\theta}_t$).*

Theorem 4.2.8 follows from Theorems 4.1.2 and 4.2.3–4.2.5.

4.3. Observations of diffusion-type processes

Consider the family of statistical experiments generated by observations ξ^t of the diffusion-type process ξ that is the solution of the differential equation (1.2.38), whose coefficients $a(\theta)$ and b satisfy the conditions of Theorem 1.2.3 with $\mathsf{P} = \mathsf{P}_\theta$ and $\widetilde{\mathsf{P}} = \mathsf{P}_y$ for any $\theta, y \in \Theta$ which ensure the local equivalence of the measures P_θ and P_y. From Theorems 1.2.3 and 1.2.4 we get the form of the local density process $z(y, \theta)$ of the measure P_y with respect to the measure P_θ defined by formulas (3.4.7) and (3.4.8). From this we obtain the expression for the normalized likelihood ratio $Z_{t,\theta}(u) = z_t(\theta + \varphi_t(\theta)u, \theta)$ to be used in this section. The theorems in Section 4.2 can be reformulated for $Z_{t,\theta}$ in a trivial way and hence are omitted. We restrict our consideration to particular cases of diffusion-type processes.

The case $a(\theta) = bs(\theta) + h$ with determinate function $s(\theta)$. Assume that $a(\theta) = h + s(\theta)b$ and the function $s(\theta) = (s_t(\theta))_{t\in\mathbf{R}_+}$ depends only on $\theta \in \Theta$ and $t \in \mathbf{R}_+$ and is independent of ξ. In addition to the conditions of Theorem 1.2.3, we assume that $b_t \neq 0$ (P_θ-a.s.) and $s^2(\theta) \circ I_t < \infty$ for all $\theta \in \Theta$ and $t \in \mathbf{R}_+$. Then $\gamma(y, \theta) = s(y) - s(\theta)$ and in order that the conditions of Theorems 4.2.1–4.2.6 and hence of Theorems 4.1.1 and 4.1.2 be fulfilled it is sufficient to impose restrictions only on the function $s(\theta)$, as in the scheme "a signal in white noise" ($h = 0$, $b = 1$).

Denote by $\mathbf{L}_2(0, T)$ the space of square integrable functions $f = (f_t)_{0\le t\le T}$ defined on $[0, T]$ and assuming values in $\mathbf{R}^m$, so that

$$\|f\|_T^2 = \int_0^T |f_t|^2\, dt < \infty.$$

Under the above assumptions, $s(\theta) \in \mathbf{L}_2(0, t)$ for all $\theta \in \Theta$ and $t \in \mathbf{R}_+$.

DEFINITION 4.3.1. A function $s(\theta)$ is said to be *differentiable with respect to θ* in $\mathbf{L}_2(0, t)$ (in other words, *differentiable in mean square*) if there exists a function $\dot{s}(\theta) = (\dot{s}_u(\theta))\colon \mathbf{R}_+ \times \Theta \to \mathbf{R}^k$ such that

$$\int_0^t \left|s_u(\theta + h) - s_u(\theta) - h'\dot{s}_u(\theta)\right|^2 du = o\left(|h|^2\right), \qquad h \to 0.$$

If the function $s(\theta)$ is differentiable with respect to θ (in the ordinary sense), then, evidently, $\dot{s}_u(\theta) = \nabla_\theta s_u(\theta)$. In what follows we assume the function to be differentiable in the ordinary sense, which justifies the notation $\dot{s}(\theta)$.

THEOREM 4.3.1. *Let the set $\Theta \subset \mathbf{R}^k$ be bounded, and let the following conditions be fulfilled:*

1) *the function $s(\theta)$ is continuously differentiable with respect to θ in $\mathbf{L}_2(0,t)$ for any $t \in \mathbf{R}_+$;*

2) $\lim_{t\to\infty} \inf_{\theta\in\Theta} \inf_{|\lambda|=0} \lambda' I_t(\theta)\lambda = \infty$, *where*

$$I_t(\theta) = \int_0^t \dot{s}_u(\theta)\dot{s}_u(\theta)'\,du;$$

3) *for some $t_0 \in \mathbf{R}_+$*

$$\sup_{t>t_0} \sup_{\theta,y\in\Theta} \left| I_t^{-1/2}(\theta) I_t(y) I_t^{-1/2}(\theta) \right| < \infty;$$

4) *for some positive function χ_t such that $\chi_t \to \infty$ as $t \to \infty$ and some $\beta > 0$ the following relations are true:*

$$\lim_{t\to\infty} \sup_{\theta\in\Theta} \sup_{\substack{|u|<\chi_t \\ u\in\Theta_{t,\theta}}} \int_0^t \left| I_t^{-1/2}(\theta) \left(\dot{s}_v \left(\theta + I_t^{-1/2}(\theta)u \right) - \dot{s}_v(\theta) \right) \right|^2 dv = 0,$$

$$\inf_{\theta\in\Theta} \inf_{\substack{|u|>\chi_t \\ u\in\Theta_{t,\theta}}} \left| I_t^{1/2}(\theta) \right|^{-\beta} \int_0^t \left| s_v \left(\theta + I_t^{-1/2}(\theta)u \right) - s_v(\theta) \right|^2 dv > 0.$$

Then the assertions of Theorems 4.1.1 and 4.1.2 with $\varphi_t(\theta) = I_t^{-1/2}(\theta)$ are valid for the maximum likelihood estimate $\hat{\theta}_t$ and Bayes estimate $\tilde{\theta}_t$ constructed from a continuous positive aligned density $q(\theta)$ and a loss function $l(I_t^{1/2}(\theta_0)x)$, where $l \in \widetilde{\mathbf{W}}_p'$ and θ_0 is any point in Θ.

PROOF. The proof reduces to the verification of conditions **K1**–**K4** and can be found in ([**40**], the proof of Theorem 3.5.1).

Some examples of the function $s(\theta)$ in which the conditions of Theorem 4.3.1 are fulfilled, can be found in ([**40**], Example 3.5.3).

REMARK 4.3.1. For $\varphi_t(\theta) = I_t^{-1/2}(\theta)$ the additional condition of Theorem 4.2.3 is trivial since $\varphi_t(\theta) g_0 g_0' \circ I_t \varphi_t(\theta) = I_k$. From Theorem 4.3.1 it follows immediately that the conditions of Theorem 4.2.3 ensuring the fulfillment of the uniform LANC are fulfilled. It is evident that in Theorem 4.2.5 $D_t(y,\theta) = \Omega$, and hence conditions (1.2.17) and (4.2.18) are trivial and conditions (4.2.15) and (4.2.16) follow from condition 4) of Theorem 4.3.1.

DEFINITION 4.3.2. A family of statistics $(\bar{\theta}_t, t \in \mathbf{R}_+)$ is called a *uniformly consistent family of estimates of the parameter θ in the set $K \subset \Theta$* if for any $\varepsilon > 0$

$$\lim_{t\to\infty} \sup_{\theta\in K} \mathsf{P}_\theta \left\{ \left|\bar{\theta}_t - \theta\right| > \varepsilon \right\} = 0, \tag{4.3.1}$$

that is, $\bar{\theta} \to \theta$ in probability P_θ as $t \to \infty$ uniformly with respect to $\theta \in K$. If (4.3.1) is fulfilled for $K = \Theta$, we shall say that $(\bar{\theta}_t, t \in \mathbf{R}_+)$ is a *uniformly consistent family of estimates*.

In the following theorems existence and nonexistence conditions for uniformly consistent estimates are formulated.

THEOREM 4.3.2. *If for some $\theta, y \in \Theta$, $\theta \neq y$,*

$$\lim_{t\to\infty} \|s(y) - s(\theta)\|_t < \infty, \tag{4.3.2}$$

then there are no uniformly consistent estimates of parameter θ. Uniformly consistent estimates can exist if and only if

$$\lim_{t\to\infty} \sup_{\theta\in\Theta} \|s(\theta)\|_t = \infty. \tag{4.3.3}$$

PROOF. See the proof in [**40**] (although in [**40**] the assertion of Theorem 4.3.1 was proved for $h = 0$ and $b = 1$, in the general case the theorem can be proved in the same way).

The example given in [**40**] shows that if condition (4.3.3) is violated, there is no uniformly consistent estimates. The existence conditions for uniformly consistent and efficient estimates were formulated in Theorem 4.3.1.

If the condition $b_t \neq 0$ (P_θ-a.s.) for all $t \in \mathbf{R}_+$ and $\theta \in \Theta$ does not hold (that is, for some $\theta \in \Theta$ the diffusion-type process ξ is degenerate), then it can happen that there is no uniformly consistent estimate valid for all functions $s(\theta)$ (and even for those satisfying condition (4.3.3)).

Indeed, let $h_t = (1/2)\xi_t$ and $b_t = \xi_t$ for all $t \in \mathbf{R}_+$, and let $\mathsf{P}_\theta\{\xi_0 = 0\} = p$ for all $\theta \in \Theta$. Then the following lemma is valid.

LEMMA 4.3.1. *For any estimate $\bar\theta_t$ of the parameter θ there exists $\varepsilon > 0$ such that for all $t \in \mathbf{R}_+$ the following inequality holds:*

$$\sup_{\theta\in\Theta} \mathsf{P}_\theta\left\{\left|\bar\theta_t - \theta\right| > \varepsilon\right\} \geq \frac{p}{2} + q\Phi\left(-\sup_{\theta\in\Theta}\|s(\theta)\|_t\right), \tag{4.3.4}$$

where $\Phi(x)$ is the distribution function of the law $\mathcal{N}(0,1)$, $q = 1 - p$.

PROOF. Let $\theta_1, \theta_2 \in \Theta$, $\theta_1 \neq \theta_2$, and let $\varepsilon > 0$ be such that $2\varepsilon < |\theta_2 - \theta_1|$. Then

(4.3.5)
$$\begin{aligned}\sup_{\theta\in\Theta} \mathsf{P}_\theta\left\{\left|\bar\theta_t - \theta\right| > \varepsilon\right\} &\leq \frac{1}{2}\mathsf{P}_{\theta_1}\left\{\left|\bar\theta_t - \theta_1\right| > \varepsilon\right\} + \frac{1}{2}\mathsf{P}_{\theta_2}\left\{\left|\bar\theta_t - \theta_2\right| > \varepsilon\right\}\\ &= \frac{1}{2}\mathsf{E}_{\theta_1}\left[\chi\left(\left|\bar\theta_t - \theta_1\right| > \varepsilon\right) + z_t(\theta_2,\theta_1)\chi\left(\left|\bar\theta_t - \theta_2\right| > \varepsilon\right)\right].\end{aligned}$$

Since $|\theta_2 - \theta_1| > 2\varepsilon$, at least one of the numbers $\chi(|\bar\theta_t - \theta_1| > \varepsilon)$ and $\chi\left(|\bar\theta_t - \theta_2| > \varepsilon\right)$ is equal to 1. Hence the right-hand side of (4.3.5) is at lest

$$\frac{1}{2}\mathsf{E}_{\theta_1}\chi(z_t(\theta_2,\theta_1) \geq 1) + \frac{1}{2}\mathsf{E}_{\theta_2}\chi(z_t(\theta_2,\theta_1) < 1)$$

and equals this value if $\bar\theta_t = \theta_t^*$, where

$$\theta_t^* = \begin{cases}\theta_1, & \text{if } z_t(\theta_2,\theta_1) < 1,\\ \theta_2, & \text{if } z_t(\theta_2,\theta_1) \geq 1.\end{cases}$$

Therefore we have the following estimate:

$$\sup_{\theta\in\Theta} \mathsf{P}_\theta\left\{\left|\bar\theta_t - \theta\right| > \varepsilon\right\} \geq \frac{1}{2}\mathsf{P}_{\theta_1}\{z_t(\theta_2,\theta_1) \geq 1\} + \frac{1}{2}\mathsf{P}_{\theta_2}\{z_t(\theta_2,\theta_1) < 1\}. \tag{4.3.6}$$

Applying formulas (3.4.45), (3.4.51), and (3.4.53), which are valid for any $p \in [0,1]$, with $\theta = \theta_1$, $y_t = \theta_2$, and $v_t = \|s(\theta_2) - s(\theta_1)\|_t$, we get

$$(4.3.7)\quad \mathsf{P}_{\theta_1}\{z_t(\theta_2,\theta_1) \geq 1\} - p = q\Phi\left(-\frac{1}{2}\|s(\theta_2) - s(\theta_1)\|_t\right) = \mathsf{P}_{\theta_2}\{z_t(\theta_2,\theta_1) < 1\}.$$

The inequality (4.3.6) and equality (4.3.7) imply the estimate

$$(4.3.8)\qquad \sup_{\theta\in\Theta} \mathsf{P}_\theta\left\{|\bar\theta_t - \theta| > \varepsilon\right\} \geq \frac{p}{2} + q\Phi\left(-\frac{1}{2}\|s(\theta_2) - s(\theta_1)\|_t\right).$$

In view of the triangle inequality we have

$$(4.3.9)\qquad \|s(\theta_2) - s(\theta_1)\|_t \leq 2\sup_{\theta\in\Theta}\|s(\theta)\|_t.$$

The inequalities (4.3.8) and (4.3.9) imply the required estimate (4.3.4). □

COROLLARY 4.3.1. *If $p = 0$ and for some $y, \theta \in \Theta$, $y \neq \theta$, condition* (4.3.2) *holds, then there are no uniformly consistent estimates of parameter θ. Uniformly consistent estimates can exist only if condition* (4.3.3) *is fulfilled.*

If $p \neq 0$, then for any functions $s(\theta)$ there are no uniformly consistent estimates of parameter θ.

PROOF. If $p = 0$, then in view of equality (3.4.41), valid also for $p = 0$, we have $\xi_t \neq 0$ (P_θ-a.s.) for all $\theta \in \Theta$ and $t \in \mathbf{R}_+$. This means that $b_t \neq 0$ (P_θ-a.s.) for all $t \in \mathbf{R}_+$ and $\theta \in \Theta$. Thus the required result follows from Theorem 4.3.2.

If $p \neq 0$, then in view of estimate (4.3.4) we have

$$\sup_{\theta\in\Theta} \mathsf{P}_\theta\left\{|\bar\theta_t - \theta| > \varepsilon\right\} \geq \frac{p}{2} > 0$$

for any estimate $\bar\theta_t$ of the parameter θ for all $t \in \mathbf{R}_+$ and all sufficiently small $\varepsilon > 0$ From this the required result follows. □

Homogeneous diffusion processes. The ergodic case. Let the coefficients of equation (1.2.38) be time-independent ($a(t,\xi,\theta) = a(\xi_t,\theta)$, $b(t,\xi) = b(\xi_t)$) and the initial value be determinate ($\xi_0 = x$, $x \in \mathbf{R}^1$). Assume that $b(x) > 0$ for all $x \in \mathbf{R}^1$ and conditions (3.4.54) are fulfilled, under which ξ is a positive recurrent process with respect to P_θ with finite invariant measure π_θ defined by formula (3.4.55). We restrict our consideration to checking the conditions of Theorem 4.2.5 since the finding of rather simple sufficient conditions of the other theorems of Section 4.2 presents no problems. The conditions of Theorems 4.2.1–4.2.3 can be checked similarly to Theorem 3.4.8. As in Theorem 3.4.8, it is necessary to assume that $I(\theta) = \int f(x,\theta)f(x,\theta)'\pi_\theta(dx)$ is a nondegenerate matrix with finite elements, where $f(x,\theta) = b^{-1}(x)\nabla_\theta a(x,\theta)$, and to put $\varphi_t(\theta) = (tI(\theta))^{-1/2}$.

Let us consider the conditions of Theorem 4.2.5. We assume K to be a one-point compact set, $K = \{\theta\}$. It is evident that

$$(4.3.10)\qquad \gamma^2(y,\theta)\circ I_t \geq \sum_{i=1}^{\nu_t}\gamma^2(y,\theta)\circ I_{\tau_i}^{\tau_{i-1}},$$

where

$$\gamma^2(y,\theta)\circ I_v^u = \int_u^v \gamma_s^2(y,\theta)\,ds$$

and the random variables τ_i and ν_t are as introduced in Section 3.4.

Let a random variable τ_1 satisfy the Cramér condition: $\mathsf{E}_\theta \exp\{\delta\tau_1\} < \infty$ for some $\delta > 0$. Then from the Cramér theorem on large deviations ([**105**], Chapter 8, Theorem 1) for $\alpha \in (0, \frac{1}{2})$ we get the following estimate as $n \to \infty$:

$$\mathsf{P}_\theta\left\{\sum_{i=1}^{n} \tau^{(i)} > na(\theta) + \sigma(\theta)n^{\alpha+1/2}\right\} \le \exp(-cn^{2\alpha}), \qquad c < \frac{1}{2}, \tag{4.3.11}$$

where $\tau^{(i)} = \tau_i - \tau_{i-1}$ are independent identically distributed random variables, $\tau^{(1)} = \tau_1$, $a(\theta) = \mathsf{E}_\theta\,\tau_1$, and $\sigma^2(\theta) = \mathrm{Var}_\theta\,\tau_1$. Now we choose a positive integer n_t so that

$$n_t a(\theta) + \sigma(\theta)n_t^{\alpha+1/2} = t(1 + o(1)), \qquad t \to \infty,$$

and use the equality

$$\{\nu_t < n\} = \left\{\sum_{i=1}^{n} \tau^{(i)} \ge t\right\}$$

and the estimate (4.3.11) to obtain

$$\mathsf{P}_\theta\{\nu_t < n_t\} \le \exp\left\{-\tilde{c}t^{2\alpha}\right\}, \qquad \tilde{c} > 0, \tag{4.3.12}$$

as $t \to \infty$.

Put $f_s = f(\theta + s\varphi_t(\theta)u)$, where $s \in [0, 1]$ and $f(\theta) = (f(\xi_t, \theta))_{t\in\mathbf{R}_+}$. Furthermore, put $y = \theta + \varphi_t(\theta)u$ and $|u| \le t^\delta$, $\delta \in (0, \frac{1}{2})$. Then, assuming that $a(x, \theta) \in \mathbf{C}_1^0(\Theta)$ with respect to the variable θ, for $t \to \infty$ we have

(4.3.13)

$$\begin{aligned}
\sum_{i=1}^{n} \gamma^2(y, \theta) \circ I_{\tau_i}^{\tau_{i-1}} &= \sum_{i=1}^{n} \left(\int_0^1 f_s\,ds, y - \theta\right)^2 \circ I_{\tau_i}^{\tau_{i-1}} \\
&= \sum_{i=1}^{n} \Bigg[\left(\int_0^1 f_s\,ds - f_0, y - \theta\right)^2 \\
&\qquad + 2\left(\int_0^1 f_s\,ds - f_0, y - \theta\right)(f_0, y - \theta) + (f_0, y - \theta)^2\Bigg] \circ I_{\tau_i}^{\tau_{i-1}} \\
&\ge \sum_{i=1}^{n} \left[(f_0, y - \theta)^2 + 2\left(\int_0^1 f_s\,ds - f_0, y - \theta\right)(f_0, y - \theta)\right] \circ I_{\tau_i}^{\tau_{i-1}} \\
&\ge \frac{|u|^2}{t}\left(\sum_{i=1}^{n} \eta_i - 2\left(\sum_{i=1}^{n} \eta_i \sum_{i=1}^{n} \zeta_i\right)^{1/2}\right),
\end{aligned}$$

where

$$\eta_t = \left(I^{-1/2}(\theta)f_0, \frac{u}{|u|}\right)^2 \circ I_{\tau_i}^{\tau_{i-1}},$$

$$\zeta_t = \sup_{|y-\theta|\le\Delta} \left|I^{-1/2}(\theta)(f(y) - f(\theta))\right|^2 \circ I_{\tau_i}^{\tau_{i-1}},$$

and Δ is so small that $\bar{a}(\theta) > 4\bar{\bar{a}}(\theta)$, $\bar{a}(\theta) = \mathsf{E}_\theta\,\eta_1$, and $\bar{\bar{a}}(\theta) = \mathsf{E}_\theta\,\zeta_1$. Note that $\eta_1, \eta_2, \ldots$, as well as $\zeta_1, \zeta_2, \ldots$, are independent identically distributed random variables.

Consider the random event

$$D_t = \left\{ \left| \sum_{i=1}^{n_t} \eta_i - n_t \bar{a}(\theta) \right| < \bar{\sigma}(\theta) n_t^{\alpha+1/2} \right\} \cap \left\{ \sum_{i=1}^{n_t} \zeta_i < n_t \bar{\bar{a}}(\theta) + \bar{\bar{\sigma}}(\theta) n_t^{\alpha+1/2} \right\},$$

where $\bar{\sigma}^2(\theta) = \mathrm{Var}_\theta\, \eta_1$ and $\bar{\bar{\sigma}}^2(\theta) = \mathrm{Var}_\theta\, \zeta_1$. Then, assuming that the random variables η_1 and ζ_1 satisfy the Cramér condition, from the theorem on large deviations for $t \to \infty$ we get

$$\mathsf{P}_\theta(D_t^c) \leq \exp\left\{-\bar{c} t^{2\alpha}\right\}, \qquad \bar{c} > 0,\ D_t^c = \Omega \setminus D_t.$$

Taking into account (4.3.10), (4.3.12), and (4.3.13), we conclude that conditions (4.2.15) and (4.2.17) of Theorem 4.2.5 are fulfilled if we put $\chi_t = t^\delta$ and $D_t(y,\theta) = \{\nu_t \geq n_t\} \cap D_t$ for $y = \theta + \varphi_t(\theta) u$ and $|u| \leq \chi_t$.

Now let $|u| > \chi_t = t^\delta$, $\delta \in (0, \frac{1}{2})$, and for some $d \in (0,1)$ and any $s \in \mathbf{R}_+$

$$\gamma^2(\theta + \varphi_t(\theta) u, \theta) \circ I_s \geq t^{-d} (g^2(\theta) \circ I_s),$$

where $g(\theta) = (g_v(\theta))_{v \in \mathbf{R}_+}$, $g_v(\theta) = g(\xi_v, \theta)$ and the function $g(x,\theta)$ is not identically zero.

Note that if $k = 1$ and $f_t(\theta) \geq \varkappa > 0$, where $\varkappa$ is a nonrandom constant, then it is enough to put $g_v(\theta) = \varkappa I^{-1/2}(\theta)$ and $d = 1 - 2\delta$. If the random variables $z_1 = g^2(\theta) \circ I_{\tau_1}$ satisfy the Cramér condition and $\mathsf{E}_\theta\, z_1 > 0$ as in the case $|u| \leq \chi_t$, then conditions (4.2.16) and (4.2.18) of Theorem 4.2.5 are fulfilled. Thus we have obtained the sufficient conditions under which Theorem 4.2.5 is true for one-point compact sets.

4.4. Observations of counting processes

Consider the parametric family of statistical experiments generated by observations ξ^t of a counting process ξ with compensator $\nu(\theta)$ with respect to the measure P_θ (see Sections 1.2. and 3.5). We assume that the conditions of Theorem 1.2.5 with $\mathsf{P} = \mathsf{P}_\theta$ and $\widetilde{\mathsf{P}} = \mathsf{P}_y$ are fulfilled for any $y, \theta \in \Theta$ which ensure the local equivalence of the measures P_y and P_θ. From Theorem 1.2.5 we obtain the form of the local density process $z(y,\theta)$ of the measure P_y with respect to the measure P_θ (see the representation (3.5.13)–(3.5.15)). The theorems of Section 4.2 can be easily reformulated for a counting process in terms of the compensator $\nu(\theta)$ and hence are omitted here. Instead, we limit ourselves to some examples illustrating the verification of the conditions of these theorems for particular counting processes.

Counting processes with determinate compensators. Let compensators $\nu(\theta)$ be determinate and time-continuous for all $\theta \in \Theta$. We assume that there exists a nonnegative function $\lambda(y,\theta) = (\lambda_t(y,\theta))$, measurable with respect to t and such that $\nu_t(y) = \lambda(y,\theta) \circ \nu(\theta)_t$ and $|\ln \lambda(y,\theta)| \circ \nu(\theta)_t < \infty$ for all $t \in \mathbf{R}_+$ and $y, \theta \in \Theta$. This assumption implies that $\mathsf{P}_y \overset{\text{loc}}{\sim} \mathsf{P}_\theta$ for all $y, \theta \in \Theta$ (see Section 1.2). Also, we assume that $\ln \lambda(y,\theta) \in \mathbf{C}_1^0(\Theta)$ with respect to the variable $y \in \Theta$ and in the notation of Section 3.5 the matrix $I_t(\theta) = l_0 l_0' \circ \nu(\theta)_t$ has finite elements and is nonnegative for all $t \in \mathbf{R}_+$. Then it suffices to assume in Theorems 4.2.1–4.2.3 that $\varphi_t(\theta) = I^{-1/2}(\theta)$ and $|I_t^{-1}(\theta)| \to 0$ as $t \to \infty$. In this case the additional condition in Theorem 4.2.3 is trivial since $\varphi_t(\theta) I_t \varphi_t(\theta) = I_k$, and the other conditions of Theorems 4.2.1–4.2.3 are easily obtainable.

Since the Hellinger process $h(y,\theta)$ of order $\frac{1}{2}$ is determinate for the measures P_y and P_θ, it is enough to put $D_t(y,\theta)=\Omega$ for all $t\in\mathbf{R}_+$ and $y,\theta\in\Theta$ in Theorem 4.2.5 and to require that only conditions (4.2.15) and (4.2.16) be fulfilled. The following lemma formulates sufficient conditions for the relations (4.2.15) and (4.2.16) to hold.

LEMMA 4.4.1. *Let Θ be a bounded set, $K\subset\Theta$ any compact set, and let the following conditions be fulfilled:*

1) $\ln\lambda(y,\theta)\in\mathbf{C}_1^0(\Theta)$ *with respect to the variable $y\in\Theta$, the matrix $I_t(\theta)=l_0l_0'\circ\nu(\theta)_t$ is nonsingular and has finite elements for all $t\in\mathbf{R}_+$ and $|I_t^{-1}(\theta)|\to 0$ as $t\to\infty$;*

2) $\varphi_t(\theta)=I_t^{-1/2}(\theta)$ *and there exists a positive function χ_t such that $\chi_t\to\infty$ and $\chi_t|\varphi_t(\theta)|\to 0$ as $t\to\infty$ for all $\theta\in\Theta$, and, in addition,*

$$\lim_{t\to\infty}\sup_{\theta\in K}\sup_{\substack{u\in\Theta_{t,\theta}\\ |u|<\chi_t}}\left|\varphi_t(\theta)\big(n(\theta+\varphi_t(\theta)u,\theta)-n(\theta,\theta)\big)\right|^2\circ\nu(\theta)_t=0,$$

where $n(y,\theta)=(1/2)l(y,\theta)\lambda^{-1/2}(y,\theta)$ and $l(y,\theta)=\nabla_y\lambda(y,\theta)$;

3) *there exists a number $b>0$ such that*

$$\inf_{\theta\in K}\inf_{\substack{u\in\Theta_{t,\theta}\\ |u|>\chi_t}}|\varphi_t^{-1}(\theta)|^{-b}\left|\lambda^{1/2}(\theta+\varphi_t(\theta)u,\theta)-1\right|^2\circ\nu(\theta)_t\ge c>0$$

for all $t\ge t_0$ for some $t_0>0$, where c is a constant, dependent possibly on K and independent of t.

Then the relations (4.2.15) *and* (4.2.16) *are true.*

PROOF. For $y=\theta+\varphi_t(\theta)u$ we have

$$\begin{aligned}\left|\lambda^{1/2}(y,\theta)-1\right|^2\circ\nu(\theta)_t&=\left(\int_0^1 n_s\,ds,y-\theta\right)^2\circ\nu(\theta)_t\\&=(n_0,y-\theta)^2\circ\nu(\theta)_t+\left(\int_0^1 n_s\,ds-n_0,y-\theta\right)^2\circ\nu(\theta)_t\\&\quad+2\left(\int_0^1 n_s\,ds-n_0,y-\theta\right)(n_0,y-\theta)\circ\nu(\theta)_t,\end{aligned}\tag{4.4.1}$$

where $n_s=n(\theta+s(y-\theta),\theta)$, $s\in[0,1]$. The absolute values of the second and third terms on the right-hand side of (4.4.1) are estimated from above as follows:

$$\begin{aligned}&|u|^2\int_0^1|\varphi_t(\theta)(n_s-n_0)|^2\,ds\circ\nu(\theta)_t\\&\quad+2\left(|u|^2\int_0^1|\varphi_t(\theta)(n_s-n_0)|^2\,ds\circ\nu(\theta)_t(n_0,y-\theta)^2\circ\nu(\theta)_t\right)^{1/2}.\end{aligned}$$

Taking into account the equality

$$\begin{aligned}4(n_0,y-\theta)^2\circ\nu(\theta)_t&=(l_0,\varphi_t(\theta)u)^2\circ\nu(\theta)_t\\&=u'\varphi_t(\theta)(l_0l_0'\circ\nu(\theta)_t)\varphi_t(\theta)u=|u|^2,\end{aligned}$$

where $l_0=l(\theta,\theta)$, and condition 2), we get the relation (4.2.15).

Relation (4.2.16) follows trivially from condition 3). $\square$

The conditions of Theorems 4.2.4. and 4.2.6 can be easily reformulated for the case where $\nu(\theta)$ is a determinate compensator.

EXAMPLE 4.4.1. Let $\nu_t(\theta) = c(\theta)t$, where $0 < c(\theta) < \infty$ for all $\theta \in \Theta \subset \mathbf{R}^k$, $k \geq 1$. We assume that $\ln c(\theta) \in \mathbf{C}_1^0(\Theta)$ and the vectors $\dot{c}(\theta)$ and $\dot{c}(\theta)/c(\theta)$ are bounded for all $\theta \in \Theta$, where $\dot{c}(\theta) = \nabla_y c(\theta)$. Then, following the arguments of Theorem 3.3.2 for $y_t = \theta + \varphi_t(\theta)u$ and $\varphi_t(\theta) = (c(\theta)t)^{-1/2}$, we get for $t \to \infty$

$$\Lambda_t(\theta + \varphi_t(\theta)u, \theta) = u'\eta_t - (1/2)u'I(\theta)u + r_t(u, \theta),$$

where

$$\begin{gathered} I(\theta) = c^{-2}(\theta)\dot{c}(\theta)\dot{c}(\theta)', \\ \mathcal{L}(\eta_t|\mathsf{P}_\theta) \xrightarrow{w} \mathcal{N}(0, I(\theta)), \qquad t \to \infty, \\ \mathsf{P}_\theta\text{-}\lim_{t\to\infty} r_t(u, \theta) = 0. \end{gathered}$$

Since for $k > 1$ the matrix $I(\theta)$ is singular, the above representation implies that in the case $\nu_t(\theta) = c(\theta)t$ the family $(P_\theta^t, \theta \in \Theta)$ does not satisfy the LANC if $k > 1$ (see Lemma 3.4.3 in the diffusion case). Therefore the available theory of parameter estimation is applicable only in the case of one-dimensional parameters ($k = 1$).

EXAMPLE 4.4.2. Let the compensators $\nu(\theta)$ be of the form

$$\nu(\theta) = \int_0^t e^{a+\theta s}\, ds, \qquad t \in \mathbf{R}_+,$$

where $a \in \mathbf{R}^1$, $\theta \in \Theta = (0, A)$, $A < \infty$ [**47**]. In this case $\lambda_t(y, \theta) = \exp\{(y - \theta)t\}$ and the conditions of Theorem 1.2.5 are fulfilled with $\mathsf{P} = \mathsf{P}_\theta$ and $\widetilde{\mathsf{P}} = \mathsf{P}_y$ for all $y, \theta \in \Theta$. Evidently, $\lambda_t(y, \theta) \in \mathbf{C}_1^0(\Theta)$ with respect to the variable $y \in \Theta$ and $l_t(y, \theta) = t\lambda_t(y, \theta)$. Hence it is enough to put $\varphi_t(\theta) = (s^2 \circ \nu(\theta)_t)^{-1/2}$ in Theorems 4.2.1–4.2.3. Then we have

$$\varphi_t^2(\theta) = \theta t^{-2} \exp(-a - \theta t)(1 + o(1)) \tag{4.4.2}$$

as $t \to \infty$. It is easy to see that condition **K2** can hold for the normalizing function $\varphi_t(\theta)$ only in the case of one-point compact sets $K \subset \Theta$ since otherwise the uniform convergence in (4.1.12) and the continuity of the function $B(y, \theta)$ are violated. For one-point compact sets K the conditions of Theorems 4.2.1–4.2.5 are fulfilled and hence conditions **K1–K4** are valid. Then, based on Theorem 4.1.1, the maximum likelihood estimates $\hat{\theta}_t$ are asymptotically normal with respect to P_θ with parameters $(\theta, \varphi_t^2(\theta))$, so that the asymptotic variance decreases at the rate (4.4.2). Moreover, the moments $\mathsf{E}_\theta\, |\varphi_t^{-1}(\theta)(\hat{\theta}_t - \theta)|^n$, $n > 0$, of any order converge and the family of estimates $(\hat{\theta}_t)$ is asymptotically efficient. Based on Theorem 4.1.2, the Bayes estimates possess similar properties.

EXAMPLE 4.4.3. Let the compensator $\nu(\theta)$ be of the form

$$\nu_t(\theta) = \int_0^t e^{\theta_1 + \theta_2 s}\, ds, \qquad \theta = (\theta_1, \theta_2)' \in \Theta = (A, B) \times (0, C),$$
$$A > -\infty, \qquad B, C < \infty$$

(see Example 3.5.6). In this case

$$\lambda_t(y,\theta) = \exp\{(y_1-\theta_1)+(y_2-\theta_2)t\}, \qquad l_t(y,\theta) = \begin{pmatrix}1\\ t\end{pmatrix}\lambda_t(y,\theta),$$

where $y=(y_1,y_2)'$.

As was shown in Example 3.5.6, the matrix $I_t(\theta) = l_0 l_0' \circ \nu(\theta)_t$ is nonsingular, and hence it is enough to put $\varphi_t(\theta) I_t(\theta)\varphi_t(\theta) = I_k$ in Theorems 4.2.1–4.2.3. Then, as was shown in Example 3.5.6, we have

$$\varphi_t(\theta) = \theta_2^{3/2} \exp\left\{-\frac{\theta_1+\theta_2 t}{2}\right\}\begin{pmatrix} t+o(t) & -1+o(1) \\ -1+o(1) & (1+\theta_2^{-1}+o(1))t^{-1}\end{pmatrix}$$

as $t\to\infty$. This means that similarly to the previous example condition **K2** can hold only for the compacta of the form $K = K_1\times K_2 \subset \Theta$, where K_1 is any compact set in (A,B) and K_2 is a one-point compact set in $(0,C)$. The validity of conditions **K1–K4** for these compact sets can be easily established.

REMARK 4.4.1. Other examples of nonhomogeneous Poisson processes satisfying conditions **K1–K4** can be found in [**47, 57, 134, 173, 190**]. In [**57**] conditions were established which in view of Remark 2 in [**81**] are even more restrictive than some of the conditions of Theorems 4.2.1–4.2.6.

Renewal processes. Let ξ be a renewal process with jump times

$$T_n = \sum_{i=1}^{n} \tau_i, \qquad i=1,2,\dots,$$

where $\tau_1,\tau_2,\dots$ are independent identically distributed positive random variables with continuous distribution function $F(t;\theta) = \mathsf{P}_\theta\{\tau_1 < t\}$, $\theta\in\Theta\subset\mathbf{R}^k$. Now we put $F(t) = F(t;\theta)$ and $\widetilde{F}(t) = F(t;y)$ for some $\theta, y\in\Theta$ and assume that the conditions of Theorem 3.5.7, which ensure the local equivalence of the measures P_y and P_θ, are fulfilled. From Theorem 3.5.7 we obtain the form of the local density $z(y,\theta)$ of the measure P_y with respect to the measure P_θ (see the representation (3.5.13)–(3.5.15)).

For simplicity, we assume that K is a one-point compact set. The sufficient conditions for the fulfillment of Theorem 4.2.3 were actually obtained in Section 3.5 while verifying the conditions of Theorem 3.5.4. Here we put $\varphi_t(\theta) = (a^{-1}(\theta)I(\theta)t)^{-1/2}$, where $a(\theta) = \mathsf{E}_\theta\,\tau_1$, and assume the matrix $I(\theta) = \mathsf{E}_\theta\, I^{(1)}(\theta)$ to be nonsingular with finite elements (the matrix $I^{(1)}(\theta)$ is defined by (3.5.98)). Clearly, the compact set K can be chosen arbitrarily. Then the validity of Theorem 4.2.3 is easily proved.

The sufficient conditions of Theorems 4.2.4 and 4.2.6 are obtained using the same reasonings as in the case of Theorem 3.5.4. Here we restrict ourselves to verifying only the conditions of Theorem 4.2.5.

Using the form of compensator $\nu(\theta)$ for a renewal process (see (3.5.47) and (3.5.48)), we get

$$\left(\lambda^{1/2}(y,\theta)-1\right)^2 \circ \nu(\theta)_t \ge \sum_{i=1}^{\xi_{t-}} \int_0^{\tau_i} \left(l^{1/2}(s;y,\theta)-1\right)^2 d\pi(s;\theta), \tag{4.4.3}$$

where $l(s;y,\theta)$ and $\pi(s,\theta)$ are the functions defined in Section 3.5.

We assume that the random variable τ_1 satisfies the Cramér condition and choose a positive integer n_t so that for $t \to \infty$

$$n_t a(\theta) + \sigma(\theta) n_t^{\alpha+1/2} = t(1 + o(1)),$$

where $a(\theta) = \mathsf{E}_\theta\, \tau_1$, $\sigma^2(\theta) = \operatorname{Var}_\theta \tau_1$, and $\alpha \in (0, \frac{1}{2})$. Exactly as it was done while deriving the estimate (4.3.12), from the Cramér theorem on large deviations [**105**] we get

$$\mathsf{P}_\theta\{\xi_{t-} < n_t\} \le \exp\{-ct^{2\alpha}\}, \qquad c > 0, \tag{4.4.4}$$

as $t \to \infty$.

Let $l^{1/2}(s; y, \theta) \in \mathbf{C}_1^0(\Theta)$ with respect to the variable $y \in \Theta$ and let $k_s(y, \theta) = \nabla_y l^{1/2}(s; y, \theta)$. Putting $y = \theta + \varphi_t(\theta) u$, $|u| \le t^\delta$, $\delta \in (0, \frac{1}{2})$, exactly as when deriving the estimate (4.3.13), we get

(4.4.5)

$$\sum_{i=1}^{n} \int_0^{\tau_i} \Big(l^{1/2}(s; y, \theta) - 1 \Big)^2 d\pi(s; \theta) \ge \frac{a(\theta)|u|^2}{t} \left(\sum_{i=1}^{n} h_i - 2\Big(\sum_{i=1}^{n} h_i \sum_{i=1}^{n} z_i \Big)^{1/2} \right)$$

as $t \to \infty$, where

$$h_i = \int_0^{\tau_i} \left(I^{-1/2}(\theta) k_s(\theta, \theta), \frac{u}{|u|} \right)^2 d\pi(s; \theta),$$

$$z_i = \sup_{|y-\theta| \le \Delta} \int_0^{\tau_i} \Big| I^{-1/2}(\theta)(k_s(y, \theta) - k_s(\theta, \theta)) \Big|^2 d\pi(s; \theta).$$

The quantity Δ is sufficiently small, so that $\tilde{a}(\theta) > 4\tilde{\tilde{a}}(\theta)$, $\tilde{a}(\theta) = \mathsf{E}_\theta\, h_1$, and $\tilde{\tilde{a}}(\theta) = \mathsf{E}_\theta\, z_1$. Consider the event

$$D_t = \left\{ \left| \sum_{i=1}^{n_t} h_i - n_t \tilde{a}(\theta) \right| < \tilde{\sigma}(\theta) n_t^{\alpha+1/2} \right\} \cap \left\{ \sum_{i=1}^{n_t} z_i < n_t \tilde{\tilde{a}}(\theta) + \tilde{\tilde{\sigma}}(\theta) n_t^{\alpha+1/2} \right\},$$

where $\tilde{\sigma}^2(\theta) = \operatorname{Var}_\theta h_1$ and $\tilde{\tilde{\sigma}}^2(\theta) = \operatorname{Var}_\theta z_1$.

Assuming that the random variables h_1 and z_1 satisfy the Cramér condition, from the Cramér large deviation theorem [**105**] we have $\mathsf{P}_\theta(D_t^c) \le \exp\{-\tilde{c} t^{2\alpha}\}$ as $t \to \infty$. Here $\tilde{c} > 0$ and $D_t^c = \Omega \setminus D_t$.

This and (4.4.3)–(4.4.5) imply the fulfillment of conditions (4.2.15) and (4.2.17), provided that $\chi_t = t^\delta$, $\delta \in (0, \frac{1}{2})$, and $D_t(y, \theta) = \{\xi_{t-} \ge n_t\} \cap D_t$ for $y = \theta + \varphi_t(\theta) u$, $|u| \le \chi_t$.

Now suppose $|u| > \chi_t = t^\delta$, $\delta \in (0, \frac{1}{2})$, and for some $d \in (0, 1)$ the following estimate holds for any $s \in \mathbf{R}_+$:

$$\int_0^s \Big(l^{1/2}(v; \theta + \varphi_t(\theta) u, \theta) - 1 \Big)^2 d\pi(v; \theta) \ge t^{-d} \int_0^s \varkappa_v^2(\theta)\, d\pi(v; \theta),$$

where the function $\varkappa_v(\theta)$ is not identically zero. Assuming that the random variable $v_1 = \int_0^{\tau_1} \varkappa_s^2(\theta)\, d\pi(s; \theta)$ satisfies the Cramér condition, we see, exactly as in the case $|u| \le \chi_t$, that conditions (4.2.16) and (4.2.18) are fulfilled.

Thus, the validity of Theorem 4.2.5 is established for one-point compact sets.

CHAPTER 5

Asymptotic Information-Theoretic Problems in Parameter Estimation

5.1. Asymptotic behavior of the Shannon information in observations with respect to an unknown parameter

Statement of the problem. Let an observation ξ^t depend on a random parameter θ taking values in $(\Theta, \mathcal{B}(\Theta))$, where Θ is an open set in $\mathbf{R}^k$. Let the parameter θ be distributed according to a law Q. Denote by P_y^t the conditional distribution of the observation ξ^t under the condition that $\theta = y$ and $y \in \Theta$. Thus the family of statistical experiments $(X^t, \mathfrak{B}^t, (P_y^t, y \in \Theta))$ is generated by the conditional distributions of the observation ξ^t, provided θ is known. We assume the measures P_y^t and P_z^t to be mutually absolutely continuous for all $y, z \in \Theta$, $t \in \mathbf{R}_+$.

Denote by $I(\xi^t, \theta)$ the *amount of Shannon information contained in the observation ξ^t with respect to the unknown parameter θ*. According to [**31, 106**]

$$I(\xi^t, \theta) = \begin{cases} \int_{X^t \times \Theta} \log \frac{dP_{(\xi^t,\theta)}}{d(P_{\xi^t} \times Q)}(x, y)\, P_{(\xi^t,\theta)}(dx, dy), & \text{if } P_{(\xi^t,\theta)} \ll P_{\xi^t} \times Q, \\ \infty, & \text{otherwise}, \end{cases}$$

where $P_{(\xi^t,\theta)}$ is the distribution of the vector (ξ^t, θ), P_{ξ^t} is the distribution of the observation ξ^t,

$$P_{\xi^t}(A) = \int_\Theta P_y^t(A)\, Q(dy), \qquad A \in \mathfrak{B}^t,$$

and $P_{\xi^t} \times Q$ is the product of the measures P_{ξ^t} and Q.

Here and later in this chapter the logarithm is taken to base 2. All other notions of information theory and their properties can be found in [**31, 106**], unless otherwise stated.

Consider the behavior of $I(\xi^t, \theta)$ as $t \to \infty$ under certain restrictions on the density of measures $\rho_t(x; y, z) = dP_y^t/dP_z^t(x)$, $x \in X^t$, and on the distribution Q.

Absolutely continuous distributions of parameter θ. Let the measure Q be Lebesgue absolutely continuous with density $q(y)$, $y \in \Theta$. For convenience, we assume that the measure Q is defined on the whole space $(\mathbf{R}^k, \mathcal{B}^k)$ and put $Q(\Theta^c) = 0$. Hence $q(y) = 0$ for $y \in \Theta^c$.

In this case the definition of the information $I(\xi^t, \theta)$ implies that

$$(5.1.1) \qquad I(\xi^t, \theta) = \mathsf{E} \log \left|\det \varphi_t^{-1}(\theta)\right| + h(\theta) - \mathsf{E} \log \int Z_t(u)\, du,$$

provided that all the quantities in (5.1.1) are well defined and finite. Here $h(\theta) = -\mathsf{E} \log q(\theta)$ is the *differential entropy* of the distribution of a vector θ, $\varphi_t(y)$ is a nonsingular (Q-a.s.) matrix for $t \in \mathbf{R}_+$ such that $|\varphi_t(y)| \to 0$ as $t \to \infty$ for all $y \in \Theta$, $\widetilde{Z}_t(u) = Z_t(u)q(\theta+\varphi_t(\theta)u)/q(\theta)$ for all $u \in \Theta_{t,\theta}$, and $\widetilde{Z}_t(u) = 0$ for $u \in \Theta^c_{t,\theta}$, where $Z_t = \rho_t(\xi^t; \theta + \varphi_t(\theta)u, \theta)$ for $u \in \Theta_{t,\theta}$ (in the notation of Section 4.1, $Z_t(u)$ is $Z_{t,\theta}(u)$). For convenience, we assume that $Z_t(u)$ is defined for all $u \in \mathbf{R}^k$ and put $Z_t(u) = 0$ for $u \in \Theta^c_{t,\theta}$.

Under certain assumptions

$$\lim_{t\to\infty} \mathsf{E} \log \int \widetilde{Z}_t(u)\,du = \mathsf{E} \log \int Z(u)\,du, \tag{5.1.2}$$

where $Z(u)$ is a random function defined for $u \in \mathbf{R}^k$.

Then if the limit in (5.1.2) is finite as $t \to \infty$, the equality (5.1.1) takes the form

$$I(\xi^t, \theta) = \mathsf{E} \log \left|\det \varphi_t^{-1}(\theta)\right| + h(\theta) - \mathsf{E} \log \int Z(u)\,du + o(1). \tag{5.1.3}$$

The proof of this formula is based on the following theorem justifying relation (5.1.2).

THEOREM 5.1.1. *Let the following conditions be fulfilled:*

1) *finite-dimensional distributions of the random function $Z_t(u)$ converge, as $t \to \infty$, to finite-dimensional distributions of the random function $Z(u)$;*

2) *for all $u, v, y \in \mathbf{R}^k$ we have*

$$\mathsf{E}_y \left|Z_t(u) - Z_t(v)\right| \le C|u - v| \left(1 + |y|^p + \left(|\varphi_t(y)u| \vee |\varphi_t(y)v|\right)^p\right), \tag{5.1.4}$$

where the constants $C > 0$ and $p \ge 0$ are independent of t, u, v, y, and $\mathsf{E}\,|\theta|^p < \infty$ and $\mathsf{E}\,|\varphi_t(\theta)|^p \le L < \infty$ for all $t \in \mathbf{R}_+$;

3) *for all $N > 0$ and $u \in \mathbf{R}^k$ we have*

$$\mathsf{P}\left\{Z_t(u) > |u|^{-N}\right\} \le C_N |u|^{-N},$$

where the constant C_N depends only on N;

4) *one of the following two conditions is fulfilled:*

4′) *for each $N > 0$ we have*

$$\mathsf{P}\text{-}\lim_{A\to\infty} A^N \sup_{|u|>A} Z_t(u) = 0$$

uniformly in t, and for some $\Delta > 0$ and $\delta > 0$ we have

$$\lim_{t\to\infty} \mathsf{P}\left\{t^\delta \left|\det \varphi_t(\theta)\right| > \Delta\right\} = 1;$$

4″) *the function $q(y)$ is bounded and*

$$\mathsf{P}\text{-}\lim_{A\to\infty} \int_{|u|>A} Z_t(u)\,du = 0$$

uniformly in t;

5) *for some positive constants* $A, B, C, a, b, c, d, g, f$, *independent of* $t \in \mathbf{R}_+$ *and* $u \in \mathbf{R}^k$, *the following inequalities are valid:*

$$\int \left|q(y + \varphi_t(y)u) - q(y)\right| dy \le At^{-b}|u|^a, \tag{5.1.5}$$

$$\int |y|^p q(y + \varphi_t(y)u)\, dy \le B\left(1 + |u|^c t^{-d}\right), \tag{5.1.6}$$

$$\int |\varphi_t(y)|^p q(y + \varphi_t(y)u)\, dy \le C\left(1 + |u|^f t^{-g}\right), \tag{5.1.7}$$

where p *is the constant of condition* 2).

Then relation (5.1.2) *is valid.*

PROOF. To prove (5.1.2), it is enough to establish the convergence

$$\mathcal{L}\Big(\int \widetilde{Z}_t(u)\, du \,\Big|\, \mathsf{P}\Big) \xrightarrow{w} \mathcal{L}\Big(\int Z(u)\, du \,\Big|\, \mathsf{P}\Big), \qquad t \to \infty, \tag{5.1.8}$$

and uniform integrability of $\log \int \widetilde{Z}_t(u)\, du$ [**7**], that is, to show that for some $t_0 \in \mathbf{R}_+$

$$\lim_{N\to\infty} \sup_{t \ge t_0} \mathsf{E}\left|\log \int \widetilde{Z}_t(u)\, du\right| \chi\left(\left|\log \int \widetilde{Z}_t(u)\, du\right| > N\right) = 0. \tag{5.1.9}$$

These relations can be proved in the same way as Lemmas 3.1–3.3 of [**15**], with appropriate modification.

In view of the Chebyshev inequality, the equality $\mathsf{E}_y\, Z_t(u) = 1$, and inequality (3.1.5), for any $\varepsilon > 0$ and $H > 0$ we have

$$\begin{aligned} &\mathsf{P}\left\{\int_{|u|\le H} \left|\widetilde{Z}_t(u) - Z_t(u)\right| du > \varepsilon\right\} \\ &\quad \le \frac{1}{\varepsilon} \int \left(\int_{|u|\le H} \left|q(y + \varphi_t(y)u) - q(y)\right| \mathsf{E}_y\, Z_t(u)\, du\right) dy \\ &\quad \le \frac{A_k H^{k+a}}{\varepsilon t^b}, \end{aligned} \tag{5.1.10}$$

where the constant A_k depends only on k.

Putting $H = t^\lambda$, $\lambda < b/(k + a)$, we see that the right-hand side of the inequality (5.1.10) vanishes as $t \to \infty$. If condition 4″) is fulfilled, then in view of the inequality

$$\int_{|u|>t^\lambda} \widetilde{Z}_t(u)\, du \le \left(\sup_y q(y)/q(\theta)\right) \int_{|u|>t^\lambda} Z_t(u)\, du$$

we get

$$\mathsf{P}\text{-}\lim_{t\to\infty} \int_{|u|>t^\lambda} \widetilde{Z}_t(u)\, du = 0. \tag{5.1.11}$$

If condition 4′) is fulfilled, then

$$\mathsf{P}\text{-}\lim_{t\to\infty}\int_{|u|>t^{\lambda}} Z_t(u)\,du = 0. \tag{5.1.12}$$

In view of the inequality

$$\begin{aligned}\int_{|u|>t^{\lambda}} \widetilde{Z}_t(u)\,du &\le \sup_{|u|>t^{\lambda}} Z_t(u)\int \frac{q(\theta+\varphi_t(\theta)u)}{q(\theta)}\,du\\ &= \sup_{|u|>t^{\lambda}} Z_t(u)\frac{\left|\det\varphi_t^{-1}(\theta)\right|}{q(\theta)} = t^{\lambda N}\sup_{|u|>t^{\lambda}} Z_t(u)\frac{\left|\det\varphi_t^{-1}(\theta)\right|}{t^{\lambda N}q(\theta)},\end{aligned}$$

condition 4′) implies relation (5.1.11). Combining (5.1.10)–(5.1.12), we get

$$\mathsf{P}\text{-}\lim_{t\to\infty}\Big(\int \widetilde{Z}_t(u)\,du - \int Z_t(u)\,du\Big) = 0. \tag{5.1.13}$$

By Theorem II.1.22 in [**40**] and conditions 1) and 2), for any $H>0$ we have

$$\mathcal{L}\bigg(\int_{|u|\le H} Z_t(u)\,du\;\bigg|\;\mathsf{P}\bigg) \xrightarrow{w} \mathcal{L}\bigg(\int_{|u|\le H} Z(u)\,du\;\bigg|\;\mathsf{P}\bigg), \qquad t\to\infty.$$

Taking into account condition 4), we get

$$\mathcal{L}\Big(\int Z_t(u)\,du\;\Big|\;\mathsf{P}\Big) \xrightarrow{w} \mathcal{L}\Big(\int Z(u)\,du\;\Big|\;\mathsf{P}\Big), \qquad t\to\infty. \tag{5.1.14}$$

Relations (5.1.13) and (5.1.14) imply (5.1.8).

Now we proceed to the proof of (5.1.9). First we prove the equality

$$\lim_{H\to\infty}\sup_{t\ge t_0}\mathsf{E}\log\int \widetilde{Z}_t(u)\,du\,\chi\Big(\log\int \widetilde{Z}_t(u)\,du > H\Big) = 0. \tag{5.1.15}$$

For any integer $H>0$ we have

$$\begin{aligned}&\mathsf{E}\log\int \widetilde{Z}_t(u)\,du\,\chi\Big(\log\int \widetilde{Z}_t(u)\,du > (a+k)H\Big)\\ &\quad= \sum_{l=H}^{\infty}\mathsf{E}\log\int \widetilde{Z}_t(u)\,du\,\chi\Big(2^{(a+k)l} < \int \widetilde{Z}_t(u)\,du \le 2^{(a+k)(l+1)}\Big)\\ &\quad\le (a+k)\sum_{l=H}^{\infty}(l+1)\mathsf{P}\Big\{\int \widetilde{Z}_t(u)\,du > 2^{(a+k)l}\Big\}\\ &\quad\le (a+k)\sum_{l=H}^{\infty}(l+1)\bigg[\mathsf{P}\bigg\{\int_{|u|\le 2^{l/2}} \widetilde{Z}_t(u)\,du > 2^{(a+k)l-1}\bigg\}\\ &\qquad\qquad + \mathsf{P}\bigg\{\int_{|u|>2^{l/2}} \widetilde{Z}_t(u)\,du > 2^{(a+k)l-1}\bigg\}\bigg],\end{aligned} \tag{5.1.16}$$

where a is the constant of condition 5).

In view of (5.1.5),

$$\int q\big(y+\varphi_t(y)u\big)\,dy \le 1 + A|u|^a \tag{5.1.17}$$

for $t \ge 1$; then, applying the Chebyshev inequality, we get

$$\begin{aligned}
\mathsf{P}\bigg\{\int_{|u|\le 2^{l/2}} \widetilde{Z}_t(u)\,du > 2^{(a+k)l-1}\bigg\} &\le 2^{1-(a+k)l}\,\mathsf{E}\int_{|u|\le 2^{l/2}} \widetilde{Z}_t(u)\,du\\
&\le 2^{1-(a+k)l}\int_{|u|\le 2^{l/2}} du \int q\big(y+\varphi_t(y)u\big)\,\mathsf{E}_y\, Z_t(u)\,dy\\
&\le 2^{1-(a+k)l}\int_{|u|\le 2^{l/2}} (1+A|u|^a)\,du \le C2^{-(a+k)l/2}.
\end{aligned}\tag{5.1.18}$$

Assume that

$$\mathsf{P}\bigg\{\int_{\Gamma_m} \widetilde{Z}_t(u)\,du > \frac{1}{m^2}\bigg\} \le \frac{C}{m^2}, \tag{5.1.19}$$

where $\Gamma_m = \{u\colon m < |u| \le m+1\}$.

Then for $l \ge 2$ we have

(5.1.20)

$$\begin{aligned}
\mathsf{P}\bigg\{\int_{|u|>2^{l/2}} \widetilde{Z}_t(u)\,du > 2^{(a+k)l-1}\bigg\} &\le \mathsf{P}\bigg\{\int_{|u|>2^{l/2}} \widetilde{Z}_t(u)\,du > 1\bigg\}\\
&\le \sum_{m=2^{l/2}}^{\infty} \mathsf{P}\bigg\{\int_{\Gamma_m} \widetilde{Z}_t(u)\,du > \frac{1}{m^2}\bigg\} \le C2^{-l/2}.
\end{aligned}$$

Obviously, relations (5.1.16), (5.1.18), and (5.1.20) imply (5.1.15) for $t_0 = 1$.

Now we prove (5.1.19). We divide the spherical layer Γ_m into M, $M \le Dmh^{1-k}$, regions $\Gamma_m^1, \Gamma_m^2, \dots, \Gamma_m^M$, each of diameter not greater than h. Choose some points $u^j \in \Gamma_m^j$, $j = 1, 2, \dots, M$. Then the following chain of inequalities is true:

$$\begin{aligned}
&\mathsf{P}\bigg\{\int_{\Gamma_m} \widetilde{Z}_t(u)\,du > \frac{1}{m^2}\bigg\}\\
&\quad\le \mathsf{P}\bigg\{\sum_{j=1}^{M}\int_{\Gamma_m^j} \frac{q\big(\theta+\varphi_t(\theta)u\big)}{q(\theta)} Z_t(u^j)\,du > \frac{1}{2m^2}\bigg\}\\
&\qquad+ \mathsf{P}\bigg\{\sum_{j=1}^{M}\int_{\Gamma_m^j} \frac{q\big(\theta+\varphi_t(\theta)u\big)}{q(\theta)} \big|Z_t(u) - Z_t\left(u^j\right)\big|\,du > \frac{1}{2m^2}\bigg\}\\
&\quad\le \sum_{j=1}^{M} \mathsf{P}\bigg\{Z_t(u^j)\int_{\Gamma_m^j} \frac{q\big(\theta+\varphi_t(\theta)u\big)}{q(\theta)}\,du > \frac{1}{2Mm^2}\bigg\}\\
&\qquad+ 2m^2\sum_{j=1}^{M} \mathsf{E}\int_{\Gamma_m^j} \frac{q\big(\theta+\varphi_t(\theta)u\big)}{q(\theta)} \big|Z_t(u) - Z_t\left(u^j\right)\big|\,du
\end{aligned}$$

$$\leq \sum_{j=1}^{M}\left[\mathsf{P}\left\{Z_t(u^j) > \frac{1}{|u^j|^N}\right\} + \mathsf{P}\left\{\int_{\Gamma_m^j} \frac{q(\theta+\varphi_t(\theta)u)}{q(\theta)}\,du > \frac{|u^j|^N}{2Mm^2}\right\}\right]$$
$$+ 2m^2\sum_{j=1}^{M}\int_{\Gamma_m^j} du \int q(y+\varphi_t(y)u)\,\mathsf{E}_y\left|Z_t(u) - Z_t\left(u^j\right)\right| dy.$$

Taking into account that $m \leq |u^j| \leq m+1$, in view of conditions 2) and 3) we get

$$\mathsf{P}\left\{\int_{\Gamma_m} \widetilde{Z}_t(u)\,du > \frac{1}{m^2}\right\}$$
$$\leq \frac{MC_N}{m^N} + \frac{2M}{m^{N-2}}\int_{\Gamma_m} du \int q(y+\varphi_t(y)u)dy$$
$$+ 2m^2Ch\int_{\Gamma_m} du \int q(y+\varphi_t(y)u)\left[1+|y|^p+(m+1)^p|\varphi_t(y)|^p\right]dy.$$

Applying the inequalities (5.1.6), (5.1.7), and (5.1.17), for $t > 1$ we have

$$\mathsf{P}\left\{\int_{\Gamma_m} \widetilde{Z}_t(u)\,du > \frac{1}{m^2}\right\} \leq \frac{MC_N}{m^N} + \frac{CM}{m^{N-2-k-a}} + Chm^{2+k+c\vee f\vee a}$$
$$\leq \frac{C_N}{m^{N-1}h^{k-1}} + \frac{C}{m^{N-3-k-a}h^{k-1}} + Chm^{2+k+c\vee f\vee a}.$$

Putting $h = m^{-4-k-c\vee f\vee a}$ and $N = 1+a+(k-1)(c\vee f\vee a)+k(k+4)$, we get the required estimate (5.1.19). Thus relation (5.1.15) is completely proved.

Now we prove the equality

$$\lim_{H\to\infty}\sup_{t\geq t_0}\mathsf{E}\left|\log\int \widetilde{Z}_t(u)\,du\right|\chi\left(\log\int \widetilde{Z}_t(u)\,du < -H\right) = 0. \tag{5.1.21}$$

For any integer $H > 0$ we have

$$\begin{aligned}
&\mathsf{E}\left|\log\int \widetilde{Z}_t(u)\,du\right|\chi\left(\log\int \widetilde{Z}_t(u)\,du < -kH\right)\\
&\quad= \sum_{l=H}^{\infty}\mathsf{E}\left|\log\int \widetilde{Z}_t(u)\,du\right|\chi\left(2^{-k(l+1)} \leq \int \widetilde{Z}_t(u)\,du < 2^{-kl}\right)\\
&\quad\leq k\sum_{l=H}^{\infty}(l+1)\mathsf{P}\left\{\int \widetilde{Z}_t(u)\,du < 2^{-kl}\right\}.
\end{aligned} \tag{5.1.22}$$

But, obviously,

$$\begin{aligned}
\mathsf{P}\left\{\int \widetilde{Z}_t(u)\,du < 2^{-kl}\right\} &\leq \mathsf{P}\left\{\int_{B_l} \widetilde{Z}_t(u)\,du < 2^{-kl}\right\}\\
&= \mathsf{P}\left\{\int_{B_l}\left(\widetilde{Z}_t(u)-1\right)du < -\left(2^k-1\right)2^{-kl}\right\}\\
&\leq \mathsf{P}\left\{\int_{B_l}\left|\widetilde{Z}_t(u)-1\right|du > \left(2^k-1\right)2^{-kl}\right\}\\
&\leq (2^k-1)^{-1}2^{kl}\int_{B_l}\mathsf{E}\left|\widetilde{Z}_t(u)-1\right|du,
\end{aligned} \tag{5.1.23}$$

where $B_l = \{u = (u_1, u_2, \ldots, u_k)\colon |u_i| \leq 2^{-l}, i = 1, 2, \ldots, k\}$.

Taking into consideration that $Z_t(0) = 1$ and using conditions 2) and 5), for $t \geq 1$ and $|u| \leq 1$ we get

$$
\begin{aligned}
\mathsf{E}\left|\widetilde{Z}_t(u) - 1\right| &\leq \int q(y)\, \mathsf{E}_y \left| \frac{q(y+\varphi_t(y)u)}{q(u)} Z_t(u) - 1 \right| dy \\
&\leq \int q\big(y+\varphi_t(y)u\big)\, \mathsf{E}_y \left|Z_t(u) - Z_t(0)\right| dy \\
&\quad + \int \left|q\big(y+\varphi_t(y)u\big) - q(y)\right| dy \\
&\leq c(|u| + |u|^a) \leq C|u|^\gamma,
\end{aligned}
\tag{5.1.24}
$$

where $C > 0$ and $\gamma > 0$ are some constants ($\gamma = a \wedge 1$).

Since $|u| \leq k^{1/2}2^{-l}$ for $u \in B_l$, from (5.1.23) and (5.1.24) for $t \geq 1$ we have

$$
\mathsf{P}\left\{ \int \widetilde{Z}_t(u)\, du < 2^{-kl} \right\} \leq C2^{-\gamma l},
$$

and hence (5.1.22) imply (5.1.21) with $t_0 = 1$.

Relations (5.1.15) and (5.1.21) imply relation (5.1.9) for $t_0 = 1$. □

From Theorem 5.1.1 and relations (5.1.1) and (5.1.2) the following theorem follows.

THEOREM 5.1.2. *If the conditions of Theorem* 5.1.1 *are fulfilled and the quantities* $\mathsf{E}\left|\log q(\theta)\right|$ *and* $\mathsf{E}\left|\log\left|\det \varphi_t(\theta)\right|\right|$ *are finite, then for* $t \to \infty$ *the asymptotic formula* (5.1.3) *is valid.*

Note that the finiteness of $\mathsf{E}\left|\log \int Z(u)\, du\right|$ follows from the convergence (5.1.8) and uniform integrability of $\log \int \widetilde{Z}(u)\, du$ with respect to the measure P [**7**].

COROLLARY 5.1.1. *If the conditions of Theorem* 5.1.2 *are fulfilled and the family of measures* $(P_y^t, y \in \Theta)$ *possesses the LAN property as* $t \to \infty$ *at all points* $y \in \Theta$ *with normalizing matrix* $\varphi_t(y)$, *then for* $t \to \infty$

$$
I(\xi^t, \theta) = \mathsf{E}\log\left|\det \varphi_t^{-1}(\theta)\right| + h(\theta) - \log(2\pi e)^{k/2} + o(1). \tag{5.1.25}
$$

PROOF. It suffices to observe that in view of the LANC

$$
Z(u) = \exp\left\{u'\eta - 2^{-1}|u|^2\right\}, \qquad \mathcal{L}(\eta|\mathsf{F}) = \mathcal{N}(0, I_k).
$$

From this it follows that

$$
\mathsf{E}\log \int Z(u)\, du = \log(2\pi e)^{k/2}, \tag{5.1.26}
$$

and hence (5.1.1) and (5.1.26) imply (5.1.25). □

An arbitrary distribution of parameter θ. Now let the measure Q be not absolutely continuous with respect to Lebesgue measure. We assume that there is at most a countable set of isolated points $\Theta' = \{y_1, y_2, \dots\} \subset \Theta$ such that the measure $\widetilde{Q}(A) = Q(A \setminus \Theta')$, $A \in \mathcal{B}^k$, is absolutely continuous with respect to Lebesgue measure and $Q(\{y_i\}) > 0$ for all $i = 1, 2, \dots$. Put

$$\mathsf{P}\{\theta \in \Theta'\} = p, \qquad \mathsf{P}\{\theta = y_j/\chi = 1\} = p_j, \quad j = 1, 2, \dots,$$
$$\mathsf{P}\{\theta \in A/\chi = 0\} = \int_A q(y)\, dy, \qquad A \in \mathcal{B}^k,$$

where $\chi = \chi(\Theta'; \theta)$ and $q(y)$ is the density of conditional distribution of the vector θ with respect to Lebesgue measure under the condition that $\chi = 0$.

By the formula of conditional information,

$$I\left(\xi^t, \theta\right) = I\left(\xi^t, \chi\right) + pI\left(\xi^t, \theta/\chi = 1\right) + qI\left(\xi^t, \theta/\chi = 0\right), \tag{5.1.27}$$

where $q = 1 - p$ and $I(\xi^t, \theta/\chi = i)$ is the *conditional information in ξ^t with respect to θ under the condition that $\chi = i$.*

Under certain assumptions for $t \to \infty$ we have

$$I\left(\xi^t, \chi\right) = H(\chi) + o(1), \tag{5.1.28}$$
$$I\left(\xi^t, \theta/\chi = 1\right) = H(\theta/\chi = 1) + o(1), \tag{5.1.29}$$

where $H(\chi)$ is the *entropy of distribution of the random variable* χ and $H(\theta \setminus \chi = 1)$ is the *conditional entropy of distribution of the random variable θ under the condition that* $\chi = 1$. Similarly to (5.1.1), we have

$$\begin{aligned} I\left(\xi^t, \theta/\chi = 0\right) = \mathsf{E}\left\{\log\left|\det \varphi_t^{-1}(\theta)\right| / \chi = 0\right\} + h(\theta/\chi = 0) \\ - \mathsf{E}\left\{\log \int \widetilde{Z}_t(u)\, du/\chi = 0\right\}, \end{aligned} \tag{5.1.30}$$

if the random variables in (5.1.30) are well defined and finite. Here $h(\theta/\chi = 0)$ is the *conditional differential entropy of distribution of the random vector θ under the condition that* $\chi = 0$. Under certain conditions,

$$\lim_{t \to \infty} \mathsf{E}\left\{\log \int \widetilde{Z}_t(u)\, du/\chi = 0\right\} = \mathsf{E}\left\{\log \int Z(u)\, du/\chi = 0\right\}. \tag{5.1.31}$$

Thus from (5.1.27)–(5.1.31) we have for $t \to \infty$

$$\begin{aligned} I\left(\xi^t, \theta\right) = q\Bigg[\mathsf{E}\left\{\log\left|\det \varphi_t^{-1}(\theta)\right| / \chi = 0\right\} \\ + h(\theta/\chi = 0) - \mathsf{E}\left\{\log \int Z(u)\, du/\chi = 0\right\}\Bigg] \\ + pH(\theta/\chi = 1) + H(\chi) + o(1). \end{aligned} \tag{5.1.32}$$

Now we prove (5.1.32). As is obvious from the previous arguments it suffices to prove (5.1.28), (5.1.29), and (5.1.31).

LEMMA 5.1.1. *Let* $H(\theta/\chi=1)<\infty$ *and* $\mathsf{E}_y\,\rho_t^{1/2}(y',y)\to 0$ *as* $t\to\infty$ *for any* y, $y'\in\Theta'$, $y\neq y'$. *Then relation* (5.1.29) *is valid.*

PROOF. From the properties of information we have

$$I\left(\xi^t,\theta/\chi=1\right)=H\left(\theta/\chi=1\right)-\mathsf{E}\,H\left(\theta/\xi^t,\chi=1\right),$$

where $\mathsf{E}\,H(\theta/\xi^t,\chi=1)$ is the *average conditional entropy of distribution of the vector* θ *under the condition that* ξ^t *is given and* $\chi=1$. Hence in order to prove (5.1.29) it is enough to show that $\mathsf{E}\,H(\theta/\xi^t,\chi=1)\to 0$ as $t\to\infty$.

First let $\Theta'=\{y_1,y_2,\dots,y_n\}$, where $n<\infty$. Then

$$\mathsf{E}\,H\left(\theta/\xi^t,\chi=1\right)=-\,\mathsf{E}\Big\{\sum_{j=1}^{n}p_j(\xi^t)\log p_j(\xi^t)/\chi=1\Big\},$$

where $p_j(x)=\mathsf{P}\{\theta=y_j/\xi^t=x,\chi=1\}$ is the a posteriori distribution of the vector θ. According to the Bayes formula,

$$p_j(x)=p_j\rho_t(x;y_j,y)\Big(\sum_{i=1}^{n}p_i\rho_t(x;y_i,y)\Big)^{-1},\qquad x\in X^t,$$

for any $y\in\Theta$. Since $0\le-\sum_{j=1}^{n}p_j(x)\log p_j(x)\le\log n$, the lemma can be proved by showing that

$$\lim_{t\to\infty}\mathsf{P}\Big\{-\sum_{j=1}^{n}p_j(\xi^t)\log p_j(\xi^t)>\varepsilon\;/\;\chi=1\Big\}=0$$

for any $\varepsilon>0$. To prove this relation, it is enough to show that

$$\mathsf{P}\left\{\varepsilon<p_j\left(\xi^t\right)<1-\varepsilon/\chi=1\right\}\to 0$$

as $t\to\infty$ for any $\varepsilon>0$ and $j=1,2,\dots,n$. We have

$$\mathsf{P}\left\{\varepsilon<p_j\left(\xi^t\right)<1-\varepsilon/\chi=1\right\}=\sum_{i=1}^{n}\mathsf{P}\left\{\varepsilon<p_j\left(\xi^t\right)<1-\varepsilon/\theta=y_i\right\}p_i. \tag{5.1.33}$$

Let us represent $p_j(x)$ in the following way:

$$p_j(x)=p_j\rho_t(x;y_j,y_i)\Big(p_i+\sum_{r\neq i}p_r\rho_t(x;y_r,y_i)\Big)^{-1}. \tag{5.1.34}$$

For any $\varepsilon>0$ the following inequalities are true:

$$\mathsf{P}_{y_i}\{\rho_t(y_j,y_i)>\varepsilon\}\le\varepsilon^{-1/2}\,\mathsf{E}_{y_i}\,\rho_t^{1/2}(y_j,y_i),\qquad j\neq i, \tag{5.1.35}$$

$$\begin{aligned}\mathsf{P}_{y_i}\Big\{\sum_{r\neq i}p_r\rho_t(y_r,y_i)>\varepsilon\Big\}&\le\sum_{r\neq i}\mathsf{P}_{y_i}\left\{\rho_t(y_r,y_i)>\frac{\varepsilon}{(n-1)p_r}\right\}\\&\le\sqrt{\frac{\varepsilon}{n-1}}\sum_{r\neq i}\sqrt{p_r}\,\mathsf{E}_{y_i}\,\rho_t^{1/2}(y_r,y_i).\end{aligned} \tag{5.1.36}$$

Since by the conditions of the lemma, $\mathsf{E}_{y_i}\,\rho_t^{1/2}(y_r, y_i) \to 0$ as $t \to \infty$ for $r \neq i$, from relations (5.1.34)–(5.1.36) we get that $\mathsf{P}_{y_i}\{p_j(\xi^t) > \varepsilon\} \to 0$ if $j \neq i$, and $\mathsf{P}_{y_i}\{p_j(\xi^t) < 1 - \varepsilon\} \to 0$ if $j = i$ for any $\varepsilon > 0$ as $t \to \infty$. Hence the equality (5.1.33) implies that $\mathsf{P}\{\varepsilon < p_j(\xi^t) < 1 - \varepsilon/\chi = 1\} \to 0$ as $t \to \infty$ for any $\varepsilon > 0$ and $j = 1, 2, \ldots, n$. Thus the assertion of the lemma is proved in the case when the set Θ' is finite.

Now let $\Theta' = \{y_1, y_2, \ldots\}$ be, in general, a countable set. Consider the random variable $\chi_n = \chi(\Theta'_n; \theta)$, where $\Theta'_n = \{y_1, y_2, \ldots, y_n\}$, $n < \infty$. By the formula for conditional information,

$$\mathsf{E}\,H\left(\theta/\xi^t, \chi = 1\right) = \mathsf{E}\,H\left(\chi_n/\xi^t, \chi = 1\right) + \mathsf{E}\,H\left(\theta/\xi^t, \chi_n, \chi = 1\right). \tag{5.1.37}$$

Since $\mathsf{P}\{\chi_n = 1/\chi = 1\} \to 1$ as $n \to \infty$, we have

$$\mathsf{E}\,H\left(\chi_n/\xi^t, \chi = 1\right) \leq H(\chi_n/\chi = 1) \to 0 \quad \text{as } n \to \infty.$$

The second term on the right-hand side of (5.1.37) can be represented in the form

$$\mathsf{E}\,H\left(\theta/\xi^t, \chi_n, \chi = 1\right) = \sum_{i=0}^{1} \mathsf{E}\,H\left(\theta/\xi^t, \chi_n = i, \chi = 1\right) \mathsf{P}\{\chi_n = i/\chi = 1\}.$$

Taking into account the inequality

$$\mathsf{E}\,H\left(\theta/\xi^t, \chi_n = 0, \chi = 1\right) \leq H(\theta/\chi_n = 0, \chi = 1)$$

and the equality

$$\begin{aligned} &H(\theta/\chi_n = 0, \chi = 1)\mathsf{P}\{\chi_n = 0/\chi = 1\} \\ &\qquad = \mathsf{P}\{\chi_n = 0/\chi = 1\} \log \mathsf{P}\{\chi_n = 0/\chi = 1\} - \sum_{l=n+1}^{\infty} p_l \log p_l, \end{aligned}$$

we get that

$$\mathsf{E}\,H\left(\theta/\xi^t, \chi_n = 0, \chi = 1\right) \mathsf{P}\{\chi_n = 0/\chi = 1\} \to 0$$

as $n \to \infty$, uniformly in t. Now using the first part of the proof of the lemma (the case $\Theta' = \Theta'_n$, $n < \infty$), for a fixed $n < \infty$ we get

$$\mathsf{E}\,H\left(\theta/\xi^t, \chi_n = 1, \chi = 1\right) \mathsf{P}\{\chi_n = 1/\chi = 1\} \to 0$$

as $t \to \infty$. Thus the second term on the right-hand side of (5.1.37) can be made arbitrarily small by choosing first n and then t sufficiently large. Therefore

$$\mathsf{E}\,H\left(\theta/\xi^t, \chi = 1\right) \to 0 \quad \text{as } t \to \infty$$

if the set Θ' is countable. □

LEMMA 5.1.2. *Let the following conditions be fulfilled:*

1) *under the condition $\chi = 0$ the conditional finite-dimensional distributions of the random field $Z_t(u)$ weakly converge, as $t \to \infty$, to the conditional finite-dimensional distributions of the random field $Z(u)$ under the condition that $\chi = 0$;*

2) *for all $y \in \Theta \setminus \Theta'$ and $u, v \in \mathbf{R}^k$ we have*

$$\mathsf{E}_y \left| Z_t(u) - Z_t(v) \right| \le C|u - v| \left[1 + |y|^p + \left(|\varphi_t(y)u| \vee |\varphi_t(y)v| \right)^p \right],$$

where the constants $C > 0$ and $p \ge 0$ are independent of t and y,

$$\mathsf{E}\{|\theta|^p / \chi = 0\} < \infty,$$

and $\mathsf{E}\{|\varphi_t(\theta)|^p / \chi = 0\} \le L < \infty$;

3) *for any $N > 0$ and all $u \in \mathbf{R}^k$,*

$$\mathsf{P}\left\{ Z_t(u) > |u|^{-N} / \chi = 0 \right\} \le C_N |u|^{-N},$$

where the constant C_N depends only on N;

4) *one of the following conditions is fulfilled:*

4′) *for any $N > 0$ and $\varepsilon > 0$*

$$\lim_{A \to \infty} \sup_t \mathsf{P}\left\{ A^N \sup_{|u| > A} Z_t(u) > \varepsilon \;/\; \chi = 0 \right\} = 0$$

and for some $\delta > 0$ and $\Delta > 0$

$$\lim_{t \to \infty} \mathsf{P}\left\{ t^{-\delta} \left| \det \varphi_t(\theta) \right| \ge \Delta / \chi = 0 \right\} = 1;$$

4″) *the function $q(y)$ is bounded and for any $\varepsilon > 0$*

$$\lim_{A \to \infty} \sup_t \mathsf{P}\left\{ \int_{|u| > A} Z_t(u)\, du > \varepsilon \;/\; \chi = 0 \right\} = 0;$$

5) *condition 5) of Theorem 5.1.1 is fulfilled.*

Then relation (5.1.31) is true.

Lemma 5.1.2 follows from Theorem 5.1.1.

LEMMA 5.1.3. *Let the following conditions be fulfilled:*

1) *the function $q(y)$ is continuous on $\Theta \setminus \Theta'$;*

2) *for any $y \in \Theta$ and some $\alpha \in (0, 1)$*

$$\mathsf{P}_y\text{-}\lim_{t \to \infty} \left| \det \varphi_t(y) \right|^{-1} \sup_{|u| > |\det \varphi_t(y)|^{\alpha}} \rho_t(y + u, y) = 0;$$

3) *for some $\beta > 0$ and $A > 0$ for all $y \in \Theta$ and $|u| \le 1$*

$$\mathsf{E}_y \left| Z_t(u) - Z_t(0) \right| \le A|u|^{\beta}.$$

Then relation (5.1.28) is true.

PROOF. Obviously, $I\left(\xi^t,\chi\right)=H(\chi)-\mathsf{E}\,H\left(\chi/\xi^t\right)$, where

$$\mathsf{E}\,H\left(\chi/\xi^t\right)=-\sum_{i=0}^{1}\mathsf{E}\,q_i\left(\xi^t\right)\log q_i\left(\xi^t\right),$$
$$q_i(x)=\mathsf{P}\left\{\chi=i/\xi^t=x\right\},\qquad x\in X^t.$$

Therefore to prove relation (5.1.28), it suffices to show that

$$\mathsf{P}\left\{\varepsilon<q_i\left(\xi^t\right)<1-\varepsilon\right\}\to 0\quad\text{as } t\to\infty$$

for all $\varepsilon>0$ and $i=0,1$. But

$$\mathsf{P}\left\{\varepsilon<q_i(\xi^t)<1-\varepsilon\right\}=\int\mathsf{P}_y\left\{\varepsilon<q_i\left(\xi^t\right)<1-\varepsilon\right\}Q(dy)$$

and hence it is enough to show that $\mathsf{P}_y\{\varepsilon<q_i(\xi^t)<1-\varepsilon\}\to 0$ as $t\to\infty$ for any $\varepsilon>0$ and Q-almost all $y\in\mathbf{R}^k$.

First we consider $q_1(\xi^t)$. By the Bayes formula, we have

$$q_1(\xi^t)=\frac{p\sum_{i=1}^{\infty}p_i\rho_t(y_i,y)}{p\sum_{i=1}^{\infty}p_i\rho_t(y_i,y)+q\int\rho_t(v,y)q(v)\,dv}\tag{5.1.38}$$

for any $y\in\Theta$. Let $y=y_r\in\Theta'$. Using the Chebyshev inequality, for any $\varepsilon>0$ and $\gamma>0$ we get

(5.1.39)
$$\begin{aligned}\mathsf{P}_{y_r}\Big\{\sum_{i\neq r}p_i\rho_t(y_i,y_r)>\varepsilon\Big\}&\le\mathsf{P}_{y_r}\Big\{\sum_{\substack{i\neq r\\|y_i-y_r|\le\gamma}}p_i\rho_t(y_i,y_r)>\frac{\varepsilon}{2}\Big\}\\&\quad+\mathsf{P}_{y_r}\Big\{\sum_{\substack{i\neq r\\|y_i-y_r|>\gamma}}p_i\rho_t(y_i,y_r)>\frac{\varepsilon}{2}\Big\}\\&\le\frac{2}{\varepsilon}\sum_{\substack{i\neq r\\|y_i-y_r|\le\gamma}}p_i+\mathsf{P}_{y_r}\left\{\sup_{|y_i-y_r|>\gamma}\rho_t(y_i,y_r)>\frac{\varepsilon}{2}\right\}.\end{aligned}$$

It is evident that the first term on the right-hand side of (5.1.39) vanishes as $\gamma\to 0$ uniformly in t and, due to condition 2), the second term also vanishes as $t\to\infty$ for any $\gamma>0$ and $\varepsilon>0$. From this it follows that

$$\mathsf{P}_{y_r}\text{-}\lim_{t\to\infty}\sum_{i\neq r}p_i\rho_t(y_i,y_r)=0.\tag{5.1.40}$$

Similarly to (5.1.39), for any $\varepsilon>0$ and $\gamma>0$ we get

$$\begin{aligned}&\mathsf{P}_{y_r}\left\{\int\rho_t(v,y_r)q(v)\,dv>\varepsilon\right\}\\&\qquad\le\frac{2}{\varepsilon}\int_{|v-y_r|\le\gamma}q(v)\,dv+\mathsf{P}_{y_r}\left\{\sup_{|v-y_r|>\gamma}\rho_t(v,y_r)>\frac{\varepsilon}{2}\right\}.\end{aligned}$$

Therefore in view of condition 2) and arbitrariness of γ we have

$$\mathsf{P}_{y_r}\text{-}\lim_{t\to\infty}\int \rho_t(v,y_r)q(v)\,dv = 0. \tag{5.1.41}$$

Relations (5.1.38), (5.1.40), and (5.1.41) imply that $\mathsf{P}_{y_r}\{q_1(\xi^t) < 1-\varepsilon\} \to 0$ as $t\to\infty$ for all $\varepsilon > 0$.

Now let $y \in (\Theta\backslash\Theta')\backslash(\{y\colon q(y)=0\}\cup N)$, where N is the set of limit points of the set Θ'. Since the set N is at most countable, we have $Q(\{y\colon q(y)=0\}\cup N) = 0$ and $q(y) > 0$ for $y \in (\Theta\backslash\Theta')\backslash(\{y\colon q(y)=0\}\cup N)$. Let us show that $\mathsf{P}_y\{q_1(\xi^t) > \varepsilon\} \to 0$ as $t\to\infty$ for any $\varepsilon > 0$. To this end it suffices to show that

$$\mathsf{P}_{y_r}\text{-}\lim_{t\to\infty}\sum_{i=1}^{\infty} p_i\rho_t(y_i,y)\Big/\int \rho_t(v,y)q(v)\,dv = 0. \tag{5.1.42}$$

Since the function $q(y)$ is continuous, for sufficiently small u we have

$$q\big(y+\varphi_t(y)u\big) > \frac{1}{2}q(y).$$

Moreover, for sufficiently small $\gamma > 0$ and sufficiently large t

$$\begin{aligned}\int \rho_t(v,y)q(v)\,dv &= |\det\varphi_t(y)|\int \rho_t\big(y+\varphi_t(y)u,y\big)q\big(y+\varphi_t(y)u\big)\,du\\ &\geq |\det\varphi_t(y)|\int_{B_\gamma} \rho_t\big(y+\varphi_t(y)u,y\big)q\big(y+\varphi_t(y)u\big)\,du\\ &\geq \frac{1}{2}q(y)\,|\det\varphi_t(y)|\int_{B_\gamma}\rho_t\big(y+\varphi_t(y)u,y\big)\,du,\end{aligned}$$

where $B_\gamma = \{u = (u_1,u_2,\dots,u_k)\colon 0\le u\le\gamma, i=1,2,\dots,k\}$.

Thus for any $\varepsilon > 0$ and sufficiently small $\gamma > 0$ and sufficiently large t we get

$$\begin{aligned}&\mathsf{P}_y\left\{\sum_{i=1}^{\infty} p_i\rho_t(y_i,y)\left(\int \rho_t(v,y)q(v)\,dv\right)^{-1} > \varepsilon\right\}\\ &\quad\le \mathsf{P}_y\left\{\sum_{i=1}^{\infty} p_i\rho_t(y_i,y)\left(\frac{1}{2}q(y)\,|\det\varphi_t(y)|\int_{B_\gamma} Z_t(u)\,du\right)^{-1} > \varepsilon\right\}\\ &\quad\le \mathsf{P}_y\left\{\int_{B_\gamma} Z_t(u)\,du < \frac{1}{2}\gamma^k\right\}\\ &\qquad+\mathsf{P}_y\left\{\sum_{i=1}^{\infty} p_i\rho_t(y_i,y) > \frac{\varepsilon}{4}q(y)\,|\det\varphi_t(y)|\,\gamma^k\right\}.\end{aligned} \tag{5.1.43}$$

In view of condition 3), analogously to (5.1.23) and (5.1.24) we have

$$\begin{aligned}\mathsf{P}_y\left\{\int_{B_\gamma} Z_t(u)\,du < \frac{1}{2}\gamma^k\right\} &\le \mathsf{P}_y\left\{\int_{B_\gamma} |Z_t(u)-Z_t(0)|\,du > \frac{1}{2}\gamma^k\right\}\\ &\le 2\gamma^{-k}\int_{B_\gamma} \mathsf{E}_y\,|Z_t(u)-Z_t(0)|\,du \le 2A\gamma^{\beta}.\end{aligned}$$

Therefore the first term on the right-hand side of (5.1.43) vanishes as $\gamma \to 0$ uniformly in t. In view of the estimate

$$\mathsf{P}_y\left\{\sum_{i=1}^{\infty} p_i\rho_t(y_i,y) > \varepsilon\right\} \le \frac{2}{\varepsilon}\sum_{|y_i-y|\le\varepsilon^\alpha} p_i + \mathsf{P}_y\left\{\sup_{|y_i-y|>\varepsilon^\alpha} \rho_t(y_i,y) > \frac{\varepsilon}{2}\right\},$$

similar to (5.1.39), and condition 2) the second term on the right-hand side of (5.1.43) also vanishes as $t \to \infty$ for fixed γ. Here we have taken into consideration that the set $\{i\colon |y_i - y| \le \varepsilon^\alpha\}$ is empty for sufficiently small ε. Therefore $\sum_{|y_i-y|\le\varepsilon^\alpha} p_i = 0$. Since γ is chosen arbitrarily, from (5.1.43) we get (5.1.42).

Thus we have proved that $\mathsf{P}\{\varepsilon < q_1(\xi^t) < 1-\varepsilon\} \to 0$ as $t \to \infty$ for any $\varepsilon > 0$.

Now we consider the quantity $q_0(\xi^t)$. By the Bayes formula for any $y \in \Theta$ we have

$$q_0(\xi^t) = \frac{q\int \rho_t(v,y)q(v)\,dv}{p\sum_{i=1}^\infty p_i\rho_t(y_i,y) + q\int\rho_t(v,y)q(v)\,dv}.$$

Then, as for $q_1(\xi^t)$, we see that $\mathsf{P}_y\{q_0(\xi^t) > \varepsilon\} \to 0$ as $t \to \infty$ for any $\varepsilon > 0$ and $y \in \Theta'$, while $\mathsf{P}_y\{q_0(\xi^t) < 1-\varepsilon\} \to 0$ as $t \to \infty$ for any $\varepsilon > 0$ and for Q-almost all $y \in \Theta \setminus \Theta'$. □

Lemmas 5.1.1–5.1.3 and relations (5.1.2)–(5.1.31) imply the following theorem.

THEOREM 5.1.3. *If* $\mathsf{E}\{\log|\det\varphi_t(\theta)|/\chi = 0\}$ *and* $h(\theta/\chi = 0)$ *are finite, and the conditions of Lemmas* 5.1.1–5.1.3 *are fulfilled, then* (5.1.32) *holds as* $t \to \infty$.

REMARK 5.1.1. It is obvious that condition 3) of Lemma 5.1.13 follows from condition 2) of Lemma 5.1.2.

REMARK 5.1.2. In the proof of Lemma 5.1.3 condition 2) was not fully used. As is easily seen from the proof, it is sufficient that for all $y \in \Theta\setminus\Theta'$ for which $q(y) \ne 0$,

$$\mathsf{P}_y\text{-}\lim_{t\to\infty} \left|\det\varphi_t^{-1}(y)\right| \sup_{\substack{v\in\Theta' \\ |v-y|>|\det\varphi_t(y)|^\alpha}} \rho_t(v,y) = 0,$$

and for any $y \in \Theta'$ and all $\gamma > 0$

$$\mathsf{P}_y\text{-}\lim_{t\to\infty}\sup_{|v-y|>\gamma}\rho_t(v,y) = 0.$$

REMARK 5.1.3. Theorem 5.1.3 can also be proved in the case where the limit points of the set Θ' belong to the set Θ'. However, the set $\Theta' \cap A$ does not have to be dense in A for every open set $A \subset \Theta$. The proofs of Lemmas 5.1.1 and 5.1.2 remain the same, except for slight changes in the proof of Lemma 5.1.3.

5.2. Lower bounds for the information about a statistical estimate of a parameter

Auxiliary notions and assumptions. Let $\bar\theta_t = \bar\theta_t(\xi^t)$ be a certain statistical estimate for a random parameter θ with distribution given by a measure Q on $(\mathbf{R}^k, \mathcal{B}^k)$ constructed from an observation ξ^t of a general nature (see Section 1.2). The probability distribution of an estimate $\bar\theta_t$ on $(\mathbf{R}^k, \mathcal{B}^k)$ is denoted by V^t, and the conditional distribution of an estimate $\bar{\theta}_t$, under the condition $\theta = y$ is denoted by V_y^t. Let $L(y,z)$, $y,z \in \mathbf{R}^k$, be a nonnegative Borel function regarded as a loss function, and $\varepsilon(\bar{\theta}_t; y) = \mathsf{E}_y\, L(\theta, \bar{\theta}_t)$, $y \in \mathbf{R}^k$, the *risk function* for the estimate $\bar{\theta}_t$.

We are going to construct lower bounds for the Shannon information $I(\bar{\theta}_t, \theta)$ about the parameter θ contained in the estimate $\bar{\theta}_t$. To do so, we use the risk function $\varepsilon(\bar{\theta}_t; y)$, $y \in \mathbf{R}^k$.

It is easy to derive from the definition of the Shannon information that

$$I\left(\bar{\theta}_t, \theta\right) = \mathsf{E}\, I\left(V_\theta^t | V^t\right), \tag{5.2.1}$$

where $I(V_y^t|V^t)$ is the entropy of the measure V_y^t with respect to the measure V^t or, in other words, the Kullback–Leibler information (see the definition in Section 2.3).

Note that

$$I\left(V_y^t|V^t\right) = \begin{cases} \int \log(dV_y^t/dV^t(z))V_y^t(dz), & \text{if } V_y^t \ll V^t, \\ \infty, & \text{otherwise.} \end{cases}$$

Moreover, for all measures G and S on $\mathcal{B}^k$ and all $y \in \mathbf{R}^k$ and $b \geq 0$ we introduce the quantity

$$J_y(S, G; b) = I(S|G) + bL(y; S) \log e, \tag{5.2.2}$$

where $L(y; S) = \int L(y,z)\, S(dz)$.

We also introduce the probability measure U_y^G on $\mathcal{B}^k$

$$U_y^G(A) = a^G(y) \int_A \exp\{-b(y)L(y,z)\}\, G(dz), \qquad A \in \mathcal{B}^k, \tag{5.2.3}$$

where $y \in \mathbf{R}^k$, $b(y)$ is a nonnegative Borel function and

$$a^G(y) = \left(\int \exp\{-b(y)L(y,z)\}\, G(dz)\right)^{-1}. \tag{5.2.4}$$

LEMMA 5.2.1. *For all probability measures U_y, $y \in \mathbf{R}^k$, any measure G on $\mathcal{B}^k$, and any nonnegative Borel function $b(y)$, $y \in \mathbf{R}^k$, the following equality holds:*

$$J_y(U_y, G; b(y)) = J_y\left(U_y^G, G; b(y)\right) + I\left(U_y | U_y^G\right) \tag{5.2.5}$$

for all $y \in \mathbf{R}^k$ satisfying the condition $J_y(U_y, G; b(y)) < \infty$.

PROOF. By (5.2.2) the condition $J_y(U_y, G; b(y)) < \infty$ implies that $I(U_y|G) < \infty$ and $b(y)L(y;G) < \infty$. Since $I(U_y|G)$ is finite, we get $U_y \ll G$. Since

$$b(y)L(y;G) < \infty,$$

we have

$$\exp\{-b(y)L(y,z)\} > 0 \quad G\text{-a.s.}$$

with respect to the variable z, whence by (5.2.3) $a^G(y) < \infty$. This, together with (5.2.3), implies that $U_y^G \sim G$. From equality (5.2.3) we have

$$\log\left(dU_y^G/dG(z)\right) = \log a^G(y) - b(y)L(y,z) \log e. \tag{5.2.6}$$

Integrating (5.2.6) with respect to the measure U_y^G, by (5.2.2) we get

$$J_y\left(U_y^G, G; b(y)\right) = \log a^G(y). \tag{5.2.7}$$

Since $U_y \ll G$ and $G \ll U_y^G$, we have $U_y \ll U_y^G$ and

$$dU_y/dU_y^G(z) = (dU_y/dG(z))\left(dG/dU_y^G(z)\right).$$

Taking the logarithm and integrating with respect to the measure U_y, we get, according to (5.2.6),

$$I\left(U_y|U_y^G\right) = I(U_y|G) - \log a^G(y) + b(y)L(y; U_y)\log e. \tag{5.2.8}$$

Combining (5.2.2), (5.2.7), and (5.2.8), we get the equality (5.2.5). □

Nonasymptotic lower bounds. We assume that $t \in \mathbf{R}_+$ is fixed. Our goal is to obtain lower bounds for the information $I(\bar{\theta}_t, \theta)$ determined by the loss function $L(y, z)$ and some auxiliary functions.

THEOREM 5.2.1. *For any estimate $\bar{\theta}_t$ of the parameter θ the following inequality holds:*

$$I\left(\bar{\theta}_t, \theta\right) \geq \mathsf{E}\left(\log a(\theta) - b(\theta)\varepsilon\left(\bar{\theta}_t; \theta\right)\log e\right) - \log \mathsf{E}\, c\left(\bar{\theta}_t\right), \tag{5.2.9}$$

where $a(y) > 0$ and $b(y) \geq 0$ (Q-a.s.) are arbitrary Borel functions such that $\mathsf{E}\, J_\theta(V_\theta^t, V^t; b(\theta)) < \infty$, $\mathsf{E}\,|\log a(\theta)| < \infty$, and

$$c(z) = \mathsf{E}\, a(\theta)\exp\{-b(\theta)L(\theta, z)\}, \qquad z \in \mathbf{R}^k. \tag{5.2.10}$$

PROOF. First we use (5.2.2) with $S = V_y^t$ and $G = V^t$ and then we apply (5.2.5) with $U_y = V_y^t$ and $G = V^t$. As a result we get

$$\begin{aligned} &\mathsf{E}\left(I\left(V_\theta^t|V^t\right) + b(\theta)L\left(\theta; V_\theta^t\right)\log e\right) = \mathsf{E}\, J_\theta\left(V_\theta^t, V^t; b(\theta)\right) \\ &\quad = \mathsf{E}\left(J_\theta\left(U_\theta^{V^t}, V^t; b(\theta)\right) + I\left(V_\theta^t|U_\theta^{V^t}\right)\right) \geq \mathsf{E}\, J_\theta\left(U_\theta^{V^t}, V^t; b(\theta)\right). \end{aligned} \tag{5.2.11}$$

Since $\mathsf{E}\log a(\theta)$ is finite, we get by (5.2.7)

$$\mathsf{E}\, J_\theta\left(U_\theta^{V^t}, V^t; b(\theta)\right) = \mathsf{E}\log a(\theta) - \mathsf{E}\log\left(a(\theta)/a^{V^t}(\theta)\right). \tag{5.2.12}$$

Applying Jensen inequality, from (5.2.4) and (5.2.10) we obtain that

$$\begin{aligned} &\mathsf{E}\log\left(a(\theta)/a^{V^t}(\theta)\right) \leq \log \mathsf{E}\left(a(\theta)/a^{V^t}(\theta)\right) \\ &\qquad = \log \mathsf{E}\, a(\theta)\int \exp\{-b(\theta)L(\theta, z)\}\, V^t(dz) = \log \mathsf{E}\, c\left(\bar{\theta}_t\right). \end{aligned} \tag{5.2.13}$$

Combining (5.2.11)–(5.2.13) and using (5.2.1) and equality $L(y; V_y^t) = \varepsilon(\bar{\theta}_t; y)$, we arrive at the inequality (5.2.9). □

REMARK 5.2.1. Inequality (5.2.9) becomes an equality if and only if $V_y^t = U_y^{V^t}$ (Q-a.s.) and $a(y) = ca^{V^t}(y)$ (Q-a.s.) for some constant c. Actually, inequality (5.2.11) becomes an equality if and only if $I(V_y^t | U_y^{V^t}) = 0$ (Q-a.s.), that is, if $V_y^t = U_y^{V^t}$ (Q-a.s.) and inequality (5.2.13) becomes an equality if and only if $a(y)/a^{V^t}(y) = \text{const}$ (Q-a.s.). Note that if $V_y^t = U_y^{V^t}$ (Q-a.s.) and $a(y) = ca^{V^t}(y)$ (Q-a.s.) for some constant c, then

$$I\left(\bar{\theta}_t, \theta\right) = \mathsf{E}\left(\log a^{V^t}(\theta) - b(\theta)\varepsilon\left(\bar{\theta}_t; \theta\right) \log e\right). \tag{5.2.14}$$

REMARK 5.2.2. It follows from (5.2.7) and (5.2.11) that

$$I\left(\bar{\theta}_t, \theta\right) \geq \mathsf{E}\left(\log a^{V^t}(\theta) - b(\theta)\varepsilon\left(\bar{\theta}_t; \theta\right) \log e\right). \tag{5.2.15}$$

Moreover, (5.2.11) implies that inequality (5.2.15) becomes an equality if and only if $V_y^t = U_y^{V^t}$ (Q-a.s.).

The following result provides a method for choosing the functions $a(y)$ and $b(y)$ based on the ε-entropy approach.

THEOREM 5.2.2. *Let $\tilde{L}(y, z)$, $y, z \in \mathbf{R}^k$, be a nonnegative Borel function such that for any fixed y the function $\tilde{L}(y, z)$ is continuous with respect to z at the point $z = y$; $\tilde{L}(y, y) = 0$, $\inf_{|z-y|>c} \tilde{L}(y, z) > 0$ for all $c > 0$, and*

$$\int \tilde{L}^j(y, z) \exp\left\{-\beta_0 \tilde{L}(y, z)\right\} dz < \infty$$

for $j = 0, 1, 2$ and some $\beta_0 \in (0, \infty)$. Then for any estimate $\bar{\theta}_t$ for the parameter θ such that for all $y \in \Theta$

$$\varepsilon(\bar{\theta}_t; y) < \int \tilde{L}(y, z) \exp\left\{-\beta_0 \tilde{L}(y, z)\right\} dz \left(\int \exp\left\{-\beta_0 \tilde{L}(y, z)\right\} dz\right)^{-1}, \tag{5.2.16}$$

$\mathsf{E}\, J_\theta\big(V_\theta^t, V^t; \beta(\theta, \varepsilon(\bar{\theta}_t; \theta))\big) < \infty$, *and* $\left|\mathsf{E} \log a^G\left(\theta, \varepsilon(\bar{\theta}_t; \theta)\right)\right| < \infty$ *the following inequality holds*:

$$I\left(\bar{\theta}_t, \theta\right) \geq \mathsf{E}\left(\log a^G\left(\theta, \varepsilon\left(\bar{\theta}_t; \theta\right)\right) - \varepsilon\left(\bar{\theta}_t; \theta\right) \beta\left(\theta, \varepsilon\left(\bar{\theta}_t; \theta\right)\right) \log e\right) - \log \mathsf{E}\, \gamma\left(\bar{\theta}_t\right), \tag{5.2.17}$$

where $\beta(y, \varepsilon)$ is the solution of the equation (with respect to β)

$$\left(\int \exp\left\{-\beta \tilde{L}(y, z)\right\} dz\right)^{-1} \int \tilde{L}(y, v) \exp\left\{-\beta \tilde{L}(y, v)\right\} dv = \varepsilon, \tag{5.2.18}$$

the functions $\alpha^G(y, \varepsilon)$ and $\gamma(z)$ are given by the formulas

$$\alpha^G(y, \varepsilon) = \left(\int \exp\left\{-\beta(y, \varepsilon)\tilde{L}(y, z)\right\} G(dz)\right)^{-1}, \tag{5.2.19}$$

$$\gamma(z) = \mathsf{E}\, \alpha^G\left(\theta, \varepsilon\left(\bar{\theta}_t; \theta\right)\right) \exp\left\{-\beta\left(\theta, \varepsilon\left(\bar{\theta}_t; \theta\right)\right) L(\theta, z)\right\}, \tag{5.2.20}$$

where G is a measure on $(\mathbf{R}^k, \mathcal{B}^k)$.

PROOF. Inequality (5.2.17) obviously follows from Theorem 5.2.1 for $a(y) = \alpha^G(y, \varepsilon(\bar{\theta}_t; y))$ and $b(y) = \beta(y, \varepsilon(\bar{\theta}_t; y))$. To complete the proof of the theorem, we must prove the existence and uniqueness of the solution of equation (5.2.18) for $\varepsilon < \varepsilon_0(y)$, where $\varepsilon_0(y)$ stands for the right-hand side of inequality (5.2.16).

Denote the left-hand side of inequality (5.2.18) by $l(\beta, y)$. The conditions posed on $\tilde{L}(y,z)$ imply that the function $l(\beta, y)$ is strictly decreasing for $\beta > \beta_0$ and continuous in β. Since $l(\beta_0, y) = \varepsilon_0(y)$, it is sufficient to prove that $\lim_{\beta\to\infty} l(\beta, y) = 0$.

For all $\delta > 0$ we have

$$l(\beta, y) \leq \sup_{|z-y|\leq\delta} \tilde{L}(y,z) + l_\delta(\beta, y), \tag{5.2.21}$$

$$l_\delta(\beta, y) = \left(\int \exp\left\{-\beta\tilde{L}(y,z)\right\} dz\right)^{-1} \int_{|z-y|>\delta} \tilde{L}(y,z)\exp\left\{-\beta\tilde{L}(y,z)\right\} dz. \tag{5.2.22}$$

The continuity of $\tilde{L}(y,z)$ with respect to the variable z at the point $z = y$ implies that

$$\lim_{\delta\to 0} \sup_{|z-y|\leq\delta} \tilde{L}(y,z) = 0. \tag{5.2.23}$$

For $\beta > \beta_0$ the following estimates hold:

$$\begin{aligned} l_\delta(\beta, y) &\leq m_{\delta'}^{-1} \exp\{\beta\psi_{\delta'}(y) - (\beta-\beta_0)\varphi_\delta(y)\} \int \tilde{L}(y,z)\exp\left\{-\beta_0\tilde{L}(y,z)\right\} dz \\ &\leq m_{\delta'}^{-1} \exp\{\beta_0\varphi_\delta(y) - \beta[\varphi_\delta(y) - \psi_{\delta'}(y)]\} \int \tilde{L}(y,z)\exp\left\{-\beta_0\tilde{L}(y,z)\right\} dz, \end{aligned} \tag{5.2.24}$$

where $\psi_{\delta'}(y) = \sup_{|z-y|\leq\delta'} \tilde{L}(y,z)$, $\varphi_\delta(y) = \inf_{|z-y|>\delta} \tilde{L}(y,z)$, $\delta' > 0$, and $m_{\delta'}$ is the Lebesgue measure of the set $\{z : |z-y| \leq \delta'\}$.

According to the properties of the function $\tilde{L}(y,z)$, for every $\delta > 0$ there exists $\delta' > 0$ such that $\varphi_\delta(y) - \psi_{\delta'}(y) > 0$. By (5.2.24) this means that $\lim_{\beta\to\infty} l_\delta(\beta, y) = 0$ for every $\delta > 0$, whence by (5.2.21) and (5.2.23) we get $\lim_{\beta\to\infty} l(\beta, y) = 0$. □

REMARK 5.2.3. Obviously, it is enough that condition (5.2.16) be fulfilled only for Q-almost all $y \in \mathbf{R}^k$.

REMARK 5.2.4. If

$$\int \tilde{L}^j(y,z)\exp\left\{-\beta\tilde{L}(y,z)\right\} dz < \infty$$

for $j = 0, 1, 2$ and all $\beta > 0$, then the statement of Theorem 5.2.2 remains true without condition (5.2.16). As is seen from the proof of Theorem 5.2.2, this follows from the fact that $\lim_{\beta\to 0} l(\beta, y) = \infty$. But this condition is obviously true since the Lebesgue measure of the set $A_N = \{z : \tilde{L}(y,z) < N\}$ is finite for all $N < \infty$ and

$$l(\beta, y) \geq \frac{\int_{A_N} \tilde{L}(y,z)\exp\{-\beta\tilde{L}(y,z)\} dz + N\int_{A_N^c} \exp\{-\beta\tilde{L}(y,z)\} dz}{\int_{A_N} \exp\{-\beta\tilde{L}(y,z)\} dz + \int_{A_N^c} \exp\{-\beta\tilde{L}(y,z)\} dz},$$

where $A_N^c = \mathbf{R}^k \setminus A_N$, whence $\underline{\lim}_{\beta\to 0} l(\beta, y) \ge N$ for all $N < \infty$.

The functions $\tilde{L}(y,z)$ and $L(y,z)$ and the measure G that generate the lower bound (5.2.17) are rather arbitrary. However, as follows from Remarks 5.2.1 and 5.2.2, estimate (5.2.17) is good only if the functions $\tilde{L}(y,z)$ and $L(y,z)$ are close to each other and the measure G is close to the measure V^t. We consider the case $\tilde{L}(y,z) = L(y,z)$, $G = Q$ in more detail. Then we study the asymptotic behavior, as $t \to \infty$, of estimate (5.2.17) for consistent estimates $\bar{\theta}_t$ (in this case Q and V^t are close to each other). We need the following version of Theorem 5.2.2.

THEOREM 5.2.3. *Let* $\tilde{L}(y,z) = l(B(y)(z-y))$, *where* $B(y)$ *is a positive definite matrix such that* $0 < \det B(y) < \infty$ (Q-*a.s.*), *while* $l(z)$, $z \in \mathbf{R}^k$, *is a nonnegative Borel function satisfying condition* (l): *the function* $l(z)$ *is continuous at the point* $z = 0$, $l(0) = 0$, $\inf_{|z|>c} l(z) > 0$ *for all* $c > 0$, *and*

$$\int l^j(z) \exp\{-\beta l(z)\}\, dz < \infty$$

for $j = 0, 1, 2$ *and all* $\beta > 0$. *Then for all estimates* $\bar{\theta}_t$ *of the parameter* θ *such that* $\left|\mathsf{E} \log \alpha^G\left(\theta, \varepsilon(\bar{\theta}_t;\theta)\right)\right| < \infty$ *and* $\mathsf{E}\, J_\theta\left(V_\theta^t, V^t; \beta\left(\varepsilon(\bar{\theta}_t;\theta)\right)\right) < \infty$ *the following inequality holds*:

$$I\left(\bar{\theta}_t, \theta\right) \ge \mathsf{E}\Big(\log \alpha^G\left(\theta, \varepsilon\left(\bar{\theta}_t;\theta\right)\right) - \varepsilon\left(\bar{\theta}_t;\theta\right) \beta\left(\varepsilon\left(\bar{\theta}_t;\theta\right)\right) \log e\Big) - \log \mathsf{E}\, \gamma\left(\bar{\theta}_t\right), \tag{5.2.25}$$

where G *is a measure on* $(\mathbf{R}^k, \mathcal{B}^k)$ *and* $\beta(\varepsilon)$ *is the solution of the equation*

$$\Big(\int \exp\{-\beta l(v)\}\, dv\Big)^{-1} \int l(v) \exp\{-\beta l(v)\}\, dv = \varepsilon, \tag{5.2.26}$$

$$\alpha^G(y,\varepsilon) = \Big(\int \exp\{-\beta(\varepsilon) l\big(B(y)(z-y)\big)\}\, G(dz)\Big)^{-1}, \tag{5.2.27}$$

$$\gamma(z) = \mathsf{E}\, \alpha^G\left(\theta, \varepsilon\left(\bar{\theta}_t;\theta\right)\right) \exp\left\{-\beta\left(\varepsilon\left(\bar{\theta}_t;\theta\right)\right) L(\theta, z)\right\}. \tag{5.2.28}$$

PROOF. The statement of the theorem follows from Theorem 5.2.2 and Remark 5.2.4. It is only necessary to take into account that in the case $\tilde{L}(y,z) = l(B(y)(z-y))$ the equation (5.2.18) takes the form (5.2.26) after the change $v = B(y)(z-y)$ and therefore $\beta(y,\varepsilon) = \beta(\varepsilon)$ for all $y \in \mathbf{R}^k$. Moreover, the functions $\alpha^G(y,\varepsilon)$ and $\gamma(z)$ given by (5.2.19) and (5.2.20) take the form (5.2.27) and (5.2.28), respectively, for $\beta(y,\varepsilon) = \beta(\varepsilon)$. □

Asymptotic lower bounds. Absolutely continuous case. Consider the asymptotic behavior of the lower bound (5.2.25) as $t \to \infty$ under the condition that the measure Q is absolutely continuous with respect to Lebesgue measure.

THEOREM 5.2.4. *Suppose* $L(y,z) = \tilde{L}(y,z) = l(\varphi_t^{-1}(y)(z-y))$, *where* $\varphi_t(y)$ *is a positive definite matrix for all* $y \in \mathbf{R}^k$, $t \in \mathbf{R}_+$, *the function* $l(z)$ *satisfies condition* (l) *of Theorem* 5.2.3 *and moreover* $|\varphi_t(y)| \to 0$ (Q-*a.s.*) *as* $t \to \infty$, $\det \varphi_t(z)\varphi_t^{-1}(z') \le K < \infty$ *and* $l(\varphi_t^{-1}(z)\varphi_t(z')v) \ge l(av)$ *for all* $t \in \mathbf{R}_+$, $z, z' \in \Theta$,

$v \in \mathbf{R}^k$, and some constants K and $a \neq 0$ independent of z, z', v, t. Further, let the measure Q be absolutely continuous with respect to Lebesgue measure and have density $q(y)$ such that $\mathsf{E}\,|\log q(\theta)| < \infty$, $\mathsf{E}\,|\log\det\varphi_t(\theta)| < \infty$ for all $t \in \mathbf{R}_+$. Finally, let $\bar\theta_t$ be a consistent estimate for the parameter θ such that $0 < \varepsilon_0 \le \varepsilon(\bar\theta_t; y) \le \varepsilon^0 < \infty$ for all $y \in \Theta$, $t \in \mathbf{R}_+$. Then, as $t \to \infty$, the following inequality holds:

$$I\left(\bar\theta_t, \theta\right) \ge \mathsf{E}\log\det\varphi_t^{-1}(\theta) + h(\theta) + \mathsf{E}\, m\left(\varepsilon\left(\bar\theta_t; \theta\right)\right) + o(1), \tag{5.2.29}$$

where

$$m(\varepsilon) = \log\alpha(\varepsilon) - \varepsilon\beta(\varepsilon)\log e, \tag{5.2.30}$$

$$\alpha(\varepsilon) = \left(\int \exp\{-\beta(\varepsilon) l(v)\}\, dv\right)^{-1}, \tag{5.2.31}$$

and $\beta(\varepsilon)$ is the solution of equation (5.2.26).

PROOF. If $I(\bar\theta_t, \theta) = \infty$, then inequality (5.2.29) is trivial. Now let $I(\bar\theta_t, \theta) < \infty$. Then, obviously, the condition $\mathsf{E}\, J_\theta\left(V_\theta^t, V^t; \beta\left(\varepsilon(\bar\theta_t; \theta)\right)\right) < \infty$ is satisfied. Moreover, we have

(5.2.31′)

$$\begin{aligned}
&\mathsf{E}\log\alpha^Q\left(\theta, \varepsilon\left(\bar\theta_t; \theta\right)\right)\\
&\quad= h(\theta) + \mathsf{E}\log\det\varphi_t^{-1}(\theta)\\
&\qquad - \mathsf{E}\log\int_{\Theta_{t,\theta}} \exp\left\{-\beta\left(\varepsilon\left(\bar\theta_t; \theta\right)\right) l(v)\right\} \frac{q(\theta + \varphi_t(\theta)v)}{q(\theta)}\, dv\\
&\quad= h(\theta) + \mathsf{E}\log\det\varphi_t^{-1}(\theta) + \mathsf{E}\log\alpha\left(\varepsilon\left(\bar\theta_t; \theta\right)\right)\\
&\qquad - \mathsf{E}\log\left[\alpha\left(\varepsilon\left(\bar\theta_t; \theta\right)\right)\int_{\Theta_{t,\theta}} \exp\left\{-\beta\left(\varepsilon\left(\bar\theta_t; \theta\right)\right) l(v)\right\} \frac{q(\theta + \varphi_t(\theta)v)}{q(\theta)}\, dv\right]\\
&\quad\ge h(\theta) + \mathsf{E}\log\det\varphi_t^{-1}(\theta) + \mathsf{E}\log\alpha\left(\varepsilon\left(\bar\theta_t; \theta\right)\right) - \mathsf{E}\,\lambda_t(\theta),
\end{aligned}$$

where $\Theta_{t,y} = \varphi_t^{-1}(y)(\Theta - y)$ and

$$\lambda_t(y) = \left|\alpha\left(\varepsilon\left(\bar\theta_t; \theta\right)\right)\int_{\Theta_{t,y}} \exp\left\{-\beta\left(\varepsilon\left(\bar\theta_t; \theta\right)\right) l(v)\right\} \frac{q(\theta + \varphi_t(\theta)v)}{q(\theta)}\, dv - 1\right|.$$

The first equality in (5.2.31′) and the conditions of the theorem imply that

$$\left|\mathsf{E}\log\alpha^Q\left(\theta, \varepsilon\left(\bar\theta_t; \theta\right)\right)\right| < \infty.$$

Thus estimate (5.2.25) for $G = Q$ and inequality (5.2.31′) imply that (5.2.29) can be obtained from the equalities

$$\lim_{t\to\infty} \mathsf{E}\,\lambda_t(\theta) = 0, \tag{5.2.32}$$

$$\lim_{t\to\infty} \mathsf{E}\,\gamma\left(\bar\theta_t\right) = 1, \tag{5.2.33}$$

where $\gamma(z)$ is the function defined by (5.2.27) and (5.2.28) for $G = Q$, $B(y) = \varphi_t^{-1}(y)$, and $L(y, z) = l\left(\varphi_t^{-1}(y)(z - y)\right)$.

First we prove (5.2.32). Let the following additional condition be fulfilled:

(*A*) The function $q(y)$ is continuous on $\Theta^0 = \{y \in \Theta\colon q(y) \neq 0\}$, $m \le q(y) \le M$ for $y \in \Theta^0$ and some $m > 0$, $M < \infty$, and $|\varphi_t(y)| \to 0$ uniformly in $y \in \Theta^0$ as $t \to \infty$.

Then for all $\delta > 0$ we have
(5.2.34)

$$\begin{aligned}\mathsf{E}\,\lambda_t(y) = \mathsf{E}\Bigg|\alpha\left(\varepsilon\left(\bar\theta_t;\theta\right)\right)\Bigg[\int_{\Theta_{t,\theta}} \exp\left\{-\beta\left(\varepsilon\left(\bar\theta_t;\theta\right)\right) l(v)\right\}\left(\frac{q(\theta+\varphi_t(\theta)v)}{q(\theta)} - 1\right) dv \\ - \int_{\Theta^c_{t,\theta}} \exp\left\{-\beta\left(\varepsilon\left(\bar\theta_t;\theta\right)\right) l(v)\right\} dv\Bigg]\Bigg| \\ \le \frac{\omega_\delta}{m} + \left(\frac{M}{m}+1\right)\mathsf{E}\,\alpha\left(\varepsilon\left(\bar\theta_t;\theta\right)\right)\int_{|\varphi_t(\theta)v|>\delta} \exp\left\{-\beta\left(\varepsilon\left(\bar\theta_t;\theta\right)\right) l(v)\right\} dv \\ + \mathsf{E}\,\alpha\left(\varepsilon\left(\bar\theta_t;\theta\right)\right)\int_{\Theta^c_{t,\theta}} \exp\left\{-\beta\left(\varepsilon\left(\bar\theta_t;\theta\right)\right) l(v)\right\} dv,\end{aligned}$$

where $\Theta^c_{t,y} = \mathbf{R}^k \setminus \Theta_{t,y}$ and $\omega_\delta = \mathsf{E}\sup_{|z|\le\delta}|q(\theta+z) - q(\theta)|$. By the Lebesgue dominant convergence theorem $\omega_\delta \to 0$ as $\delta \to 0$, since $q(y)$ is continuous and bounded. The following estimates hold:

$$\alpha\left(\varepsilon\left(\bar\theta_t;\theta\right)\right) \le m^{-1}_{t,\delta'}(\theta)\exp\left\{\beta\left(\varepsilon\left(\bar\theta_t;\theta\right)\right)\psi_{t,\delta'}(\theta)\right\}, \tag{5.2.35}$$

$$\begin{aligned}&\int_{|\varphi_t(\theta)v|>\delta} \exp\left\{-\beta\left(\varepsilon\left(\bar\theta_t;\theta\right)\right) l(v)\right\} dv \\ &\quad\le \exp\left\{-\left[\beta\left(\varepsilon\left(\bar\theta_t;\theta\right)\right) - \beta(\tilde\varepsilon)\right]\varphi_{t,\delta}(\theta)\right\}\int \exp\{-\beta(\tilde\varepsilon)l(v)\}\,dv,\end{aligned}\tag{5.2.36}$$

where

$$\psi_{t,\delta'}(\theta) = \sup_{|\varphi_t(\theta)v|\le\delta'} l(v), \qquad \varphi_{t,\delta}(\theta) = \inf_{|\varphi_t(\theta)v|>\delta} l(v),$$

$m_{t,\delta'}(\theta)$ is the Lebesgue measure of the set $\{v\colon |\varphi_t(\theta)v| \le \delta'\}$ and $\tilde\varepsilon > \varepsilon^0$ and $\delta' > 0$ are constants.

From (5.2.35) and (5.2.36) we obtain
(5.2.37)

$$\begin{aligned}&\mathsf{E}\,\alpha\left(\varepsilon\left(\bar\theta_t;\theta\right)\right)\int_{|\varphi_t(\theta)v|>\delta} \exp\left\{-\beta\left(\varepsilon\left(\bar\theta_t;\theta\right)\right) l(v)\right\} dv \\ &\quad\le \int \exp\{-\beta(\tilde\varepsilon)l(v)\}\,dv \\ &\qquad\times \mathsf{E}\,m^{-1}_{t,\delta'}(\theta)\exp\left\{-\beta\left(\varepsilon\left(\bar\theta_t;\theta\right)\right)\left[\varphi_{t,\delta}(\theta) - \psi_{t,\delta'}(\theta)\right] + \beta(\tilde\varepsilon)\varphi_{t,\delta}(\theta)\right\}.\end{aligned}$$

Since $|\varphi_t(y)| \to 0$ uniformly in $y \in \Theta^0$ as $t \to \infty$, we can choose the constant δ' so small as to satisfy the equality $\psi_{t,\delta'}(y) = o(\varphi_{t,\delta}(y))$ uniformly in $y \in \Theta^0$ as $t \to \infty$ and for fixed δ and δ'. Taking into account $\varepsilon(\bar\theta_t; y) \le \varepsilon^0$, from (5.2.37) we get, as $t \to \infty$,
(5.2.38)

$$\begin{aligned}&\mathsf{E}\,\alpha\left(\varepsilon\left(\bar\theta_t;\theta\right)\right)\int_{|\varphi_t(\theta)v|>\delta} \exp\left\{-\beta\left(\varepsilon\left(\bar\theta_t;\theta\right)\right) l(v)\right\} dv \\ &\quad\le \int \exp\{-\beta(\tilde\varepsilon)l(v)\}\,dv\,\mathsf{E}\,m^{-1}_{t,\delta'}(\theta)\exp\left\{-\left[\beta\left(\varepsilon^0\right) - \beta(\tilde\varepsilon)\right]\varphi_{t,\delta}(\theta)(1+o(1))\right\},\end{aligned}$$

where $o(1) \to 0$ uniformly in $y \in \Theta^0$ as $t \to \infty$. Condition (A) implies that $\varphi_{t,\delta}(y) \to \infty$ and $m_{t,\delta'}(y) \to \infty$ uniformly in $y \in \Theta^0$ as $t \to \infty$ for fixed δ, δ', and

$\beta(\varepsilon^0) - \beta(\tilde{\varepsilon}) > 0$, since the function $\beta(\varepsilon)$ is monotone. Therefore (5.2.38) implies that the second term on the right-hand side of (5.2.34) approaches zero as $t \to \infty$ and $\delta > 0$ is fixed.

Let $\Theta^{0,\varepsilon} \subset \Theta^0$ be such that $\inf\{|z-y|; y \in \Theta^{0,\varepsilon}, z \in \Theta^c\} \geq \varepsilon > 0$. Then the following estimate holds:
(5.2.39)

$$\begin{aligned}
&\mathsf{E}\alpha\left(\varepsilon\left(\bar{\theta}_t;\theta\right)\right) \int_{\Theta^c_{t,\theta}} \exp\left\{-\beta\left(\varepsilon\left(\bar{\theta}_t;\theta\right)\right) l(v)\right\} dv \\
&\qquad \leq Q\left(\Theta \setminus \Theta^{0,\varepsilon}\right) + \int_{\Theta^{0,\varepsilon}} \alpha\left(\varepsilon\left(\bar{\theta}_t;y\right)\right) \int_{\Theta^c_{t,y}} \exp\left\{-\beta\left(\varepsilon\left(\bar{\theta}_t;y\right)\right) l(v)\right\} dv\, Q(dy),
\end{aligned}$$

where the first term on the right-hand side can be taken as small as necessary by choosing $\Theta^{0,\varepsilon}$ small enough (this can be done by choosing ε small enough). It is obvious that $|\varphi_t(y)v| > \varepsilon$ for $y \in \Theta^{0,\varepsilon}$ and $v \in \Theta^c_{t,y}$. Thus applying estimates (5.2.35) and (5.2.36) and reasoning in the same way as in the case of estimate (5.2.38), we get that the second term on the right-hand side of (5.2.39) approaches zero as $t \to \infty$ and $\Theta^{0,\varepsilon}$ is fixed. Hence (5.2.39) implies that the third term on the right-hand side of (5.2.34) approaches zero uniformly in δ as $t \to \infty$. So, (5.2.32) is proved under the additional condition (A).

Now we turn to the proof of (5.2.33). Let $V^t = \overline{V}^t + \overline{\overline{V}}^t$, where $\overline{V}^t \ll Q$ with $\zeta_t(y) = d\overline{V}^t/dQ(y)$ and $\overline{\overline{V}}^t \perp Q$. By $\overline{\Theta}_t$ and $\overline{\overline{\Theta}}_t$ we denote the supports of the measures $\overline{V}^t$ and $\overline{\overline{V}}^t$, respectively. It is clear that $\overline{\Theta}_t \subset \Theta^0$ up to a set of zero $Q + \overline{V}^t$-measure. Since the estimate $\bar{\theta}_t$ is consistent, $\overline{\overline{V}}^t(\overline{\overline{\Theta}}_t) \to 0$ and $\zeta_t(y) \to 1$ (Q-a.s.) as $t \to \infty$. To prove equality (5.2.23) we assume that besides condition (A) the following condition is satisfied:

(B) $\inf\{|z-y|; y \in \Theta^0, z \in \overline{\overline{\Theta}}_t\} \geq \delta > 0$ for all $t \in \mathbf{R}_+$ and $\zeta_t(y) \to 1$ uniformly in $y \in \Theta^0$ as $t \to \infty$.

Then by the definition of the function $\gamma(z)$ we have
(5.2.40)

$$\begin{aligned}
&\left|\mathsf{E}\gamma\left(\bar{\theta}_t\right) - 1\right| \leq \mathsf{E}\alpha^Q\left(\varepsilon\left(\bar{\theta}_t;\theta\right)\right) \\
&\qquad\qquad \times \left[\int_{\Theta^0} \exp\left\{-\beta\left(\varepsilon\left(\bar{\theta}_t;\theta\right)\right) l\left(\varphi_t^{-1}(\theta)(z-\theta)\right)\right\} |\zeta_t(z) - 1|\, Q(dz)\right. \\
&\qquad\qquad\qquad \left. + \int_{\overline{\overline{\Theta}}_t} \exp\left\{-\beta\left(\varepsilon\left(\bar{\theta}_t;\theta\right)\right) l\left(\varphi_t^{-1}(\theta)(z-\theta)\right)\right\} \overline{\overline{V}}^t(dz)\right] \\
&\qquad \leq \sup_{z \in \Theta^0} |\zeta_t(z) - 1| \\
&\qquad\qquad + m^{-1}\overline{\overline{V}}^t\left(\overline{\overline{\Theta}}_t\right) \mathsf{E} m_{t,\delta'}^{-1}(\theta) \exp\left\{-\beta\left(\varepsilon\left(\bar{\theta}_t;\theta\right)\right)\left[\varphi_{t,\delta}(\theta) - \psi_{t,\delta'}(\theta)\right]\right\},
\end{aligned}$$

where $m_{t,\delta'}(\theta)$, $\varphi_{t,\delta}(\theta)$, and $\psi_{t,\delta'}(\theta)$ are random variables introduced in the proof of inequalities (5.2.35) and (5.2.36). Choosing δ' small enough, from (5.2.40) and conditions (A) and (B) we derive (5.2.33) similarly to the case of the second term on the right-hand side of (5.2.34).

So, relations (5.2.32) and (5.2.33), as well as relation (5.2.29), are proved under the additional conditions (A) and (B).

Now we prove (5.2.29) in the general case. We introduce auxiliary random variables. Let $\chi_m^M = \chi(\Theta_m^M; \theta)$, where $\Theta_m^M = \{y \in \Theta; m \leq q(y) \leq M\}$, $m > 0$, and $M < \infty$. It is obvious that $\mathsf{P}\{\chi_m^M = 1\} \to 1$ as $m \to 0$ and $M \to \infty$. By Luzin's theorem [**102**], for all $\delta > 0$ there exists a set $\Theta_\delta \subset \Theta^0$ such that $Q(\Theta^0 \setminus \Theta_\delta) < \delta$ and the function $q(y)$ is continuous on Θ_δ. Putting $\chi_\delta = \chi(\Theta_\delta; \theta)$, we get $\mathsf{P}\{\chi_\delta = 1\} \to 1$ as $\delta \to 0$. By Egorov's theorem [**102**], for all $\varepsilon > 0$ there exists a set $\Theta^\varepsilon \subset \Theta^0$ such that $Q(\Theta^0 \setminus \Theta^\varepsilon) < \varepsilon$ and $|\varphi_t(y)| \to 0$ uniformly in $y \in \Theta^\varepsilon$ as $t \to \infty$. Thus for the variable $\chi^\varepsilon = \chi(\Theta^\varepsilon; \theta)$ we have $\mathsf{P}\{\chi^\varepsilon = 1\} \to 1$ as $\varepsilon \to 0$. If the event $\{\chi_m^M = 1\} \cap \{\chi_\delta = 1\} \cap \{\chi^\varepsilon = 1\}$ occurs, then condition (A) holds. By Egorov's theorem [**102**], for all $\sigma > 0$ there exists a set $\widehat{\Theta}^\sigma \subset \Theta^0$ such that $Q(\Theta^0 \setminus \widehat{\Theta}^\sigma) < \sigma$ and $\zeta_t(y) \to 1$ uniformly in $y \in \widehat{\Theta}^\sigma$ as $t \to \infty$. It is obvious that one can choose, for small σ, a set $\overline{\Theta}^\sigma \subset \widehat{\Theta}^\sigma$ such that $\inf\{|z - y|; y \in \overline{\Theta}^\sigma, z \in \overline{\overline{\Theta}}_t\} \geq \sigma$ and $Q(\Theta^0 \setminus \overline{\Theta}^\sigma) < c\sigma$ for some constant $c > 0$. Putting $\overline{\chi}^\sigma = \chi(\overline{\Theta}^\sigma; \theta)$, we get $\mathsf{P}\{\overline{\chi}^\sigma = 1\} \to 1$ as $\sigma \to 0$.

Now let $\widetilde{\chi}_\nu = \chi(\widetilde{\Theta}_\nu; \theta)$, where $\nu = (m, M, \delta, \varepsilon, \sigma)$ and $\widetilde{\Theta}_\nu = \Theta_m^M \cap \Theta_\delta \cap \Theta^\varepsilon \cap \overline{\Theta}^\sigma$. It is evident that $\mathsf{P}\{\widetilde{\chi}_\nu = 1\} \to 1$ as $\nu \to \nu_0 = (0, \infty, 0, 0, 0)$. It is easy to see that the conditional distribution of the vector θ, given $\widetilde{\chi}_\nu = 1$, satisfies conditions (A) and (B).

Applying equality $I\left(\bar{\theta}_t, (\theta, \widetilde{\chi}_\nu)\right) = I\left(\bar{\theta}_t, \theta\right)$ and conditional information formula [**31, 106**], we get

$$(5.2.41)\qquad I\left(\bar{\theta}_t, \theta\right) = I\left(\bar{\theta}_t, \widetilde{\chi}_\nu\right) + \mathsf{E}\, I\left(\bar{\theta}_t, \theta/\widetilde{\chi}_\nu\right) \geq \sum_{i=0}^{1} I\left(\bar{\theta}_t, \theta/\widetilde{\chi}_\nu = i\right) \mathsf{P}\{\widetilde{\chi}_\nu = i\}.$$

By inequality (5.2.29) proved under the additional conditions (A) and (B) we have

$$(5.2.42)\qquad \begin{aligned} I\left(\bar{\theta}_t, \theta/\widetilde{\chi}_\nu = 1\right) &\geq \mathsf{E}\left\{m\left(\varepsilon\left(\bar{\theta}_t; \theta\right)\right)/\widetilde{\chi}_\nu = 1\right\} \\ &\quad + h\left(\theta/\widetilde{\chi}_\nu = 1\right) + \mathsf{E}\left\{\log\det \varphi_t^{-1}(\theta)/\widetilde{\chi}_\nu = 1\right\} + r_t(\nu), \end{aligned}$$

where $r_t(\nu) \to 0$ as $t \to \infty$ for a fixed vector ν. Using Theorem 5.2.3 for

$$\alpha^G(y, \varepsilon) = \alpha(y, \varepsilon) q^{-1}(y) \mathsf{P}\{\widetilde{\chi}_\nu = 0\},$$

where

$$\alpha(y, \varepsilon) = \left(\int \exp\left\{-\beta(\varepsilon) l\left(\varphi_t^{-1}(y)(z - y)\right)\right\} dz\right)^{-1},$$

we obtain the estimate

$$(5.2.43)\qquad \begin{aligned} I\left(\bar{\theta}_t, \theta/\widetilde{\chi}_\nu = 0\right) &\geq \mathsf{E}\left\{m\left(\varepsilon\left(\bar{\theta}_t; \theta\right)\right)/\widetilde{\chi}_\nu = 0\right\} + h\left(\theta/\widetilde{\chi}_\nu = 0\right) \\ &\quad + \mathsf{E}\left\{\log\det \varphi_t^{-1}(\theta)/\widetilde{\chi}_\nu = 0\right\} - \log \mathsf{E}\left\{\widetilde{\gamma}\left(\bar{\theta}_t\right)/\widetilde{\chi}_\nu = 0\right\}, \end{aligned}$$

where

$$(5.2.44)\qquad \widetilde{\gamma}(z) = \int_{C_\nu} \frac{\exp\left\{-\beta\left(\varepsilon\left(\bar{\theta}_t; y\right)\right) l\left(\varphi_t^{-1}(y)(z - y)\right)\right\}}{\int \exp\left\{-\beta\left(\varepsilon\left(\bar{\theta}_t; y\right)\right) l\left(\varphi_t^{-1}(y)(x - y)\right)\right\} dx}\, dy$$

and $C_\nu = \Theta \setminus \widetilde{\Theta}_\nu$.

Combining (5.2.41)–(5.2.43) and accounting for the equality

$$h(\theta) - \mathsf{E}\, h(\theta/\widetilde{\chi}_\nu) = I(\theta, \widetilde{\chi}_\nu) = H(\widetilde{\chi}_\nu),$$

we have

$$I\left(\bar{\theta}_t, \theta\right) \geq \mathsf{E}\, m\left(\varepsilon\left(\bar{\theta}_t; \theta\right)\right) + h(\theta) + \mathsf{E} \log \det \varphi_t^{-1}(\theta) - H(\widetilde{\chi}_\nu) + r_t(\nu)\mathsf{P}\{\widetilde{\chi}_\nu = 1\} - \mathsf{P}\{\widetilde{\chi}_\nu = 0\} \log \mathsf{E}\left\{\widetilde{\gamma}\left(\bar{\theta}_t\right) / \widetilde{\chi}_\nu = 0\right\}. \tag{5.2.45}$$

Since $H(\widetilde{\chi}_\nu) \to 0$ as $\nu \to \nu_0$ and $r_t(\nu) \to 0$ for any fixed ν as $t \to \infty$, to complete the proof of the theorem we need, in view of (5.2.45), to show that

$$\mathsf{P}\{\widetilde{\chi}_\nu = 0\} \log \mathsf{E}\left\{\widetilde{\gamma}\left(\bar{\theta}_t\right) / \widetilde{\chi}_\nu = 0\right\} \to 0$$

uniformly in t as $\nu \to \nu_0$.

Making first the change $v = \varphi_t^{-1}(y)(x-y)$ in the inner integral in (5.2.44) and then the change $w = \varphi_t^{-1}(z)(y-z)$ in the outer integral, from the inequalities $\det \varphi_t(z)\varphi_t^{-1}(z') \leq K < \infty$ we get the estimate

$$\widetilde{\gamma}(z) \leq K \int_{C_\nu^{t,z}} \frac{\exp\left\{-\beta\left(\varepsilon\left(\bar{\theta}_t; z + \varphi_t(z)w\right)\right) l\left(-\varphi_t^{-1}(z + \varphi_t(z)w)\varphi_t(z)w\right)\right\}}{\int \exp\left\{-\beta\left(\varepsilon\left(\bar{\theta}_t; z + \varphi_t(z)w\right)\right) l(v)\right\} dv}\, dw, \tag{5.2.46}$$

where $C_\nu^{t,z} = \varphi_t^{-1}(z)(C_\nu - z)$.

Since $\varepsilon(\bar{\theta}_t; y) \geq \varepsilon_0$, for $\bar{z} = z + \varphi_t(z)w$ we have

$$\int \exp\left\{-\beta\left(\varepsilon\left(\bar{\theta}_t; \bar{z}\right)\right) l(v)\right\} dv \geq \int \exp\{-\beta(\varepsilon_0) l(v)\}\, dv = L > 0. \tag{5.2.47}$$

Taking into account that $l\left(-\varphi_t^{-1}(\bar{z})\varphi_t(z)w\right) \geq l(-aw)$ and $\varepsilon(\bar{\theta}_t; y) \leq \varepsilon^0$, we get

$$\int_{C_\nu^{t,z}} \exp\left\{-\beta\left(\varepsilon\left(\bar{\theta}_t; z + \varphi_t(z)w\right)\right) l\left(-\varphi_t^{-1}(z + \varphi_t(z)w)\varphi_t(z)w\right)\right\} dw \leq \int \exp\left\{-\beta\left(\varepsilon^0\right) l(aw)\right\} dw = L' < \infty. \tag{5.2.48}$$

Therefore (5.2.46)–(5.2.48) imply that $\mathsf{P}\{\widetilde{\chi}_\nu = 0\}\, \mathsf{E}\left\{\widetilde{\gamma}(\bar{\theta}_t)/\widetilde{\chi}_\nu = 0\right\} \to 0$ uniformly in t as $\nu \to \nu_0$. □

Asymptotic lower bounds. General case. Assume that there exists at most a countable set of isolated points $\Theta' = \{y_1, y_2, \dots\} \subset \Theta$ such that the measure $\widetilde{Q}$, $\widetilde{Q}(A) = Q(A \setminus \Theta')$, $A \in \mathbf{R}^k$, is absolutely continuous with respect to Lebesgue measure and $Q(\{y\}) > 0$ for all $y \in \Theta'$. Put

$$\mathsf{P}\{\theta \in \Theta'\} = p, \qquad \mathsf{P}\{\theta \in A/\chi = 0\} = \int_A q(y)\, dy, \quad A \in \mathcal{B}^k,$$

where $\chi = \chi(\Theta'; \theta)$ and $q(y)$ is the density of the conditional distribution of θ, given $\chi = 0$, with respect to Lebesgue measure.

By the complete information formula we have $(q = 1 - p)$

$$I\left(\bar{\theta}_t, \theta\right) = I\left(\bar{\theta}_t, \chi\right) + pI\left(\bar{\theta}_t, \theta/\chi = 1\right) + qI\left(\bar{\theta}_t, \theta/\chi = 0\right). \tag{5.2.49}$$

Since the estimate $\bar{\theta}_t$ is consistent and the information is semicontinuous [**106**], we have

$$\varliminf_{t \to \infty} I\left(\bar{\theta}_t, \chi\right) \geq I(\theta, \chi) = H(\chi),$$

whence by $I(\bar{\theta}_t, \chi) \leq H(\chi)$ we get

$$\lim_{t\to\infty} I\left(\bar{\theta}_t, \chi\right) = H(\chi). \tag{5.2.50}$$

In a similar way, from the semicontinuity of information for consistent estimates $\bar{\theta}_t$ we get that

$$\lim_{t\to\infty} I\left(\bar{\theta}_t, \theta/\chi = 1\right) = H(\theta/\chi = 1). \tag{5.2.51}$$

Therefore relations (5.2.49)–(5.2.51) and Theorem 5.2.4 applied to $I(\bar{\theta}_t, \theta/\chi = 0)$ imply the following result.

THEOREM 5.2.5. *Let $L(y,z) = \widetilde{L}(y,z) = l\left(\varphi_t^{-1}(y)(z-y)\right)$, where the functions $l(z)$ and $\varphi_t(y)$ satisfy the conditions of Theorem 5.2.4 and the measure Q is such that*

$$H(\theta/\chi = 1) < \infty, \qquad \mathsf{E}\{|\log q(\theta)|\,/\chi = 0\} < \infty,$$
$$\mathsf{E}\{|\log \det \varphi_t(\theta)|\,/\chi = 0\} < \infty.$$

Further, let $\bar{\theta}_t$ be a consistent estimate for a parameter θ such that $0 < \varepsilon_0 \leq \varepsilon(\bar{\theta}_t; y) \leq \varepsilon^0 < \infty$ for all $y \in \Theta \setminus \Theta'$, $t \in \mathbf{R}_+$. Then, as $t \to \infty$, the following inequality holds:

$$\begin{aligned} I(\bar{\theta}_t, \theta) \geq q\left[\mathsf{E}\left\{\left|\log \det \varphi_t^{-1}(\theta)\right|/\chi = 0\right\} + h(\theta/\chi = 0) + \mathsf{E}\left\{m\left(\varepsilon\left(\bar{\theta}_t; \theta\right)\right)/\chi = 0\right\}\right] \\ + pH(\theta/\chi = 1) + H(\chi) + o(1), \end{aligned} \tag{5.2.52}$$

where $m(\varepsilon)$ is the function defined by (5.2.30) *and* (5.2.31).

It is clear from Theorem 5.2.5 that for $y \in \Theta'$ the risk function $\varepsilon(\bar{\theta}_t; y)$ occurs in the terms of order $o(1)$ in the case where $H(\theta/\chi = 1) < \infty$. However, if

$$H(\theta/\chi = 1) = \infty,$$

this is not the case. Nevertheless in this case an analog of inequality (5.2.52) can also be obtained. To do this, one can use the same line of reasoning as in the proof of Theorem 5.2.5, with the exception that instead of (5.2.51), we must use the obvious inequality

$$I\left(\bar{\theta}_t, \theta/\chi = 1\right) \geq H_{\varepsilon(\bar{\theta}_t)}(\theta/\chi = 1) = \inf I(\eta, \theta/\chi = 1), \tag{5.2.53}$$

where the infimum is taken over all random vectors η assuming values in $\mathbf{R}^k$ and satisfying the inequality

$$\mathsf{E}\left\{l\left(\varphi_t^{-1}(\theta)(\eta - \theta)\right)/\chi = 1\right\} \leq \varepsilon\left(\bar{\theta}_t\right) = \mathsf{E}\left\{l\left(\varphi_t^{-1}(\theta)\left(\bar{\theta}_t - \theta\right)\right)/\chi = 1\right\}. \tag{5.2.54}$$

So, we obtained the following generalization of Theorem 5.2.5.

THEOREM 5.2.6. *Let all the conditions of Theorem* 5.2.5 *be satisfied except for* $H(\theta/\chi = 1) < \infty$. *Then the following inequality holds*:
(5.2.55)
$$I(\bar{\theta}_t, \theta) \geq q\Big[\mathsf{E}\left\{\left|\log\det\varphi_t^{-1}(\theta)\right|/\chi = 0\right\} + h(\theta/\chi = 0) + \mathsf{E}\left\{m\left(\varepsilon\left(\bar{\theta}_t;\theta\right)\right)/\chi = 0\right\}\Big] + pH_{\varepsilon(\bar{\theta}_t)}(\theta/\chi = 1) + H(\chi) + o(1),$$

where $H_{\varepsilon(\bar{\theta}_t)}(\theta/\chi = 1)$ *is defined by relations* (5.2.53) *and* (5.2.54) *and the function* $m(\varepsilon)$ *is defined by relations* (5.2.30) *and* (5.2.31).

REMARK 5.2.5. In information theory $H_\varepsilon(\theta/\chi = 1)$ is called the *ε-entropy of the conditional distribution of* θ, provided that $\chi = 1$ and

$$\mathsf{E}\left\{l\left(\varphi_t^{-1}(\theta)(\eta - \theta)\right)/\chi = 1\right\} \leq \varepsilon.$$

If $H(\theta/\chi = 1) = \infty$, a formula for $H_\varepsilon(\theta/\chi = 1)$ in a special case of the loss function $l(y)$ can be found in [**59**]. In the case of arbitrary functions $l(y)$ the calculation of the ε-entropy $H_\varepsilon(\theta/\chi = 1)$ is a complicated analytical problem and we shall not discuss it in this book.

5.3. Bounds for risk functions of consistent estimates

We get asymptotic information inequalities of the Cramér–Rao and Hájek type for consistent estimates as $t \to \infty$, using asymptotic relations for the information function $I(\xi^t, \theta)$ and lower bounds for $I(\bar{\theta}_t, \theta)$.

An auxiliary result. Below we need the following lemma on properties of the function $m(\varepsilon)$ introduced in Theorem 5.2.4. The lemma plays an important role in the theory of ε-entropy [**59**].

LEMMA 5.3.1. *Let* $l(v)$, $v \in \mathbf{R}^k$, *be a nonnegative Borel function satisfying conditions* (l) *of Theorem* 5.3.3, *and let* $m(\varepsilon) = \log\alpha(\varepsilon) - \varepsilon\beta(\varepsilon)\log e$, $\varepsilon > 0$, *where* $\beta(\varepsilon)$ *is the solution of equation* (5.2.26) *and* $\alpha(\varepsilon)$ *is given by equality* (5.2.31). *Then* $m(\varepsilon)$ *is a strictly decreasing convex twice differentiable function for* $\varepsilon \in (0, \infty)$.

PROOF. According to Theorem 5.2.2 and Remark 5.2.4, condition (l) implies the existence and uniqueness of a solution of equation (5.2.26) for any $\varepsilon \in (0, \infty)$. Therefore the functions $\alpha(\varepsilon)$, $\beta(\varepsilon)$, and $m(\varepsilon)$ are well defined for all $\varepsilon \in (0, \infty)$. Condition ($l$) also implies that the function

$$\omega(\beta) = \left(\int \exp\{-\beta l(v)\}\,dv\right)^{-1}$$

is strictly increasing and differentiable in $\beta \in (0, \infty)$. Also, by condition (l) we get, after differentiation, that

$$\mathfrak{z}(\beta) = \omega(\beta)\int l(v)\exp\{-\beta l\}\,dv$$

is a strictly decreasing and differentiable function in $\beta \in (0, \infty)$. Based on the well-known results on inverse functions, we obtain that the functions $\alpha(\varepsilon)$ and $\beta(\varepsilon)$ are strictly monotone and differentiable in $\varepsilon \in (0, \infty)$. This implies that the function $m(\varepsilon)$ is differentiable for $\varepsilon \in (0, \infty)$.

Differentiating the equality $\alpha(\varepsilon)\int \exp\{-\beta(\varepsilon)l(v)\}\,dv = 1$, we get

$$\alpha'(\varepsilon)\int \exp\{-\beta(\varepsilon)l(v)\}\,dv - \alpha(\varepsilon)\beta'(\varepsilon)\int l(v)\exp\{-\beta(\varepsilon)l(v)\}\,dv = 0.$$

Dividing this equality by $\int \exp\{-\beta(\varepsilon)l(v)\}\,dv$ and taking into account that

$$\mathfrak{z}(\beta(\varepsilon)) = \varepsilon,$$

we have $\alpha'(\varepsilon) - \varepsilon\alpha(\varepsilon)\beta'(\varepsilon) = 0$, whence we get

$$m'(\varepsilon) = \frac{\log e}{\alpha(\varepsilon)}\left[\alpha'(\varepsilon) - \alpha(\varepsilon)\beta(\varepsilon) - \varepsilon\alpha(\varepsilon)\beta'(\varepsilon)\right] = -\beta(\varepsilon)\log e < 0 \tag{5.3.1}$$

for all $\varepsilon \in (0,\infty)$ since $\beta(\varepsilon) > 0$ for all $\varepsilon \in (0,\infty)$. This implies that the function $m(\varepsilon)$ is strictly decreasing. Since the function $\beta(\varepsilon)$ is differentiable, from (5.3.1) it follows that the second derivative $m''(\varepsilon)$ exists and $m''(\varepsilon) = -\beta'(\varepsilon)\log e > 0$ by the properties of the function $\mathfrak{z}(\beta)$. Therefore the function $m(\varepsilon)$ is convex. □

Lower bounds of an integral type. The following theorem provides a lower bound for the risk functions $\varepsilon(\overline{\theta}_t; y)$ in the integral form. This bound is similar to the integral analogs of the Cramér–Rao inequalities called *information inequalities* [**38**]. This term is natural in the case under consideration, since all the bounds of this section are obtained from information-theoretic considerations. Below we keep the notation and assumptions of Sections 5.1 and 5.2.

THEOREM 5.3.1. *If the conditions of Theorem* 5.2.5 *hold and* $p \neq 1$, *then*

$$\begin{aligned} \mathsf{E}\left\{m\left(\varepsilon\left(\overline{\theta}_t;\theta\right)\right)/\chi = 0\right\} \le \frac{1}{q}&\left[I\left(\xi^t,\theta\right) - pH\left(\theta/\chi = 1\right) - H(\chi)\right] \\ &- \mathsf{E}\left\{\log\left|\varphi_t^{-1}(\theta)\right|/\chi = 0\right\} - h(\theta/\chi = 0) + o(1) \end{aligned} \tag{5.3.2}$$

as $t \to \infty$. *If additionally the conditions of Theorem* 5.1.3 *are satisfied, then*

$$\mathsf{E}\left\{m\left(\varepsilon\left(\overline{\theta}_t;\theta\right)\right)/\chi = 0\right\} \le -\mathsf{E}\left\{\log\int Z(u)\,du/\chi = 0\right\} + o(1) \tag{5.3.3}$$

as $t \to \infty$ *and moreover*

$$\mathsf{E}\left\{\varepsilon\left(\overline{\theta}_t;\theta\right)/\chi = 0\right\} \ge m^{-1}\left(-\mathsf{E}\left\{\log\int Z(u)\,du/\chi = 0\right\}\right) + o(1), \tag{5.3.4}$$

where $m^{-1}(z)$ *is the inverse function to* $m(\varepsilon)$.

PROOF. Inequality (5.3.2) follows from Theorem 5.2.5 together with the inequality $I(\xi^t,\theta) \ge I(\overline{\theta}_t,\theta)$, which holds for all estimates $\overline{\theta}_t$. Inequality (5.3.3) follows from (5.3.2) and Theorem 5.1.3. Moreover, the function $m(\varepsilon)$ is convex by Lemma 5.3.1. Thus from inequality (5.3.3) we get that

$$m\left(\mathsf{E}\left\{\varepsilon\left(\overline{\theta}_t;\theta\right)/\chi = 0\right\}\right) \le -\mathsf{E}\left\{\log\int Z(u)\,du/\chi = 0\right\} + o(1) \tag{5.3.5}$$

as $t \to \infty$. Moreover, the function $m(\varepsilon)$ is continuous and strictly decreasing by Lemma 5.3.1. Hence there exists the inverse function $m^{-1}(\varepsilon)$, which is also continuous and strictly decreasing. Therefore by (5.3.5) we have

$$\mathsf{E}\left\{\varepsilon\left(\overline{\theta}_t;\theta\right)/\chi = 0\right\} \ge m^{-1}\left(-\mathsf{E}\left\{\log\int Z(u)\,du/\chi = 0\right\}\right) + o(1),$$

whence we get (5.3.4), since $m^{-1}(z)$ is continuous. □

COROLLARY 5.3.1. *If the conditions of Theorem* 5.2.4 *hold, then*

$$\mathsf{E}\, m\left(\varepsilon\left(\overline{\theta}_t;\theta\right)\right) \le I\left(\xi^t,\theta\right) - \mathsf{E}\log\left|\det\varphi_t^{-1}(\theta)\right| - h(\theta) + o(1) \tag{5.3.6}$$

as $t\to\infty$. *If additionally the conditions of Theorem* 5.1.2 *are satisfied, then*

$$\mathsf{E}\, m\left(\varepsilon\left(\overline{\theta}_t;\theta\right)\right) \le -\mathsf{E}\log\int Z(u)\,du + o(1) \tag{5.3.7}$$

as $t\to\infty$, *and moreover*

$$\mathsf{E}\,\varepsilon(\overline{\theta}_t;\theta) \ge m^{-1}\left(-\mathsf{E}\log\int Z(u)\,du\right) + o(1). \tag{5.3.8}$$

PROOF. The corollary follows from Theorem 5.3.1 for $p=0$. Besides, it should be noted that Theorems 5.2.4 and 5.1.2 are particular cases of Theorems 5.2.5 and 5.1.3, respectively, for $p=0$ and $\mathsf{E}\{\cdot/\chi=0\}=\mathsf{E}\{\cdot\}$. □

COROLLARY 5.3.2. *Let* $L(y,z) = |\varphi_y^{-1}(y)(z-y)|^2$, $y,z\in\Theta$, *where* $\varphi_t(y)$, $t\in\mathbf{R}_+$, *are positive definite matrices such that* (Q-*a.s.*)

$$|\varphi_t(y)|\to 0, \qquad t\to\infty,$$

and

$$\left|\varphi_t(z)\varphi^{-1}(y)\right| \le C, \qquad z,y\in\Theta,\ t\in\mathbf{R}_+,$$

where the constant C *is independent of* z, y, *and* t. *Further, let the measure* Q *be absolutely continuous with respect to Lebesgue measure and have density* $q(y)$, $\mathsf{E}\,|\log q(\theta)|<\infty$. *Assume that* $\overline{\theta}_t$ *is a consistent estimate for the parameter* θ *such that*

$$0<\varepsilon_0\le\varepsilon\left(\overline{\theta}_t;y\right)\le\varepsilon^0<\infty$$

for all $y\in\Theta$ *and* $t\in\mathbf{R}_+$. *Then*

$$\mathsf{E}\log\varepsilon\left(\overline{\theta}_t;\theta\right) \ge \frac{2}{k}\left[\mathsf{E}\log\left|\det\varphi_t^{-1}(\theta)\right| + h(\theta) - I\left(\xi^t,\theta\right)\right] + \log\frac{k}{2\pi e} + o(1) \tag{5.3.9}$$

as $t\to\infty$. *Additionally, if the conditions of Theorem* 5.1.2 *hold, then*

$$\mathsf{E}\log\varepsilon\left(\overline{\theta}_t;\theta\right) \ge \frac{2}{k}\,\mathsf{E}\log\int Z(u)\,du + \log\frac{k}{2\pi e} + o(1) \tag{5.3.10}$$

as $t\to\infty$. *In particular, if the family* $(\mathsf{P}_y^t, y\in\Theta)$ *possesses the LAN property at* Q-*almost all points* $y\in\Theta$, *then*

$$\mathsf{E}\log\varepsilon\left(\overline{\theta}_t;\theta\right) \ge \log k + o(1) \tag{5.3.11}$$

as $t\to\infty$.

PROOF. It is obvious that the function $l(v) = |v|^2$ satisfies condition (l). It is not hard to show that in this case the solution of equation (5.2.26) is of the form $\beta(\varepsilon) = k(2\varepsilon)^{-1}$ and therefore $\alpha(\varepsilon) = (k/(2\pi\varepsilon))^{k/2}$. Consequently, $m(\varepsilon) = (k/2)\log(k/(2\pi e\varepsilon))$. Now we use Theorem 5.2.4 with $l(v) = |v|^2$ and get (5.3.9) from (5.3.6). To apply Theorem 5.2.4 in the case under consideration, we need to show that

$$\left|\det \varphi_t(z)\varphi_t^{-1}(z')\right| \le K \quad \text{and} \quad \left|\varphi_t^{-1}(z)\varphi_t(z')v\right|^2 \ge |av|^2$$

for all $z, z' \in \Theta$, $t \in \mathbf{R}_+$, and $v \in \mathbf{R}^k$, where the constants $K > 0$ and $a \neq 0$ are independent of z, z', t, v and $\mathsf{E}\big|\log|\det\varphi_t(\theta)|\big| < \infty$ for all $t \in \mathbf{R}_+$.

Let $\lambda_{1,t} \le \lambda_{2,t} \le \cdots \le \lambda_{k,t}$ be the eigenvalues of the symmetric positive definite matrix $G_t = A_t A_t'$, where $A_t = \varphi_t(z)\varphi_t^{-1}(z')$. Then [**15**]

$$\prod_{i=1}^{k} \lambda_{i,t} = \det G_t = \det A_t \det A_t' = (\det A_t)^2, \tag{5.3.12}$$

$$\sum_{i=1}^{k} \lambda_{i,t} = \operatorname{Sp} G_t = \operatorname{Sp} A_t A_t' = |A_t|^2 \le c^2. \tag{5.3.13}$$

Since

$$\left(\prod_{i=1}^{k} \lambda_{i,t}\right)^{1/k} \le \frac{1}{k}\sum_{i=1}^{k} \lambda_{i,t},$$

from (5.3.12) and (5.3.13) we obtain

$$|\det A_t| = \left(\prod_{i=1}^{k} \lambda_{i,t}\right)^{1/2} \le \left(\frac{1}{k}\sum_{i=1}^{k} \lambda_{i,t}\right)^{k/2} \le \left(\frac{c^2}{k}\right)^{k/2}.$$

Therefore $\left|\det \varphi_t(z)\varphi_t^{-1}(z')\right| \le K < \infty$ for all $z, z' \in \Theta$, $t \in \mathbf{R}_+$, where $K = c^k k^{-k/2}$. It is obvious that $\lambda_{1,t}^{-1} \ge \lambda_{2,t}^{-1} \ge \cdots \ge \lambda_{k,t}^{-1}$ are the eigenvalues of the matrix G_t^{-1}. From (5.3.13) it follows that $\lambda_{k,t}^{-1} \ge c^{-2} > 0$, whence

$$\left|A_t^{-1}v\right|^2 = v'G_t^{-1}v \ge \lambda_{k,t}^{-1}|v|^2 \ge c^{-2}|v|^2.$$

Since $A_t^{-1} = \varphi_t^{-1}(z)\varphi_t(z')$, we get $|\varphi_t^{-1}(z)\varphi_t(z')v|^2 \ge |av|^2$ for $a = c^{-1} \neq 0$ and all $z, z' \in \Theta$, $t \in \mathbf{R}_+$, and $v \in \mathbf{R}^k$. Let $y_0 = y_0(t) \in \Theta$ be such that $0 < |\det \varphi_t(y_0)| < \infty$. Since $\left|\det \varphi_t(z)\varphi_t^{-1}(z')\right| \le K < \infty$ for all $z, z' \in \Theta$, we have

$$\begin{aligned} \mathsf{E}\big|\log|\det\varphi_t(\theta)|\big| &= \mathsf{E}\Big|\log\left|\det\varphi_t(\theta)\varphi_t^{-1}(y_0)\right| + \log|\det\varphi_t(y_0)|\Big| \\ &\le |\log K| + \big|\log|\det\varphi_t(y_0)|\big| < \infty. \end{aligned} \tag{5.3.14}$$

Therefore all the conditions of Theorem 5.2.4 are satisfied and inequality (5.3.9) is proved.

Relation (5.3.10) follows from inequality (5.3.9), estimate $I(\xi^t, \theta) \ge I(\overline{\theta}_t, \theta)$, and asymptotics (5.1.3). Relation (5.3.11) follows from (5.3.10) and asymptotics (5.1.25). □

Based on the work [**38**], Theorem 5.2, the following inequality can be obtained from asymptotic properties of Bayes estimates.

PROPOSITION 5.3.1. *Let conditions* **K1–K4** *of Section* 4.1 *hold for a positive definite normalizing matrix* $\varphi_t(y)$ *for* Q*-almost all single-point compact sets* $\{y\} \subset \Theta$, *and let the measure* Q *be absolutely continuous with respect to Lebesgue measure and possess a continuous and positive density* $q(y)$. *Assume that the loss function is of the form* $L(y,z) = l\left(\varphi_t^{-1}(\theta_0)(z-y)\right)$, *where* $l \in \widetilde{\mathbf{W}}'_p$ *and* θ_0 *is an arbitrary fixed point of* Θ. *Then for all estimates* $\overline{\theta}_t$ *we have*

$$\mathsf{E}\,\varepsilon\left(\overline{\theta}_t;\theta\right) \geq \mathsf{E}\,l(B(\theta_0,\theta)\eta) + o(1) \tag{5.3.15}$$

as $t \to \infty$, *where* $\mathcal{L}(\eta|\mathsf{P}_y) = \mathcal{N}(0, I_k)$ *for all* $y \in \Theta$ *and* $B(\theta_0,\theta)$ *is a nonsingular matrix in condition* **K2**.

PROOF. Let $\widetilde{\theta}_t$ be a Bayes estimate with respect to the a priori density $q(y)$ and $l\left(\varphi_t^{-1}(\theta_0)(z-y)\right)$ be the loss function. Taking into account condition **K2**, by Theorem 4.1.2 we have

$$\lim_{t\to\infty} \varepsilon\left(\widetilde{\theta}_t; y\right) = \mathsf{E}_y\, l(B(\theta_0,\theta)\eta) \tag{5.3.16}$$

for Q-almost all $y \in \Theta$. Since $\mathsf{E}\,\varepsilon(\overline{\theta}_t;\theta) \geq \mathsf{E}\,\varepsilon(\widetilde{\theta}_t;\theta)$ for any estimate $\overline{\theta}_t$, by the Fatou lemma we obtain (5.3.15) from (5.3.16). □

REMARK 5.3.1. Putting

$$L(y,z) = \widetilde{L}(y,z) = l\left(\varphi_t^{-1}(\theta_0)(z-y)\right)$$

and repeating the proof of Theorem 5.4.2, we get the estimate

$$I\left(\overline{\theta}_t;\theta\right) \geq \log\left|\det \varphi_t^{-1}(\theta_0)\right| + h(\theta) + \mathsf{E}\,m\left(\varepsilon\left(\overline{\theta}_t;\theta\right)\right) + o(1). \tag{5.3.17}$$

In so doing, we do not use the conditions

$$\left|\det \varphi_t(z)\varphi_t^{-1}(z')\right| \leq K < \infty \quad \text{and} \quad l\left(\varphi_t^{-1}(z)\varphi_t(z')v\right) \geq l(av) \quad \text{for } a \neq 0.$$

This and Theorem 5.1.2 imply that

$$\mathsf{E}\,m\left(\varepsilon\left(\overline{\theta}_t;\theta\right)\right) \leq -\mathsf{E}\log|\det B(\theta_0,\theta)| \int Z(u)\,du + o(1). \tag{5.3.18}$$

If the LAN condition is satisfied and $l(v) = |v|^2$, then (5.3.18) takes the form

$$\mathsf{E}\log \varepsilon\left(\overline{\theta}_t;\theta\right) \geq \mathsf{E}\log k\,|\det B(\theta_0,\theta)|^{2/k} + o(1) \tag{5.3.19}$$

and (5.3.15) becomes the inequality

$$\mathsf{E}\,\varepsilon\left(\overline{\theta}_t;\theta\right) \geq \mathsf{E}\,|\det B(\theta_0,\theta)|^2 + o(1). \tag{5.3.20}$$

Note that neither continuity nor positivity of $q(y)$ were required when proving inequality (5.3.19). Moreover, in the case of the loss function $l\left(\varphi_t^{-1}(\theta_0)(z-y)\right)$ the known Shannon lower bound for the ε-entropy leads to the estimate

$$I\left(\overline{\theta}_t;\theta\right) \geq \log\left|\det \varphi_t^{-1}(\theta_0)\right| + h(\theta) + m\left(\mathsf{E}\,\varepsilon\left(\overline{\theta}_t;\theta\right)\right)$$

([**60**], Theorem 1), which is a somewhat rougher bound as compared to (5.3.17). Now we can get bounds similar to (5.3.18) and (5.3.19) but slightly rougher than (5.3.18) and (5.3.19).

Lower bounds of nonintegral type. Inequalities (5.3.2), (5.3.3), (5.3.6), and (5.3.7) were derived for integrated risk functions $\varepsilon(\overline{\theta}_t; y)$. From these inequalities one can obtain usual Cramér–Rao inequalities (lower estimates for risk functions $\varepsilon(\overline{\theta}_t; y)$ for every $y \in \Theta$). Namely, the following theorem holds.

THEOREM 5.3.2. *Let* $L(y, z) = l\left(\varphi_t^{-1}(y)(z-y)\right)$, $z \in \mathbf{R}^k$, *where the function* $l(v)$ *satisfies conditions* (l) *and* $\varphi_t(y)$ *are positive definite matrices such that*

$$\begin{gathered} 0 < |\det \varphi_t(y)| < \infty \quad \text{for all } t \in \mathbf{R}_+, \\ |\varphi_t(y)| \to 0 \quad \text{as } t \to \infty \text{ for almost all } y \in \Theta, \\ \left|\det \varphi_t(z)\varphi_t^{-1}(z')\right| \le K < \infty, \\ l\left(\varphi_t^{-1}(z)\varphi_t(z')v\right) \ge l(av) \quad \text{for all } z, z' \in \Theta,\ t \in \mathbf{R}_+,\ v \in \mathbf{R}^k, \end{gathered}$$

where the constants $K > 0$ *and* $a \neq 0$ *are independent of* z, z', t, *and* v. *Further, let* $\overline{\theta}_t$ *be a consistent estimate for the parameter* θ *such that for every compact set* $K \subset \Theta$ *there exist constants* $\varepsilon_0 = \varepsilon_0(K) > 0$ *and* $\varepsilon^0 = \varepsilon^0(K) < \infty$, *where* $\varepsilon_0 \le \varepsilon(\overline{\theta}_t; y) \le \varepsilon^0$ *for all* $y \in K$ *and* $t \in \mathbf{R}_+$. *Assume that the conditions of Theorem* 5.1.2 *hold for all vectors* θ *with values in the compact set* $K \subset \Theta$ *and having uniform distribution on* K. *Then for almost all* $y \in \Theta$ *the following inequality holds*:

$$\liminf_{t\to\infty} \varepsilon\left(\overline{\theta}_t; y\right) \ge m^{-1}\left(-\mathsf{E}_y \log \int Z(u)\, du\right), \tag{5.3.21}$$

where m^{-1} *is the inverse function to* $m(\varepsilon)$ *defined in Theorem* 5.2.4.

PROOF. Let K be an arbitrary fixed compact set of Θ with $\operatorname{meas}(K) > 0$, and let a random vector θ be uniformly distributed on K (here $\operatorname{meas}(K)$ stands for the Lebesgue measure of the compact set K). Under the above assumptions, the conditions of Theorem 5.2.4 are satisfied. This follows from the equality $h(\theta) = \log \operatorname{meas}(K)$ and estimate (5.3.14). Therefore by Theorems 5.1.2 and 5.2.4 and Corollary 5.3.1 we have

$$\int_K m\left(\varepsilon\left(\overline{\theta}_t; y\right)\right) dy \le -\int_K \mathsf{E}_y \log \int Z(u)\, du\, dy + o(1)$$

as $t \to \infty$. By the Fatou lemma we get

$$\int_K \limsup_{t\to\infty} m\left(\varepsilon\left(\overline{\theta}_t; y\right)\right) dy \le -\int_K \mathsf{E}_y \log \int Z(u)\, du\, dy. \tag{5.3.22}$$

Since the compact set K is arbitrary, relation (5.3.22) implies that

$$\limsup_{t\to\infty} m\left(\varepsilon\left(\overline{\theta}_t; y\right)\right) \le -\mathsf{E}_y \log \int Z(u)\, du \tag{5.3.23}$$

for almost all $y \in \Theta$. From Lemma 5.3.1 it follows that the function $m(\varepsilon)$ is monotone and continuous and thus

$$\limsup_{t\to\infty} m\left(\varepsilon\left(\overline{\theta}_t; y\right)\right) = m\left(\liminf_{t\to\infty} \varepsilon\left(\overline{\theta}_t; y\right)\right). \tag{5.3.24}$$

Therefore (5.3.21) follows from (5.3.23) and (5.3.24). □

COROLLARY 5.3.3. *Let* $L(y,z) = \left|\varphi_t^{-1}(y)(z-y)\right|^2$, $z, y \in \Theta$, *where* $\varphi_t(y)$ *are positive definite matrices such that*

$$0 < |\det \varphi_t(y)| < \infty \quad \textit{for all } t \in \mathbf{R}_+,$$
$$|\varphi_t(y)| \to 0 \quad \textit{as } t \to \infty \textit{ for almost all } y \in \Theta,$$
$$\left|\varphi_t(z)\varphi_t^{-1}(z')\right| \le c < \infty \quad \textit{for all } z, z' \in \Theta,\ t \in \mathbf{R}_+,$$

where the constant c *does not depend on* z, z', *and* t. *Further, let* $\overline{\theta}_t$ *be a consistent estimate of the parameter* θ *such that for all compact sets* $K \subset \Theta$ *there exist constants* $\varepsilon_0 = \varepsilon_0(K) > 0$ *and* $\varepsilon^0 = \varepsilon^0(K) < \infty$ *such that* $\varepsilon_0 \le \varepsilon(\overline{\theta}_t; y) \le \varepsilon^0$ *for all* $y \in K$, $t \in \mathbf{R}_+$. *Assume that the conditions of Theorem* 5.1.2 *hold for all random vectors* θ *taking values in* K *and having uniform distribution on* K. *Moreover, let the family* $(\mathsf{P}_y^t, y \in \Theta)$ *possess the LAN property for almost all* $y \in \Theta$. *Then for almost all* $y \in \Theta$ *the following bound holds*:

$$\liminf_{t\to\infty} \varepsilon\left(\overline{\theta}_t; y\right) \ge k. \tag{5.3.25}$$

PROOF. In the proof of Theorem 5.3.2 it was shown that if $l(v) = |v|^2$, $v \in \mathbf{R}^k$, then all conditions of Theorem 5.2.4 are satisfied and $m(\varepsilon) = (k/2)\log(k/(2\pi e \varepsilon))$. Now (5.3.25) follows from (5.3.2) and equality (5.1.26). □

Inequalities (5.3.2), (5.3.3), (5.3.6), and (5.3.7) imply an asymptotically minimax bound of the Hájek type [**40, 147**]. Namely, the following theorem holds.

THEOREM 5.3.3. *Let* $L(y,z) = l\left(\varphi_t^{-1}(y)(z-y)\right)$, $z, y \in \mathbf{R}^k$, *where the function* $l(v)$ *and matrices* $\varphi_t(y)$ *satisfy the conditions of Theorem* 5.3.2 *and* $\overline{\theta}_t$ *is a consistent estimate for the parameter* θ *such that* $0 < \varepsilon_0 \le \varepsilon\left(\overline{\theta}_t; y\right) \le \varepsilon^0 < \infty$ *for all* $y \in S_\delta^k(\theta_0) = \{y \in \mathbf{R}^k : |y - \theta_0| \le \delta\}$, $t \in \mathbf{R}_+$, *where* $\delta > 0$ *and* $\theta_0 \in \Theta$ *are such that* $S_\delta^k(\theta_0) \subset \Theta$. *In general the constants* ε_0 *and* ε^0 *can depend on* θ_0 *and* δ. *Further, let the conditions of Theorem* 5.1.2 *hold for a random vector* θ *taking values in* $S_\delta^k(\theta_0)$ *and having uniform distribution on* S_δ^k. *Then*

$$\liminf_{t\to\infty} \sup_{|y-t_0|<\delta} \varepsilon\left(\overline{\theta}_t; y\right) \ge m^{-1}\left(-\inf_{|y-t_0|<\delta} \mathsf{E}_y \log \int Z(u)\,du\right). \tag{5.3.26}$$

Moreover, if the above conditions are satisfied for all sufficiently small $\delta > 0$ *and the function* $\mathsf{E}_y \log \int Z(u)\,du$ *is continuous in* y, *then*

$$\lim_{\delta\to 0} \liminf_{t\to\infty} \sup_{|y-\theta_0|<\delta} \varepsilon\left(\hat{\theta}_t; y\right) \ge m^{-1}\left(\mathsf{E}_{\theta_0} \log \int Z(u)\,du\right). \tag{5.3.27}$$

PROOF. From Theorems 5.2.4 and 5.1.2 and Corollary 5.3.1 it follows that

$$\frac{1}{m_\delta}\int_{S_\delta^k(\theta_0)} m\left(\varepsilon\left(\overline{\theta}_t; y\right)\right) dy \le -\frac{1}{m_\delta}\int_{S_\delta^k(\theta_0)} \mathsf{E}_y \log \int Z(u)\,du\,dy + o(1) \tag{5.3.28}$$

as $t \to \infty$, where m_δ is the Lebesgue measure of the set $S_\delta^k(\theta_0)$. By the properties of the function $m(\varepsilon)$,

$$\inf_{y\in S_\delta^k(\theta_0)} m\left(\varepsilon\left(\overline{\theta}_t; y\right)\right) = m\left(\sup_{y\in S_\delta^k(\theta_0)} \varepsilon\left(\overline{\theta}_t; y\right)\right),$$

whence, by (5.3.28), we obtain

$$m\left(\sup_{y\in S_\delta^k(\theta_0)} \varepsilon\left(\overline{\theta}_t; y\right)\right) \le -\inf_{y\in S_\delta^k(\theta_0)} \mathsf{E}_y \log \int Z(u)\, du + o(1)$$

as $t \to \infty$. Therefore

$$\sup_{y\in S_\delta^k(\theta_0)} \varepsilon\left(\overline{\theta}_t; y\right) \ge m^{-1}\left(-\inf_{y\in S_\delta^k(\theta_0)} \mathsf{E}_y \log \int Z(u)\, du + o(1)\right).$$

Inequality (5.3.26) follows from the continuity of the function $m^{-1}(z)$. Relation (5.3.27) follows immediately from (5.3.26) and the continuity of the function $\mathsf{E}_y \log \int Z(u)\, du$. □

COROLLARY 5.3.4. *If under the conditions of Theorem* 5.3.3 *the family* $(\mathsf{P}_y^t,\ y \in \Theta)$ *possesses the LAN property in a neighborhood of a point* $\theta_0 \in \Theta$ *with normalizing matrix* $\varphi_t(y)$, *then*

$$\lim_{\delta\to 0} \liminf_{t\to\infty} \sup_{|y-\theta_0|<\delta} \varepsilon\left(\overline{\theta}_t; y\right) \ge m^{-1}\left(\log(2\pi e)^{-k/2}\right). \tag{5.3.29}$$

In particular, if $l(v) = |v|^2$, *then*

$$\lim_{\delta\to 0} \liminf_{t\to\infty} \sup_{|y-\theta_0|<\delta} \varepsilon\left(\overline{\theta}_t; y\right) \ge k. \tag{5.3.30}$$

PROOF. To prove inequalities (5.3.29), it is sufficient to apply (5.3.27) and (5.1.26). Inequality (5.3.30) follows from (5.3.29), since in the case where $l(v) = |v|^2$ the function $m(\varepsilon)$ is of the form $m(\varepsilon) = \log(k/(2\pi e\varepsilon))^{k/2}$. □

REMARK 5.3.2. Under the LAN condition the following Hájek lower bound is well known [**40, 147**]: for all $\delta > 0$

$$\liminf_{t\to\infty} \sup_{|y-\theta_0|<\delta} \varepsilon\left(\overline{\theta}_t; y\right) \ge \mathsf{E}\, l(\eta), \tag{5.3.31}$$

where $\mathcal{L}(\eta|\mathsf{P}_y) = \mathcal{N}(0, I_k)$ for all $y \in \mathbf{R}^k$ and the loss function $L(y,z)$ is of the form $L(y,z) = l\left(\varphi_t^{-1} l(\theta_0)(z-y)\right)$. Using estimate (5.3.17) and arguing as in the proof of Theorem 5.3.3, in the case where $L(y,z) = l\left(\varphi_t^{-1}(\theta_0)(z-y)\right)$, we get the following inequality:

$$\liminf_{t\to\infty} \sup_{|y-\theta_0|<\delta} \varepsilon\left(\overline{\theta}_t; y\right) \ge m^{-1}\left(-\inf_{|y-\theta_0|<\delta} \mathsf{E}_y \log |\det B(\theta_0, y)| \int Z(u)\, du\right). \tag{5.3.32}$$

Passing to the limit in inequalities (5.3.31) and (5.3.32), we get inequality (5.3.30) under the LAN condition as $t \to \infty$ and $l(v) = |v|^2$.

Relationship to information sufficiency. The information inequalities obtained in this section are closely related to the notion of information sufficiency or, more precisely, to the notion of asymptotic information sufficiency [**107**].

Let $(Y, \mathfrak{T})$ be a measurable space and $T_t = T_t(x)$, $x \in X^t$, a measurable mapping from $(X^t, \mathfrak{B}^t)$ into $(Y, \mathfrak{T})$, usually called a statistic.

DEFINITION. A statistic T_t is called *informationally sufficient* for a parametric family $(\mathsf{P}_y^t, y \in \Theta)$ generated by observations ξ^t if $I(\xi^t, \theta) = I(T_t, \theta)$. If $I(\xi^t, \theta) - I(T_t, \theta) \to 0$ as $t \to \infty$, then the family of statistics T_t, $t \in \mathbf{R}_+$, is called *asymptotically informationally sufficient.*

From Section 4.1 it is not hard to derive conditions that guarantee the assertion of Theorem 5.1.2 which gives the asymptotics of the information $I(\xi^t, \theta)$, which takes the form (5.1.25) under condition **K1**. From Section 4.1 it is also possible to derive conditions under which, for example, inequality (5.3.11) transforms into an equality for the maximal likelihood estimates $\widehat{\theta}_t$. In view of (5.3.11) this means that $I(\xi^t, \theta) - I(\widehat{\theta}_t, \theta) \to 0$ as $t \to \infty$, that is, the estimates $\widehat{\theta}_t$, $t \in \mathbf{R}_+$, are asymptotically informationally sufficient. Similar assertions hold for Bayes estimates $\widetilde{\theta}_t$, $t \in \mathbf{R}_+$.

5.4. Observations of semimartingales

Consider a family of statistical experiments $(\mathbf{D}^t, \mathfrak{B}^t, (P_y^t, y \in \Theta))$ generated by conditional distributions of observations ξ^t of a semimartingale ξ for a given value of an unknown random parameter θ (see Sections 1.2 and 5.1). Note that in what follows the parameter θ is treated as a random vector that assumes values in an open set $\Theta \subset \mathbf{R}^k$, $k \geq 1$, and has a distribution Q. As in Section 1.2, the semimartingale ξ has a triple of predictable characteristics $(\alpha(\theta), \langle m(\theta) \rangle, \nu(\theta))$ on a stochastic basis $(\Omega, \mathcal{F}, \mathbf{F}, \mathsf{P}_\theta)$. For all fixed $\theta \in \Theta$ and $y \in \Theta$ we consider two probability measures P_θ and P_y and assume that the conditions of Theorems 1.2.1 and 1.2.2 hold for $\mathsf{P} = \mathsf{P}_\theta$ and $\widetilde{\mathsf{P}} = \mathsf{P}_y$. These conditions assure that the measures P_θ and P_y are locally equivalent. Theorems 1.2.1 and 1.2.2 give the form of the process of the local density $z(y, \theta)$ of the measure P_y with respect to the measure P_θ (see Sections 1.2 and 3.3, relations (3.3.1)–(3.3.3)). Based on representation (3.3.1)–(3.3.3) for the process $z(y, \theta)$, we get the random function $Z_t(u) = z_t(\theta + \varphi_t(\theta)u, \theta)$ introduced in Section 5.1 and used troughout this chapter.

From the results of Sections 3.3 and 4.2 it is easy to derive sufficient conditions allowing one to apply assertions of Sections 5.1–5.3 to the case of observations of semimartingales.

The following auxiliary result provides sufficient conditions for condition 1) of Theorem 5.1.1 to hold (we use the notation of Section 3.3 with appropriate changes).

LEMMA 5.4.1. *Let the conditions of Theorem* 3.3.1 *hold for* $\Delta_t = \varphi_t(\theta)u$ *and for Q-almost all* $\theta \in \Theta$ *and* $u \in \mathbf{R}^k$, *where* $\varphi_t(\theta)$ *is a matrix in condition* 2) *of Theorem* 3.3.1. *Then condition* 1) *of Theorem* 5.1.1 *holds and*

$$Z(u) = \exp\left\{u'\eta - \frac{1}{2}|u|^2\right\}, \qquad \mathcal{L}(\eta|\mathsf{P}) = \mathcal{N}(0, I_k).$$

PROOF. Due to the Lebesgue dominant convergence theorem we need to prove that conditional finite-dimensional distributions of the field $Z_t(u)$ weakly converge for $\theta = y$ to the corresponding conditional distributions of the field $Z(u)$ for Q-almost all $y \in \Theta$ as $t \to \infty$. In turn, this follows from the weak convergence, as $t \to \infty$, of conditional finite-dimensional distributions of the field $\ln Z_t(u)$ at $\theta = y$ to the corresponding conditional finite-dimensional distributions of the field $\ln Z(u)$ for Q-almost all $y \in \Theta$. According to the Cramér–Wold method [**17**] it is necessary to show that, as $t \to \infty$,

$$\mathcal{L}\Big(\sum_{i=1}^{n} \lambda_i \ln Z_t(u_i) \mid \mathsf{P}_y\Big) \xrightarrow{w} \mathcal{L}\Big(\sum_{i=1}^{n} \lambda_i \ln Z(u_i) \mid \mathsf{P}\Big) \tag{5.4.1}$$

for all $n \in \mathbf{N}$, $\lambda_1, \lambda_2, \dots, \lambda_n \in \mathbf{R}^1$, $u_1, u_2, \dots, u_n \in \mathbf{R}^k$, and Q-almost all $y \in \Theta$. Using representation (3.3.5)–(3.3.8) for $\Lambda_t(\theta + \varphi_t(\theta)u_i, \theta)$ at $\theta = y$, $i = 1, 2, \dots, n$, we obtain relation (5.4.1) by the same argument as in the proof of Theorem 3.3.1. □

REMARK 5.4.1. Lemma 5.4.1 can also be proved under the conditions of Theorem 3.3.2 instead of those of Theorem 3.3.1.

The following lemma provides sufficient conditions for condition 2) of Theorem 5.1.1 to hold (we use notation of Sections 3.3 and 4.2).

LEMMA 5.4.2. *Let condition* 1) *of Theorem* 4.2.4 *be satisfied for all* $\theta \in \Theta$. *Assume that for some constants* $p \geq 0$, $C > 0$, *and for all* $t \in \mathbf{R}_+$, $y \in \Theta$, *and* $u, v \in \Theta_{t,y}$ *the following inequality holds*:

$$\begin{aligned} \mathsf{E}_{y+\varphi_t(y)u}\Big\{ &\big|\varphi_t(y) g\big(y + \varphi_t(y)v, y + \varphi_t(y)u\big)\big|^2 \circ \langle m(y)\rangle_t \\ &+ \big|\varphi_t(y) n\big(y + \varphi_t(y)v, y + \varphi_t(y)u\big)\big|^2 * \nu\big(y + \varphi_t(y)u\big)_t\Big\} \\ &\leq C\big(1 + |y|^p + (|\varphi_t(y)u| \vee |\varphi_t(y)v|)^p\big), \end{aligned}$$

where $\varphi_t(y)$ *is a matrix of condition* 2) *of Theorem* 3.3.1, *the function* $g(y, z)$ *is defined in Section* 3.3, *and the function* $n(y, z)$ *is defined in Theorem* 4.2.4. *Then inequality* (5.1.4) *holds.*

PROOF. Repeating the proof of Theorem 4.2.4, we get

$$\mathsf{E}_y \left| Z_t^{1/2}(u) - Z_t^{1/2}(v) \right|^2 \leq \widetilde{C}\left(1 + |y|^p + \big(|\varphi_t(y)u| \vee |\varphi_t(y)v|\big)^p\right) |u - v|^2, \tag{5.4.2}$$

where $\widetilde{C}$ is a constant.

Now we use the obvious inequality

$$\mathsf{E}_y \left| Z_t(u) - Z_t(v) \right| \leq \left(\mathsf{E}_y \left| Z_t^{1/2}(u) - Z_t^{1/2}(v) \right|^2 \mathsf{E}_y \left| Z_t^{1/2}(u) + Z_t^{1/2}(v) \right|^2 \right)^2$$

and derive inequality (5.1.4) from estimate (5.4.2). □

To prove that condition 3) of Theorem 5.1.1 is satisfied, we use Theorem 4.2.5, with Θ instead of K, as in the proof of Lemma 5.4.2. Then we use the known results [**151**] to derive condition 4) of Theorem 5.1.1. Condition 5) of Theorem

5.1.1 depends on the matrix $\varphi_t(y)$ and holds for many densities $q(y)$ if the matrix $\varphi_t(y)$ is "good" enough.

In order that the theorems of Section 5.2 be true, it is necessary that the risk function of the estimate $\overline{\theta}_t$ satisfy the inequality $\varepsilon_0 \le \varepsilon\left(\overline{\theta}_t; y\right) \le \varepsilon^0$ for all $y \in \Theta$ for some constants ε_0 and ε^0 such that $0 < \varepsilon_0 < \varepsilon^0 < \infty$. The fulfillment of this condition follows from Section 4.1. For observation of semimartingales the fulfillment of this condition follows from Section 4.2. Moreover, let the conditions of Section 4.2 that ensure the fulfillment of conditions **K1**–**K4** of Section 4.1 hold and also the above conditions (Lemmas 5.4.1 and 5.4.2 and the remarks) that yield the conditions of Theorem 5.1.1. Then the information inequality (5.3.11) transforms into the asymptotic equality for the maximum likelihood estimates $\widetilde{\theta}_t$ and the Bayes estimates constructed from observations of semimartinales. This implies that the families of estimates $(\widehat{\theta}_t)$ and $(\widetilde{\theta}_t)$ are asymptotically informationally sufficient statistics (see the definition in Section 5.3).

REMARK 5.4.2. The results of Sections 5.1–5.3 are easy to check in the special cases of observations of diffusion processes and counting processes. To this end it is enough to consider the examples of Sections 3.4, 3.5, 4.3, and 4.4 and to use Lemmas 5.4.1 and 5.4.2 and the remarks in this section.

Bibliographical Notes

Chapter 1

1.1. In this section, we discuss notions and results of the general theory of stochastic processes and stochastic integration needed for the further presentation. More details on this subject are given in [**30**], [**138**], [**139**], [**153**], [**156**], [**96**], [**98**], [**128**], [**20**], [**21**], [**100**], [**207**]; also see [**1**], [**46**].

1.2. We follow Ibragimov and Khas'minskii [**40**] in describing general statistical experiments. In [**4**] and [**115**], statistical experiments are called *statistical structures*. Lemma 1.2.1 plays the crucial role in absolute continuity of probability measures. This lemma was proved by Girsanov [**17**] for Wiener processes; many other authors have later obtained this result for different types of stochastic processes; see [**25**], [**46**], [**96**], [**98**], [**113**], [**139**], [**153**], and [**156**]. Theorem 1.2.1 without the additional condition VIII) was proved in [**46**], [**153**], [**98**], and [**156**]. The local density was studied in many special cases; it was obtained in [**113**] for Markov processes, in [**96**] for diffusion type processes, in [**25**] for Markov type processes, and in [**45**] for counting processes. Similar results related to the exponential representation of distributions of stochastic processes are obtained in [**166**], [**231**]. Statistical experiments generated by stochastic processes and their properties are described in [**40**], [**146**], [**158**], and [**171**].

1.3. Theorems 1.3.1 and 1.3.2 were proved in [**84**]. A proof of Theorem 1.3.3 was given in [**83**]. Theorems 1.3.4–1.3.6 are new; for close results see [**76**], [**80**], and [**81**]. More details about semimartingales are given in [**156**], [**154**], [**27**]. Limit theorems for special cases of semimartingales can be found in [**51**], [**56**], [**71**], [**73**], [**77**], [**116**], [**119**], and [**149**].

Chapter 2

2.1. In this section we discuss some notions and results needed for the further presentation. More details are given in [**9**] and [**58**].

2.2. In this section we follow [**85**] and [**89**]. The complete group of types of the asymptotic distinguishability between families of hypotheses was introduced by the author [**85**]. The early paper [**70**] should also be mentioned, where types $\mathbf{a}_0$, $\mathbf{a}_1$, and $\mathbf{e}$ of the asymptotic distinguishability were introduced for similar diffusion processes. These types also form a complete group. Various definitions of the asymptotic distinguishability were considered by other authors: Krafft [**65**] and Ingster [**43**] gave a definition for families of hypotheses, and Liptser, Pukel'sheim, and Shiryaev [**95**], Eagleson [**141**], Eagleson and Mémin [**142**], and Le Cam and Traxler [**170**] introduced the same notion for the complete asymptotic discrimination of families of measures. Mutual contiguity for families of measures was introduced by Le Cam [**169**] (note that this notion is called simply *continuity* in that paper). Le Cam [**169**] and Roussas [**110**] gave various characterizations for the mutual contiguity. The contiguity of a family of measures to another family was defined by Hájek and Šidák [**14**]. Several characterizations for this kind of contiguity were given by Liptser, Pukel'sheim, and Shiryaev [**95**], Eagleson and Mémin [**142**], Hall and Loynes [**148**], and F. Liese [**236**]. These questions were also discussed in [**146**], [**171**], [**156**], [**174**] and [**127**].

2.3. Here we follow [**88**]. The results of this section can be found in [**87**] for the case of equivalent measures P^t and $\widetilde{P}^t$. Kullback and Leibler [**200**] introduced the relative entropy $I(P^t|\widetilde{P}^t)$ under the name *information for distinguishability*. Sanov [**201**] also used this notion in problems on large deviations for polynomial schemes (see also [**202**]–[**205**]). Theorem 2.3.1 is a generalization of a classical Stein lemma (see [**91**], [**198**], and [**219**] for other generalizations). The original proof of Stein's lemma is based on a result on large deviations. Later, Kullback [**53**] suggested another method using the law of large numbers (see also [**124**]). Theorems 2.3.2 and 2.3.3 belong to Krafft and Plachky [**164**] for the case of sequences of independent identically distributed random variables. A generalization of their results to the minimax risk in the case of composite hypotheses was given in [**206**]. Another proof of Theorem 2.3.2 was obtained by Kolomiets [**49**]. Theorems 2.3.3 and Corollary 2.3.4 are related to results on large deviations for Λ_t as $t \to \infty$. If the condition $\lim_{t\to\infty} \chi_t^{-1} \ln \alpha_t = -a$ is assumed for $a > 0$ instead of $\alpha 1'$ and condition Λ^* is satisfied, then, using Lemma 2.3.4, one can show that $\lim_{t\to\infty} \chi_t^{-1} \ln \beta(\delta_t^{+,\alpha_t}) = -b(a)$, where $b(a)$ is a certain function [**208**]. The first results on large deviations were obtained by Khintchine [**209**], Smirnoff [**210**], Cramér [**211**], and Chernoff [**212**], who studdied the case of sums of independent identically distributed random variables. For further developments see [**201**], [**205**], [**213**], [**189**], [**191**], [**143**], [**144**], [**214**]. The function $b(a)$ was studied in [**215**] and [**216**] for observations of independent identically distributed random variables and in [**217**] and [**218**] for general binary statistical experiments.

2.4. In this section we present results of [**88**] and [**92**] in a somewhat different form. Results similar to Theorems 2.4.1 and 2.4.2 were obtained by Basawa and Scott [**131**], Hornik [**150**], and Janssen [**157**]. The idea of the proof of Theorems 2.4.3 and 2.4.4 is similar to that of Theorems 2.4.1 and 2.4.2. A close result was proved in [**197**]. Relation (2.4.25) was obtained in [**197**] for independent identically distributed random variables.

2.5. Here we follow [**92**] and [**93**]. Similar results with the Gaussian law in condition $\Lambda 6$ can be found in [**88**]. Close results were obtained by Roussas [**110**] and Hall and Loynes [**148**]. The main Lemma 2.3.1 was proved for the Gaussian law $\mathcal{S} = \mathcal{N}(a, \sigma^2)$ in [**110**] with $a = -\sigma^2/2$ and in [**88**] with $a < -\sigma^2/2$. Lemma 2.5.1 was proved in [**93**] for a general law (see also [**92**]). Theorem 2.5.4 is new. General results on relative compactness and tightness can be found in [**7**], [**127**], [**110**], and [**156**].

2.6. In this section, we present results of [**88**] and [**92**] in a somewhat modified form (see also [**89**], [**90**], [**176**], [**177**]). The asymptotic decomposition of the likelihood ratio in condition $\Lambda 8^*$ was obtained in [**70**] for unbounded vectors u_t. Condition $\Lambda 8^*$ with bounded vectors u_t for measures P^t and $\widetilde{P}^t$ belonging to a parametric family is known as the *local asymptotic normality condition* and was presented in various books, mainly devoted to the theory of parametric estimation; see [**169**], [**110**], [**40**], [**29**], [**57**], [**146**], [**9**], [**184**]. Condition $\Lambda 8'$ with bounded vectors u_t for measures P^t and $\widetilde{P}^t$ belonging to a parametric family is closely related to the notion of *local asymptotic mixed normality.*

2.7. The reductions of hypotheses testing considered in this section were studied in [**88**]. In presenting them, we follow [**92**] and [**177**]. More detail and special cases as well as further results can be found in [**88**], [**92**], and [**177**].

Chapter 3

3.1. In this section, we follow [**156**] and [**49**]. For properties and applications of Hellinger processes, see [**160**], [**174**], [**178**], [**179**], [**181**]. Representation (3.1.9) for a Hellinger process of order ε was obtained in [**86**] using the multiplicative decomposition of the process $Y(\varepsilon)$. If $\varepsilon_-^t = \inf\{\varepsilon\colon H_t(\varepsilon; \widetilde{\mathsf{P}}, \mathsf{P}) < \infty\}$ and $\varepsilon_+^t = \sup\{\varepsilon\colon H_t(\varepsilon; \widetilde{\mathsf{P}}, \mathsf{P}) < \infty\}$, then $\varepsilon_-^t \uparrow \varepsilon_- \leq 0$ and $\varepsilon_+^t \downarrow \varepsilon_+ \geq 1$ as $t \to \infty$; see [**223**], [**224**]. Therefore the Hellinger process $h(\varepsilon) = h(\varepsilon; \widetilde{\mathsf{P}}, \mathsf{P})$ is well defined by Theorem 3.1.1. This allows one to get assertions of this section for all $\varepsilon \in (\varepsilon_-, \varepsilon_+)$. More details are given in [**223**]–[**225**].

3.2. The law of large numbers was proved in [**84**]. Theorem 3.2.2 is obtained by Kolomiets [**49**]. Theorem 3.2.3 is based on Theorems 1–3 in [**86**]. The idea to split the space Ω by the Hellinger process was applied for the first time to diffusion processes in [**67**]; the same method was later used in [**81**] to estimate parameters of counting processes. Theorem 3.2.4 was proved in [**92**], [**176**]. Further references can be found in [**92**] and [**176**].

3.3. This section is based on [**83**] and [**84**]. Asymptotic relation (3.3.5) was obtained in [**75**] for diffusion type processes and in [**78**] for counting processes. This relation was discussed in [**77**] for Markov processes and in [**226**] for semimartingales with determinate triplets of predictable characteristics.

3.4. General limit theorems for diffusion processes were obtained in [**66**], [**70**], and [**71**]. Types $\mathbf{a}_0$, $\mathbf{a}_1$, and $\mathbf{e}$ of the asymptotic distinguishability between families of hypotheses were introduced in [**70**] in the case of similar hypotheses. Some results for the scheme "a signal in white noise" can be found in [**82**]. Example 3.4.2 is taken from [**92**]. Results for null recurrent processes were discussed in [**87**]. Similar asymptotic results for the minimax risk belong to Gushchin [**28**]. Sequential tests for distinguishing diffusion processes were given in [**126**] and [**96**]. Properties of homogeneous diffusion processes were described in more detail in [**19**]–[**20**], [**114**], [**122**]. Corresponding results for (and applications of) Hellinger integrals and processes of order ε can be found in [**28**], [**219**], [**237**], and [**240**].

3.5. Here we present some general limit theorems of [**71**] and [**78**]. There is an extensive literature on the problem of distinguishing counting processes with determinate compensators; see, for example, [**47**], [**134**], [**72**], and [**194**]. Results discussed in this section were obtained in [**94**] for the case of determinate compensators and renewal processes. The idea of the proof of Theorem 3.5.9 was introduced in [**86**]. Generalizations of Theorems 3.5.7, 3.5.8, 3.5.10, and 3.5.11 were obtained in [**228**] for renewal processes with discontinuous compensators. Theorems 3.5.1–3.5.4 were proved in [**227**] for counting processes with discontinuous compensators; see also the survey papers [**225**] and [**230**].

Chapter 4

4.1. This section contains results on properties of statistical estimates $\hat{\theta}_t$ and $\tilde{\theta}_t$ due to Ibragimov and Khas'minskii [**40**].

4.2. Theorems 4.2.1–4.2.6 are taken from [**83**], [**84**]; see also [**175**], [**176**]. Properties of the likelihood ratio for semimartingales with determinate triplets of predictable characteristics were studied in [**226**] (the case of discontinuous characteristics was also treated in [**226**]).

4.3. An extensive literature is devoted to the problem of estimation of parameters for diffusion processes. The Gaussian processes were the first class to be studied in the literature; see, for example, [**3**], [**39**], [**50**], [**96**], [**123**], [**167**] and the references in [**40**], [**57**], [**96**]. This case was treated in detail in Arató's book [**2**]. The problem becomes complicate if the diffusion process is not Gaussian. Nevertheless, the asymptotic behavior of statistical estimates was also studied for a linear parameter by using limit theorems for stochastic integrals over the Wiener process [**52**], [**61**], [**62**], [**103**], [**116**], [**123**], [**133**]. The next step is to study the case of a nonlinear parameter, and this requires a method based on the LAN condition [**40**]. The main problem arising in this approach is to prove condition **K3** for $\beta > k$ and **K4**. To prove condition **K4**, some authors (see [**5**], [**57**], [**76**]) pose restrictions on the shift in the following form:

$$\int_0^t (a(s,x,y) - a(s,x,\theta))^2\, ds \geq \kappa |y-\theta|^2, \qquad \kappa > 0.$$

Note that for a wide class of processes no conditions of this kind are satisfied, and this does not allow one to study such processes. The truncation method for the space of trajectories used in the proof of Theorem 4.2.5 solves this problem (this method was proposed in [**64**]; see also [**68**]). The proof of condition **K3** for $\beta > k$ is based on the idea of the proof of Lemma 3.3.2 in [**40**]. This idea instead of Lemma 3.5.2 in [**40**] works in the Gaussian case (see Lemma 4 in [**76**] and Theorem 4.2.6). We mention also [**111**], where LAN condition was established for diffusion processes with periodic coefficients depending on the process. Methods of estimation were described in [**229**], [**230**], [**183**], [**238**], [**239**] for diffusion processes.

4.4. The parameter estimation for counting processes was treated in a rather great number of papers. Most of these papers deal only with the case of Poisson processes; see [**47**], [**57**], [**72**], [**134**], [**173**], [**190**]. The application of the general asymptotic method based on the LAN property [**4**] encounters the same problems as in the case of diffusion processes. The LAN property was obtained almost simultaneously for Poisson processes [**55**], Poisson type processes [**56**], and general counting processes [**71**]. The proof of condition **K3** for $\beta > k$ is simpler in this case due to an idea used for the first time in the proof of Lemma 4 in [**76**], where the case of diffusion type processes was considered; the proof for counting processes can be found in [**81**]. Condition **K4** was also proved in [**81**] by using the truncation method introduced in [**64**] for diffusion processes. Further results on properties of the likelihood ratio and parameter estimation can be found in [**229**], [**230**], [**183**], [**238**], [**239**].

Chapter 5

5.1. Theorems 5.1.1–5.1.3 are taken from [**74**] and [**79**]. Theorem 5.1.1 was proved in [**68**] for $k = 1$. Relation (5.1.3) was obtained earlier for sequences of independent random variables in [**68**] and [**117**], purely discontinuous processes with independent increments in [**73**], and Markov processes in [**77**]. Relation (5.1.32) for sequences of independent random variables was proved in [**63**]; with terms of higher orders in the asymptotic expansion it was also proved in [**187**] and [**193**] for $p = 1$ and Θ' finite. Relation (5.1.32) with terms of higher orders in the asymptotic expansion was obtained in [**242**] for $p = 1$ and the scheme of general statistical observations ξ^t. Upper and lower bounds as well as the asymptotic behavior of the information $I(\xi^t, \theta)$ up to terms of order $O(1)$ can be found in [**246**] for the case of a continuous parameter θ and observations of independent identically distributed random variables.

5.2. Theorems 5.2.1–5.2.4 are taken from [**65**], [**6**], [**74**] and given here in a somewhat different form. The idea of the proof of Lemma 5.2.1 was used for the first time in [**243**]. Inequality (5.2.29) was obtained in [**34**] and [**69**] for the case $l(v) = |v|^2$ and sequences of independent random variables. The rate of convergence in asymptotic relation (5.2.51) was studied in the case of $p = 1$ and a finite Θ' in [**187**] and [**245**] for observations of independent random variables, and in [**242**] and [**244**] for the scheme of general statistical observations.

5.3. Corollary 5.3.1 and Theorem 5.3.3 are taken from [**74**]. Theorem 5.3.2 was proved in [**65**] and [**69**] for special classes of functions $L(y, z)$ and sequences of independent random variables. The equivalence between the sufficiency and information sufficiency was proved in Linnik's book [**241**] for the Shannon information and in Kullback's book [**53**] for the Kullback information. More details on the asymptotic information sufficiency were given in [**107**], [**65**], [**69**], and [**74**].

References

1. S. V. Anulova, A. Yu. Veretennikov, N. V. Krylov, R. Sh. Liptser, and A. N. Shiryaev, *The stochastic calculus*, Itogi Nauki i Tekhniki, Sovremennye Problemy Matematiki, vol. 45, VINITI, Moscow, 1989, pp. 5–260; English transl. in *Encyclopaedia of Math. Sciences*, vol. 45, Springer-Verlag, Berlin–Heidelberg–New York, 1998.
2. M. Arató, *Linear stochastic systems with constant coefficients. A statistical approach*, Springer-Verlag, Berlin–Heidelberg–New York, 1982.
3. M. Arató, A. N. Kolmogorov, and Ya. G. Sinai, *On an estimate of parameters of a complex stationary Gaussian process*, Dokl. Akad. Nauk SSSR **146** (1962), no. 4, 747–750; English transl. in Soviet Math. Dokl. **3** (1962).
4. Jean-René Barra, *Notions fondamentales de statistique mathématique*, Dunod, Paris, 1971.
5. K. Bauer, *On asymptotic properties of estimators of drift parameter of a diffusion process*, Teor. Veroyatnost. i Primenen. **25** (1980), no. 2, 437–439; English transl. in Theory Probab. Appl. **25** (1980).
6. A. V. Bernstein, *Asymptotically similar tests*, Itogi Nauki i Tekhniki, Sovremennye Problemy Matematiki, vol. 17, VINITI, Moscow, 1979, pp. 5–260. (Russian)
7. P. Billingsley, *Convergence of probability measures*, Wiley, New York–London–Sydney–Toronto, 1968.
8. A. A. Borovkov, *Asymptotically optimal tests for testing composite hypotheses*, Teor. Veroyatnost. i Primenen. **20** (1975), no. 3, 463–487; English transl. in Theory Probab. Appl. **20** (1975).
9. ———, *Mathematical statistics*, "Nauka", Moscow, 1984; English transl., Gordon and Breach, New York, 1998.
10. ———, *Probability theory*, "Nauka", Moscow, 1986; English transl., Gordon and Breach, New York, 1998.
11. M. V. Burnashev, *On minimax detection of inaccurately known signal on a background white Gaussian noise*, Teor. Veroyatnost. i Primenen. **24** (1979), no. 1, 106–118; English transl. in Theory Probab. Appl. **24** (1979).

12a. H. L. Van Trees, *Detection, estimation and modulation theory. Part 1. Detection, estimation and linear modulation theory*, Wiley, New York–London–Sydney, 1968.

12b. ———, *Detection, estimation and modulation theory. Part 2. Nonlinear modulation theory*, Wiley, New York–London–Sydney–Toronto, 1971.

12c. ———, *Detection, estimation and modulation theory. Part 3. Radar-sonar signal processing and Gaussian signals in noise*, Wiley, New York–London–Sydney–Toronto, 1971.

13. L. G. Vetrov, *On the filtration, interpolation and extrapolation of semimartingales*, Teor. Veroyatnost. i Primenen. **27** (1982), no. 1, 24–35; English transl. in Theory Probab. Appl. **27** (1982).
14. J. Hájek and Z. Šidák, *Theory of rank tests*, Academia, Prague, 1967.
15. F. R. Gantmakher, *Theory of matrices*, "Nauka", Moscow, 1966; English transl., Chelsea, New York, 1989–1990.
16. I. M. Gel'fand and A. M. Yaglom, *Calculation of the amount of information about a random function contained in another such function*, Uspekhi Mat. Nauk **12** (1957), no. 1, 3–42; English transl. in Amer. Math. Soc. Transl. (2) **12** (1959).
17. I. V. Girsanov, *On transforming a certain class of stochastic processes by absolutely continuous substitution of measures*, Teor. Veroyatnost. i Primenen. **5** (1960), no. 3, 314–330; English transl., Theory Probab. Appl. **5** (1960), 285–301.
18. I. I. Gikhman and A. V. Skorokhod, *Introduction to the theory of random processes*, "Nauka", Moscow, 1965; English transl., Saunders, Philadelphia–London–Toronto, 1969.

19. ———, *Stochastic differential equations*, "Naukova Dumka", Kiev, 1968; English transl., Springer-Verlag, Heidelberg–New York, 1972.
20. ———, *Stochastic differential equations and their applications*, "Naukova Dumka", Kiev, 1982. (Russian)
21. ———, *The theory of stochastic processes*, vol. 3, "Nauka", Moscow, 1975; English transl., Springer-Verlag, Berlin–Heidelberg–New York, 1979.
22. B. V. Gnedenko and I. N. Kovalenko, *Introduction to queuing theory*, Second edition, "Nauka", Moscow, 1987; English transl., Birkhäuser, Boston–Basel–Berlin, 1989.
23. B. V. Gnedenko and A. N. Kolmogorov, *Limit distributions for sums of independent random variables*, Gostekhizdat, Leningrad–Moscow, 1949; English transl., Addison-Wesley, Gambridge, MA, 1954.
24. U. Grenander, *Stochastic processes and statistical inference*, Arkiv för Matematik **1** (1950), no. 3, 195–277.
25. B. I. Grigelionis, *On an absolute continuous substitution of a measure and Markov's property*, Litovsk. Mat. Sb. **9** (1969), no. 1, 57–71. (Russian)
26. ———, *On a martingale characterization of stochastic processes with independent increments*, Litovsk. Mat. Sb. **17** (1977), no. 1, 75–86. (Russian)
27. B. I. Grigelionis, K. Kubilius and R. A. Mikulevicius, *A martingale approach to functional limit theorems*, Uspekhi Mat. Nauk **37** (1982), no. 6, 39–51; English transl. in Soviet Math. Surveys **37** (1982).
28. A. A. Gushchin, *A remark on an asymptotic behavior of a minimax risk in testing diffusion type processes*, Statistics and control of stochastic processes, "Nauka", Moscow, 1989, pp. 48–55. (Russian)
29. K. O. Dzaparidze, *Parameter estimation and hypotheses testing in spectral analysis of stationary time series*, Izdat. Tbilis. Univ., Tbilisi, 1981; English transl., Springer-Verlag, Berlin–Heidelberg–New York, 1985.
30. C. Dellacherie, *Capacités et processus stochastiques*, Springer-Verlag, Berlin–Heidelberg–New York, 1972.
31. R. L. Dobrushin, *A general formulation of the basic Shannon theorem in information theory*, Uspekhi Mat. Nauk **14** (1959), no. 6, 3–104; English transl. in Russian Math. Surveys **14** (1959).
32. A. Ya. Dorogovtsev, *The theory of parameter estimation for stochastic processes*, "Vyshcha Shkola", Kiev, 1982. (Russian)
33. M. P. Ershov, *The sequential estimation of diffusion processes*, Teor. Veroyatnost. i Primenen. **15** (1970), no. 4, 705–717; English transl. in Theory Probab. Appl. **15** (1970).
34. S. Yu. Efroimovich, *The information contained in a sequence of observations*, Probl. Peredachi Inform. **15** (1979), no. 3, 24–39; English transl. in Problems Inform. Transmission **15** (1979).
35. V. M. Zvonkin, *The transformation of the phase space of a diffusion process that removes the drift*, Mat. Sb. **93** (1974), no. 1, 129–149; English transl. in Math. USSR Sb. **22** (1974).
36. V. M. Zolotarev, *One-dimensional stable distributions*, "Nauka", Moscow, 1983; English transl., Amer. Math. Soc., Providence, RI, 1986.
37. I. A. Ibragimov and Yu. A. Rozanov, *Gaussian random processes*, "Nauka", Moscow, 1970; English transl. Srpinger-Verlag, Berlin–Heidelberg–New York, 1978.
38. I. A. Ibragimov and R. Z. Has'minskii, *On information inequalities and superefficient estimators*, Probl. Peredachi Inform. **9** (1973), no. 3, 53–67; English transl. in Problems Inform. Transmission **9** (1973).
39. ———, *An estimator for a signal parameter in a Gaussian white noise*, Probl. Peredachi Inform. **10** (1974), no. 1, 39–59; English transl. in Problems Inform. Transmission **10** (1974).
40. ———, *Statistical estimation. Asymptotic theory*, "Nauka", Moscow, 1979; English transl., Srpinger-Verlag, Berlin–Heidelberg–New York, 1981.
41. I. Sh. Ibramkhalilov and A. V. Skorokhod, *Consistent estimates of parameters of stochastic processes*, "Naukova Dumka", Kiev, 1980. (Russian)
42. Yu. I. Ingster, *On minimax nonparametric detection of signals in a Gaussian white noise*, Probl. Peredachi Inform. **18** (1982), no. 2, 61–73; English transl. in Problems Inform. Transmission **18** (1982).

43. ______, *On minimax distinguishability of families of nonparametric hypotheses*, Dokl. Akad. Nauk SSSR **267** (1982), no. 3, 536–539; English transl. in Soviet Math. Dokl. **26** (1982).
44. ______, *Asymptotically optimal Bayes tests in testing composite hypotheses*, Teor. Veroyatnost. i Primenen. **28** (1983), no. 4, 738–757; English transl. in Theory Probab. Appl. **28** (1983).
45. Yu. M. Kabanov, R. Sh. Liptser, and A. N. Shiryaev, *Martingale methods in the theory of point processes*, Proc. School-Seminar on the Theory of Stochastic Processes (Druskininkai, 25–28 November 1974), Part 2, Inst. Fiz. i Mat. Akad. Nauk Litovsk. SSR, Vilnius, 1975, pp. 269–354. (Russian)
46a. ______, *Absolute continuity and singularity of locally absolutely continuous probability distributions. II*, Mat. Sb. **107** (1978), no. 3, 364–416; English transl. in Math. USSR Sb. **35** (1979), no. 5, 631–680.
46b. ______, *Absolute continuity and singularity of locally absolutely continuous probability distributions. I*, Mat. Sb. **108** (1979), no. 1, 32–61; English transl. in Math. USSR Sb. **36** (1980), no. 1, 31–58.
47. D. R. Cox and P. A. W. Lewis, *The statistical analysis of series of events*, Wiley, New York, 1966.
48a. D. R. Cox, *Renewal theory*, Wiley, New York, 1961.
48b. W. L. Smith, *Renewal theory and its ramifications*, J. Royal Statist. Society, Ser. B **20** (1958), no. 2, 243–302.
49. E. I. Kolomiets, *On asymptotical behavior of probabilities of the second type errors for Neyman-Pearson test*, Teor. Veroyanost. i Primenen. **32** (1987), no. 3, 503–522; English transl. in Theory Probab. Appl. **32** (1987).
50. V. A. Kotel'nikov, *The theory of the potential noise stability*, Gosenergoizdat, Leningrad–Moscow, 1956. (Russian)
51. K. Kubilius, *Asymptotic behavior of distributions of martingales with a continuous parameter*, Litovsk. Mat. Sb. **19** (1979), no. 4, 129–143. (Russian)
52. G. L. Kulinich, *On an estimate of a drift parameter for a stochastic diffusion equation*, Teor. Veroyatnost. i Primenen. **20** (1975), no. 2, 393–397; English transl. in Theory Probab. Appl. **20** (1975).
53. S. Kullback, *Information theory and statistics*, Wiley, New York; Chapman and Hall, London, 1959.
54. Yu. A. Kutoyants, *On a property of an estimate of a parameter in the drift coefficient*, Izv. Akad. Nauk Armyan. SSR Mat. **12** (1977), no. 4, 245–251. (Russian)
55. ______, *The estimation of an intensity parameter of a nonhomogeneous Poisson process*, Control Problems Inform. Theory **8** (1979), no. 2, 137–149. (Russian)
56. ______, *The local asymptotical normality for Poisson type processes*, Izv. Akad. Nauk Armyan. SSR Mat. **14** (1979), no. 1, 3–20. (Russian)
57. ______, *Parameter estimation for stochastic processes*, Izdat. Akad. Nauk Armyan. SSR, Erevan, 1980; English transl., Heldermann-Verlag, Berlin, 1984.
58. E. L. Lehmann, *Testing statistical hypotheses*, Wiley, New York; Chapman and Hall, London, 1959.
59. Yu. N. Lin'kov, *Epsilon-entropy of random variables*, Teor. Sluch. Prots. **1** (1973), 68–86; English transl. in Theory Random Processes **1** (1974).
60. ______, *Estimates of the epsilon-entropy of random variables with restrictions on the additive noise*, Teor. Sluch. Prots. **2** (1974), 65–73; English transl. in Theory Random Processes **2** (1974).
61. ______, *Asymptotic behavior of Bayes estimates of parameters in the drift of diffusion processes*, Teor. Sluch. Prots. **3** (1975), 50–54. (Russian)
62. ______, *Generalized Bayes estimates of the drift parameters of a diffusion process*, Teor. Veroyatnost. i Mat. Statist. **13** (1975), 92–99; English transl. in Theory Probab. and Math. Statist. **13** (1977), 99–107.
63. ______, *On the amount of information in observations relative to an unknown parameter*, Teor. Veroyatnost. i Mat. Statist. **15** (1976), 122–131; English transl., Theory Probab. and Math. Statist. **15** (1978), 126–136.
64. ______, *On statistical estimates of parameters of diffusion processes*, Second Vilnius Conf. on Probab. Theory and Math. Statist. (June 28–July 3, 1977, Vilnius), Abstracts of Commun., vol. 1, Izdat. Inst. Mat. i Kibern. Akad. Nauk Litovsk. SSR, Vilnius, 1977, pp. 240–241. (Russian)

65. ———, *On the asymptotic sufficiency and risk function of estimates of unknown parameters*, Dokl. Akad. Nauk Ukrain. SSR, Ser A (1977), no. 3, 203–206. (Russian)
66. ———, *On the asymptotic behavior of error probabilities in testing diffusion processes*, System behavior in random media, Institut Kibernetiki, Akad. Nauk Ukrain. SSR, Kiev, 1977, pp. 40–46. (Russian)
67. ———, *Estimates of nonlinear drift parameters for diffusion processes*, Teor. Sluch. Prots. **6** (1978), 68–78. (Russian)
68. ———, *On information about a parameter contained in a diffusion process*, Teor. Sluch. Prots. **6** (1978), 78–85. (Russian)
69. ———, *On Cramér-Rao's type inequalities and asymptotic sufficiency of statistical estimators*, Teor. Veroyatnost. i Mat. Statist. **19** (1978), 92–101; English transl., Theory Probab. and Math. Statist. **19** (1980), 105–115.
70. ———, *On probabilities of errors in testing similar hypotheses for diffusion processes*, Teor. Sluch. Prots. **7** (1979), 62–71. (Russian)
71. ———, *On asymptotic behavior of probabilities of errors for the likelihood ratio test in testing point processes with continuous compensators*, System behavior in random media, Institut Kibernetiki, Akad. Nauk Ukrain. SSR, Kiev, 1979, pp. 62–70. (Russian)
72. ———, *Asymptotic normality of the log-likelihood ratio and hypotheses testing for nonhomogeneous Poisson processes*, Teor. Sluch. Prots. **8** (1980), 91–100. (Russian)
73. ———, *The asymptotic normality of stochastic integrals and the statistics of discontinuous proceses*, Teor. Veroyatnost. i Mat. Statist. **23** (1980), 91–101; English transl., Theory Probab. and Math. Statist. **23** (1981), 97–107.
74. ———, *An information contained in observations and statistical estimates with respect to an unknown parameter*, VIII USSR Conference on Signal Coding and Information Transmission, Nauch. Sovet Kompl. Probl. "Kibernetika" Akad. Nauk SSSR, Moscow–Kuibyshev, 1981, pp. 54–58. (Russian)
75. ———, *On the asymptotic power of statistical tests for diffusion type processes*, Teor. Sluch. Prots. **9** (1981), 61–71. (Russian)
76. ———, *On estimates of parameters for diffusion type processes*, Teor. Sluch. Prots. **9** (1981), 71–78. (Russian)
77. ———, *Asymptotic properties of statistical estimators and tests for Markov processes*, Teor. Veroyatnost. i Mat. Statist. **25** (1981), 76–91; English transl., Theory Probab. and Math. Statist. **25** (1982), 83–98.
78. ———, *On the asymptotic power of statistical tests for counting processes*, Probl. Peredachi Inform. **17** (1981), no. 3, 69–80; English transl. in Problems Inform. Transmission **17** (1981).
79. ———, *Asymptotic behavior of Shannon information in observations of a relatively unknown parameter*, Teor. Sluch. Prots. **10** (1982), 42–50. (Russian)
80. ———, *The asymptotic normality of locally square-integrable martingales in a scheme of series*, Teor. Veroyatnost. i Mat. Statist. **27** (1982), 88–96; English transl., Theory Probab. and Math. Statist. **27** (1983), 95–103.
81. ———, *On estimates of parameters of counting processes*, Probl. Peredachi Inform. **18** (1982), no. 1, 78–93; English transl. in Problems Inform. Transmission **18** (1982).
82. ———, *On the power of tests with null asymptotic level for noncontigual alternatives*, Teor. Sluch. Prots. **11** (1983), 62–69. (Russian)
83. ———, *On the asymptotic behavior of the likelihood ratio for some statistical problems for semimartingales*, Teor. Sluch. Prots. **12** (1984), 40–48. (Russian)
84. ———, *Local densities of measures generated by semimartingales and their properties*, Teor. Sluch. Prots. **13** (1985), 43–50. (Russian)
85. ———, *Types of the asymptotic distinguishability of families of hypotheses and their characterization*, Teor. Veroyatnost. i Mat. Statist. **33** (1985), 57–67; English transl., Theory Probab. and Math. Statist. **33** (1986), 65–74.
86. ———, *Asymptotic properties of local densities of measures generated by simimartingales*, Teor. Sluch. Prots. **14** (1986), 48–55; English transl., J. Soviet Math. **53** (1991), no. 1, 49–55.
87. ———, *Asymptotic properties of the Neyman-Pearson test in the case of completely asymptotically distinguishable hypotheses*, Teor. Veroyatnost. i Mat. Statist. **35** (1986), 60–69; English transl., Theory Probab. and Math. Statist. **35** (1987), 65–74.

88. ______, *Asymptotic distinguishing between two simple statistical hypotheses*, Preprint no. 86.45, Inst. Mat. Akad. Nauk Ukrain. SSR, Kiev, 1986. (Russian)
89. ______, *Characterization of types of asymptotic distinguishability of families of hypotheses*, Teor. Sluch. Prots. **15** (1987), 64–71; English transl., J. Soviet Math. **53** (1991), 409–415.
90. ______, *Asymptotic problems of statistics for semimartingales*, The first World Congress of the Bernoulli Society, Abstracts, vol. 1, Mat. Inst. Akad. Nauk SSSR, Moscow, 1988, pp. 202–206. (Russian)
91. ______, *The generalization of Stein's lemma*, Statistics and control of stochastic processes, "Nauka", Moscow, 1989, pp. 115–117. (Russian)
92. ______, *The methods for solving asymptotic problems on testing two simple statistical hypotheses*, Preprint, no. 89.05, Inst. Prikl. Mat. i Mekh. Akad. Nauk Ukrain SSR, Donetsk, 1989. (Russian)
93. ______, *Asymptotic distinguishing between two simple statistical hypotheses*, Theory of stochastic processes and its applications, "Naukova Dumka", Kiev, 1990, pp. 89–98. (Russian)
94. Yu. N. Lin'kov and Munir Al'Shakhf, *Asymptotic properties of the likelihood ratio for counting processes*, Preprint, no. 91.02, Inst. Prikl. Mat. i Mekh. Akad. Nauk Ukrain. SSR, Donetsk, 1991. (Russian)
95. R. Sh. Liptser, F. Pukel'sheim, and A. N. Shiryaev, *On necessary and sufficient conditions for contiguity and entire asymptotic separation of probability measures*, Uspekhi Mat. Nauk **37** (1982), no. 6, 97–124; English transl. in Russian Math. Surveys **37** (1982).
96. R. Sh. Liptser and A. N. Shiryaev, *Statistics of stochastic processes*, "Nauka", Moscow, 1974; English transl., Springer-Verlag, Berlin–Heidelberg–New York, 1977.
97. ______, *The functional central limit theorems for semimartingales*, Teor. Veroyatnost. i Primenen. **25** (1980), no. 4, 683–703; English transl. in Theory Probab. Appl. **25** (1980).
98. ______, *Theory of martingales*, "Nauka", Moscow, 1986; English transl., Kluwer, Dordrecht, 1989.
99. ______, *On weak convergence of semimartingales to stochastically continuous processes with independent and conditionally independent increments*, Mat. Sb. **116** (1981), no. 3, 331–358; English transl., Math. USSR Sb. **44** (1982), no. 3, 299–323.
100. P. A. Meyer, *Probability and potential*, Blaisdell Publ. Company, A Division of Ginn and Company Waltham, Massachusetts–Toronto–London, 1966.
101. R. M. Meshcherskii, *The analysis of neuron activity*, "Nauka", Moscow, 1972. (Russian)
102. I. P. Natanson, *Theory of functions of a real variable*, Gostekhizdat, Moscow, 1957. (Russian)
103. A. A. Novikov, *On estimates of parameters of diffusion processes*, Studia Sci. Math. Hung. **7** (1972), no. 1–2, 201–209. (Russian)
104. ______, *On discontinuous martingales*, Teor. Veroyatnost. i Primenen. **20** (1975), no. 1, 13–28; English transl. in Theory Probab. Appl. **20** (1975).
105. V. V. Petrov, *Sums of independent random variables*, "Nauka", Moscow, 1972; English transl., Springer-Verlag, Berlin–Heidelberg–New York, 1975.
106. M. S. Pinsker, *Information and information stability of random variables and processes*, Izdat. Akad. Nauk SSSR, Moscow, 1960; English transl., Holden-Day, San Francisco, 1964.
107. ______, *Information contained in observations, and asymptotically sufficient statistics*, Probl. Peredachi Inform. **8** (1972), no. 1, 45–61; English transl. in Problems Inform. Transmission **8** (1972).
108. B. Ramachandran, *Advanced theory of characteristic functions*, Statist. Publ. Society, Calcutta, 1967.
109. C. R. Rao, *Linear statistical inference and its applications*, Wiley, New York–London–Sydney, 1965.
110. G. G. Roussas, *Contiguity of probability measures. Some applications in statistics*, Cambridge University Press, Cambrridge, 1972.
111. O. A. Safonova, *On asymptotic behavior of integral functionals of diffusion processes with periodic coefficients*, Ukrain. Mat. Zh. **44** (1992), no. 2, 245–252; English transl. in Ukrainian Math. J. **44** (1992).
112. D. S. Sil'vestrov, *Limit theorems for composite random fuctions*, Kiev University, Kiev, 1974. (Russian)
113. A. V. Skorokhod, *Studies in the theory of random processes*, Kiev University, Kiev, 1961; English transl., Addison-Wesley, Reading, 1965.

114. ———, *Asymptotic methods of the theory of stochastic differential equations*, "Naukova Dumka", Kiev, 1987; English transl., Amer. Math. Soc., Providence, RI, 1994.
115. Jean-Louis Soler, *Notion de liberté en statistique mathématique*, Thèse de Docteur de Troiseime Cycle, Université de Grenoble, 1970.
116. A. F. Taraskin, *On the asymptotic normality of stochastic integrals and estimates of a drift parameter for a diffusion process*, Mat. Fiz. **8** (1970), 149–163. (Russian)
117. ———, *A relationship between Shannon's and Fisher's information in diffusion processes*, Probl. Peredachi Inform. **15** (1979), no. 1, 14–26; Englich transl. in Problems Inform. Transmission **15** (1979).
118. ———, *Central limit theorem for stochastic intergrals*, Teor. Sluch. Prots. **12** (1984), 81–90. (Russian)
119. ———, *On the behavior of the likelihood ratio for semimartingales*, Teor. Veroyatnost. i Primenen. **29** (1984), no. 3, 440–451; English transl. in Theory Probab. Appl. **29** (1984).
120. G. L. Turin, *Notes on digital communication*, Van Nostrand, New York–Cincinnati–Toronto–London–Melbourne, 1969.
121. W. Feller, *An introduction to probability theory and its applications*, Second Edition, vol. 1, Wiley, New York; Chapman and Hall, London, 1958.
122. R. Z. Has'minskii, *Stochastic stability of differential equations*, "Nauka", Moscow, 1969; English transl., Sijthoff and Noordhoff, Alphen, 1980.
123. A. S. Holevo, *Estimation of the drift parameters for a diffusion process by the method of stochastic approximation*, Studies in the theory of self-controlled systems, Izdat. Vych. Tsentr. Akad. Nauk SSSR, Moscow, 1967, pp. 179–200. (Russian)
124. N. N. Chentsov, *Statistical decision rules and optimal inference*, "Nauka", Moscow, 1972; English transl., Amer. Math. Soc., Providence, RI, 1980.
125. D. M. Chibisov, *A theorem on admissible tests and its application to an asymptotic problem of hypothesis testing*, Teor. Veroyatnost. i Primenen. **12** (1967), no. 1, 96–111; English transl. in Theory Probab. Appl. **12** (1967).
126. A. N. Shiryaev, *Statistical sequential analysis*, "Nauka", Moscow, 1969; English transl., Amer. Math. Soc., Providence, RI, 1973.
127. ———, *Probability*, Second edition, "Nauka", Moscow, 1989; English transl., Graduate Texts in Mathematics, vol. 95, Springer-Verlag, Berlin–Heidelberg–New York, 1996.
128. R. J. Elliott, *Stochastic calculus and applications*, Springer-Verlag, New York–Heidelberg–Berlin, 1982.
129. M. G. Akritas, *Asymptotic theory for estimating the parameters of a Lévy process*, Ann. Inst. Statist. Math. **34** (1982), 259–280.
130. R. R. Bahadur, *Some limit theorems in statistics*, SIAM, Philadelphia, 1971.
131. I. V. Basawa and D. J. Scott, *Asymptotical optimal inference for nonergodic models*, Lecture Notes Satist., vol. 17, Springer-Verlag, Berlin–Heidelberg–New York, 1983.
132. P. Billingsley, *Statistical inference for Markov processes*, Univ. Chicago Press, Chicago, 1961.
133. B. M. Brown and J. I. Hewitt, *Asymptotic likelihood theory for diffusion processes*, J. Appl. Probab. **12** (1975), no. 2, 228–238.
134. M. Brown, *Statistical analysis of non-homogeneous Poisson processes*, Stochastic point processes: Statistical analysis, theory, and applications, Wiley, New York–London, 1972, pp. 67–89.
135. H. Chernoff, *Large sample theory: parametric case*, Ann. Math. Statist. **27** (1956), no. 1, 1–22.
136. J. T. Cox and D. Griffeath, *Large deviations for Poisson systems of independent random walks*, Z. Wahrscheinlichkeitstheor. und verw. Geb. **66** (1984), no. 4, 543–558.
137. ———, *Occupation times for critical branching Brownian motions*, Ann. Probab. **13** (1985), no. 4, 1108–1132.
138. C. Dellacherie and P. A. Meyer, *Probabilités et potentiel*, vol. I, Hermann, Paris, 1975.
139. ———, *Probabilités et potentiel*, vol. II, Hermann, Paris, 1980.
140. C. Doléans, *Intégrales stochastiques dépendant d'un paramètre*, Publ. Inst. Statist. Univ. Paris **16** (1967), 23–34.
141. G. K. Eagleson, *An extended dichotomy theorem for sequences of pairs of Gaussian measures*, Ann. Probab. **9** (1981), no. 3, 453–459.

142. G. K. Eagleson and J. Mémin, *Sur la contiguïté de deux suites de mesures: généralisation d'un théorème de Kabanov-Liptser-Shiryaev*, Lecture Notes Math., vol. 920, Springer-Verlag, Berlin–Heidelberg–New York, 1982, pp. 319–337.
143. R. S. Ellis, *Large deviations for a general class of random vectors*, Ann. Probab. **12** (1984), no. 1, 1–12.
144. ———, *Entropy, large deviations and statistical mechanics*, Springer-Verlag, Berlin–Heidelberg–New York–Tokyo, 1985.
145. W. Feller, *Fluctuation theory of recurrent events*, Trans. Amer. Math. Soc. **67** (1949), no. 1, 98–119.
146. P. E. Greenwood and A. N. Shiryaev, *Contiguity and the statistical invariance principle*, Gordon and Breach, New York, 1985.
147. J. Hájek, *Local asymptotic minimax and admissibility in estimation*, Proc. Sixth Berkley Symp. Math. Statist. and Probab., vol. 1, Univ. Calif. Press., Berkeley–Los Angeles, 1972, pp. 175–194.
148. W. J. Hall and R. M. Loynes, *On the concept of contiguity*, Ann. Probab. **5** (1977), no. 2, 278–282.
149. P. Hall and C. C. Heyde, *Martingale limit theory and its application*, Academic Press, New York, 1980.
150. K. Hornik, *On the behavior of type II errors*, Statist. Decisions **6** (1988), no. 4, 379–383.
151. I. A. Ibragimov and R. Z. Has'minskii, *On the information in a sample about a parameter*, II Int. Symp. Inform. Theory, Tsahkadsor, 1971, Akademiai Kiado, Budapest, 1973, pp. 295–309.
152. J. Jacod, *Multivariate point processes: predictable projection, Radon-Nikodym derivatives, representation of martingales*, Z. Wahrscheinlichkeitstheor. und verw. Geb. **31** (1975), no. 3, 235–253.
153. ———, *Calcul stochastique et problèmes de martingales*, Lecture Notes Math., vol. 714, Springer-Verlag, Berlin–Heidelberg–New York, 1979.
154. J. Jacod, A. Klopotowski, and J. Mémin, *Théorème de la limite centrale et convergence fonctionelle vers un processus à accroissements indépendants: la méthode des martingales*, Ann. Inst. H. Poincare. **B 28** (1982), no. 1, 1–45.
155. J. Jacod and J. Mémin, *Caractéristiques locales et conditions de continuité absolue pour les semi-martingales*, Z. Wahrscheinlichkeitstheor. und verw. Geb. **35** (1976), no. 1, 1–37.
156. J. Jacod and A. N. Shiryaev, *Limit theorems for stochastic processes*, Springer-Verlag, Berlin–Heidelberg–New York–London–Tokyo, 1987.
157. A. Janssen, *Asymptotic properties of Neyman–Pearson tests for infinite Kullback-Leibler information*, Ann. Statist. **14** (1986), no. 3, 1068–1079.
158. A. Janssen, H. Milbrodt and H. Strasser, *Infinitely divisible statistical experiments*, Lecture Notes Statist., vol. 27, Springer-Verlag, Berlin–Heidelberg–New York, 1985.
159. Yu. M. Kabanov, R. Sh. Liptser, and A. N. Shiryaev, *On absolute continuity of probability measures for Markov–Itô processes*, Lecture Notes Contr. and Inform. Sci., vol. 25, Springer-Verlag, Berlin–Heidelberg–New York, 1980, pp. 114–128.
160. ———, *On the variation distance for probability measures defined on filtered space*, Probab. Theory and Rel. Fields. **71** (1986), no. 1, 19–35.
161. T. Kailath, A. Segall, and M. Zakai, *Fubini-type theorems for stochastic integrals*, Sankhya **A 40** (1978), no. 2, 138–143.
162. N. Keiding, *Maximum likelihood estimation in the birth-and-death process*, Ann. Statist. **3** (1975), no. 2, 363–372.
163. T. Komatsu, *Statistics of stochastic processes with jumps*, Lecture Notes Math., vol. 550, Springer-Verlag, Berlin–Heidelberg–New York, 1976, pp. 276–289.
164. O. Krafft and D. Plachky, *Bounds for the power of likelihood ratio test and their asymptotic properties*, Ann. Math. Statist. **41** (1970), no. 5, 1646–1654.
165. Ch. Kraft, *Some conditions for consistency and uniform consistency of statistical procedures*, Univ. Calif. Publ. Statist. **2** (1955), no. 6, 125–142.
166. U. Küchler and M. Sørensen, *Exponential families of stochastic processes: a unifying semimartingale approach*, Int. Statist. Rev. **57** (1989), no. 2, 123–144.
167. A. Le Breton, *Estimation des paramètres d'une équation différentielle stochastique vectorielle linéaire*, C. R. Acad. Sci. **A 279** (1974), no. 8, 289–292.
168. L. Le Cam, *On some asymptotic properties of maximum likelihood estimates and related Bayes estimates*, Univ. Calif. Publ. Statist. **1** (1953), no. 11, 277–330.

169. ———, *Locally asymptotically normal families of distributions*, Univ. Calif. Publ. Statist. **3** (1960), no. 2, 37–98.
170. L. Le Cam and R. Traxler, *On the asymptotic behavior of mixtures of Poisson distributions*, Z. Wahrscheinlichkeitstheor. und verw. Geb. **44** (1978), no. 1, 1–45.
171. L. Le Cam, *Asymptotic methods in statistical decision theory*, Springer-Verlag, New York–Berlin–Heidelberg–London–Paris–Tokyo, 1986.
172. E. Lenglart, *Relation de domination entre deux processus*, Ann. Inst. H. Poincaré **B 13** (1977), no. 2, 171–179.
173. P. A. W. Lewis, *Recent results in the statistical analysis of univariate point process*, Stochastic point processes: Statistical analysis, theory, and applications, Wiley, New York–London, 1972, pp. 1–52.
174. F. Liese and I. Vajda, *Convex statictical distances*, Teubner, Leipzig, 1987.
175. Yu. N. Lin'kov, *Asymptotical properties of a local density of measures for semimartingales and their applications*, Probab. Theory and Math. Statist.: Proc. Fourth. Vilnius Conf., Vilnius, 24–29 June 1985, vol. 2, VNU Sci. Press, Utrecht, 1989, pp. 217–233.
176. ———, *Asymptotical methods of statistics for semimartingales*, Math. Res. **54** (1989), 115–139.
177. ———, *Asymptotical questions of simple hypotheses testing*, Probab. and Math. Statist. **12** (1991), no. 2, 217–243.
178. R. Sh. Liptser and A. N. Shiryaev, *On the problem of "predictable" criteria of contiguity*, Lecture Notes Math., vol. 1021, Springer-Verlag, Berlin–Heidelberg–New York, 1983, pp. 386–418.
179. ———, *On contiguity of probability measures correstonding to semimartingales*, Anal. Math. **11** (1985), no. 2, 93–124.
180. H. Luschgy, *Second order behavior of Neyman-Pearson tests for stochastic processes*, Angew. Math. und Inform., Preprint., Univ. Münster., Münster, 1990.
181. J. Mémin and A. N. Shiryaev, *Distance de Hellinger-Kakutani des lois correspondant à deux processus à accroissements indépendants*, Z. Wahrscheinlichkeitstheor. und verw. Geb. **70** (1985), no. 1, 67–89.
182. M. Metivier, *Semimartingales*, Walter de Gruyter, Berlin–New York, 1982.
183. Y. Ogata, *The asymptotic behavior of maximum likelihood estimators for stationary point processes*, Ann. Inst. Statist. Math. **30** (1978), no. 2, 243–261.
184. B. L. S. Prakasa Rao, *Asymptotic theory of statistical inference*, Wiley, New York–London, 1987.
185. C. R. Rao, *Efficient estimates and optimum inference procedures in large samples*, J. Roy. Statist. Soc. **B 24** (1962), no. 1, 46–72.
186. R. Rebolledo, *Sur les applications de la théorie des martingales à l'étude statistique d'une famille de processus ponctuels*, Lecture Notes Math., vol. 636, Springer-Verlag, Berlin–Heidelberg–New York, 1978, pp. 27–70.
187. A. Rényi, *On some problems of statistics from the point of view of information theory*, Proc. Colloq. Inform. Theory, vol. 2, Budapest, 1968, pp. 343–357.
188. W. Richter, *Übertragung von Grenzaussagen für Folgen von zufälligen Grössen auf Folgen min zufälliggen Indizes*, Teor. Veroyatnost. i Primenen. **10** (1965), no. 1, 82–94.
189. G. L. Sievers, *On the probability of large deviations and exact slopes*, Ann. Math. Statist. **40** (1969), no. 6, 1908–1921.
190. D. L. Snyder, *Random point processes*, Wiley, New York–London, 1975.
191. J. Steinebach, *Convergence rates of large deviation probabilities in the multidimensional case*, Ann. Probab. **6** (1978), no. 5, 751–759.
192. C. Stricker and M. Yor, *Calcul stochastique dépendant d'un paramètre*, Z. Wahrscheinlichkeitstheor. und verw. Geb. **45** (1978), no. 2, 109–133.
193. I. Vajda, *On the convergence of information contained in a sequence of observations*, Proc. Colloq. Inform. Theory, vol. 2, Budapest, 1968, pp. 489–501.
194. S. Verdu, *Asymptotic error probability of binary hypothesis testing for Poisson point-process observations*, IEEE Trans. Inform. Theory **32** (1986), no. 1, 113–115.
195. A. Wald, *Asymptotically most powerful tests of statistical hypotheses*, Ann. Math. Statist. **12** (1941), no. 1, 1–19.
196. ———, *Tests of statistical hypotheses concerning several parameters when the number of observations is large*, Trans. Amer. Math. Soc. **54** (1943), no. 3, 426–482.

References added for the English edition

197. B. Efron, *The power of the likelihood ratio test*, Ann. Math. Statist. **38** (1967), no. 3, 802–806.
198. I. Vajda, *Generalization of discrimination-rate theorems of Chernoff and Stein*, Kybernetika **26** (1990), no. 4, 273–288.
199. M. G. Akritas and R. A. Johnson, *Asymptotic inference in Lévy processes of the discontinuous type*, Ann. Statist. **9** (1981), no. 3, 604–614.
200. S. Kullback and R. A. Leibler, *On information and sufficiency*, Ann. Math. Statist. **22** (1951), no. 1, 79–86.
201. I. N. Sanov, *On the probability of large deviations of random variables*, Mat. Sb. **42** (1957), no. 1, 11–44; English transl. in Selected Transl. Math. Stat. and Probab. **1** (1961), 213–244.
202. R. R. Bahadur, *On optimal property of the likelihood ratio*, Proc. of the Fifth Berkeley Symposium on Math. Statist. and Probability (California, June 21–July 18, 1965; December 27, 1965–January 7, 1966) (L. Le Cam and J. Neyman, eds.), vol. 1, Univ. California Press, Berkeley–Los Angeles, 1967, pp. 13–26.
203. R. R. Bahadur, *Rates of convergence of estimates and test statistics*, Ann. Math. Statist. **38** (1967), no. 2, 303–324.
204. W. Hoeffding, *Asymptotically optimal test for multinomial distributions*, Ann. Math. Statist. **36** (1965), no. 2, 369–408.
205. ———, *On probabilities of large deviations*, Proc. of the Fifth Berkeley Symposium on Math. Statist. and Probability (California, June 21–July 18, 1965; December 27, 1965–January 7, 1966) (L. Le Cam and J. Neyman, eds.), vol. 1, Univ. California Press, Berkeley–Los Angeles, 1967, pp. 203–219.
206. D. Plachky and J. Steinebach, *A generalization of a result of Chernoff in large sample theory*, Math. Operationsforsch. Statist., Ser. Statistics **8** (1977), no. 3, 375–379.
207. M. Métivier and J. Pellaumail, *Stochastic integration*, Walter de Gruyter, Berlin–New York, 1980.
208. Yu. N. Lin'kov, *Large deviations in the problem of distinguishing counting processes*, Ukrain. Mat. Zh. **45** (1993), no. 11, 1514–1521; English transl., Ukr. Math. J. **45** (1993), 1703–1712.
209. A. A. Khintchine, *Über einen neuen Grenzwertsatz der Wahrscheinlichkeitsrechnung*, Math. Annalen **101** (1929), 745–752.
210. N. Smirnoff, *Über Wahrscheinlichkeiten grosser Abweichungen*, Rec. Soc. Math. Moscou **40** (1933), 441–455.
211. H. Cramèr, *Sur un nouveau théorème-limite de la théorie des probabilités*, Actualités Scientifiques et Industrielles **736** (1938), 5–23.
212. H. Chernoff, *A measure of asymptotic efficiency for tests of hypothesis based on the sum of observations*, Ann. Math. Stat. **23** (1952), 493–507.
213. R. R. Bahadur and R. Ranga Rao, *On deviations of the sample mean*, Ann. Math. Statist. **31** (1960), 1015–1027.
214. J. D. Deuschel and D. W. Stroock, *Large deviations*, Academic Press, Boston, 1989.
215. A. A. Borovkov and A. A. Mogul'skii, *Large deviations and statistical hypothesis testing*, "Nauka", Novosibirsk, 1992. (Russian)
216. L. Birgé, *Vitesses maximales de décroissance des erreurs et tests optimaux associés*, Z. Wahrscheinlichkeitstheorie und verw. Gebiete **55** (1981), 261–273.
217. Yu. N. Lin'kov, *Limit theorems for the likelihood ratio in the hypotheses testing problems*, Random Operators and Stochastic Equations **3** (1995), 23–40.
218. ———, *Large deviation theorems in the hypotheses testing problems*, Exploring stochastic laws: Festschrift in honour of the 70th birthday of academician V. S. Korolyuk (A. V. Skorokhod and Yu. V. Borovskikh, eds.), VSP, Utrecht, 1995, pp. 263–273.
219. I. Vajda, *Theory of statistical inference and information*, Kluwer, Dordrecht, 1989.
220. P. Jeganathan, *An extension of a result of L. Le Cam concerning asymptotic normality*, Sankhya. Indian J. Statistics. Ser. A **42** (1980), no. 3-4, 146–160.
221. ———, *On the asymptotic theory of estimation when the limit of the log-likelihood ratios is mixed normal*, Sankhya. Indian J. Statistics. Ser. A **44** (1982), no. 2, 173–212.
222. L. Le Cam, *Sur l'approximation de familles de mesures par des familles gaussiennes*, Ann. Inst. Henri Poincaré. Ser. B **21** (1985), no. 3, 225–287.
223. Yu. N. Lin'kov and Yu. A. Shevlyakov, *Large deviation theorems in the hypotheses testing problems for processes with independent increments*, Theory Stochastic Processes **2(18)** (1996), no. 3–4, 133–144.

224. ———, *Properties of the likelihood ratio for processes with independent increments*, Random Operators and Stochastic Equations **5** (1997), no. 3, 237–252.
225. ———, *Large deviation theorems in the hypotheses testing problems for processes with independent increments*, Theory Stochastic Processes **4(20)** (1998), no. 1–2, 198–210.
226. ———, *Properties of the likelihood ratio for semimartingales with determinate triplets in parametric case*, Ukrain. Mat. Zh. **51** (1999), no. 9, 1172–1180. (Russian)
227. Yu. N. Lin'kov, *Asymptotic discrimination of counting processes*, Ukrain. Mat. Zh. **45** (1993), no. 7, 972–979; English transl., Ukr. Math. J. **45** (1993), no. 7, 1077–1085.
228. Yu. N. Lin'kov and Munir Al'Shakhf, *Asymptotic discrimination of renewal processes*, Ukrain. Mat. Zh. **44** (1992), no. 10, 1382–1388; English transl., Ukr. Math. J. **44** (1992), no. 10, 1268–1275.
229. Yu. N. Lin'kov, *Limit theorems for the local density of measures in the hypotheses testing problems for counting processes*, Proc. of the Sixth. Vilnius Conf. on Probab. Theory and Math. Statist. (Vilnius, June 28–July 2, 1993) (B. Grigelionis et al., eds.), TEV, Vilnius/VSP, Utrecht, 1994, pp. 497–515.
230. ———, *Limit theorems for the local density of measures of counting processes and some statistical applications*, New trends in probability and statistics: Proc. of the second Ukrainian–Hungarian conference (Mukachevo, Ukraine, September 25–October 1, 1992) (M. Arató and M. Yadrenko, eds.), TBiMC, Kiev, 1995, pp. 143–161.
231. I. Küchler and U. Küchler, *An analytical treatment of exponential families of stochastic processes with independent stationary insrements*, Math. Nachr. **103** (1981), 21–30.
232. B. L. S. Prakasa Rao, *Estimation of the drift for diffusion process*, Statistics **16** (1985), no. 2, 263–275.
233. B. M. Bibby and M. Sørensen, *On estimation for discretely observed diffusions: A review*, Theory Stochastic Processes **2(18)** (1996), no. 1–2, 49–56.
234. M. Sørensen, *Likelihood methods for diffusions with jumps*, Statistical inference in stochastic processes (N. U. Prabhu and I. V. Basawa, eds.), Marcel Dekker, New York, 1991, pp. 67–105.
235. A. R. Pedersen, *A new approach to maximum likelihood estimation for stochastic differential equations based on discrete observations*, Scand. J. Statist. **2** (1995), 55–71.
236. F. Liese, *Hellinger integrals, contiguity and entire separation*, Kybernetika **23** (1987), no. 2, 104–123.
237. ———, *Hellinger integrals of diffusion processes*, Statistics **17** (1986), no. 1, 63–78.
238. Y. Ogata and H. Akaike, *On linear intensity models for mixed doubly stochastic Poisson and self-exciting point processes*, J. Royal Statist. Society. Ser. B **44** (1982), no. 1, 102–107.
239. Y. Ogata and D. Vere-Jones, *Inference for earthquake models: A self-correcting model*, Stochastic Processes and Appl. **17** (1984), 337–347.
240. I. Vajda, *Rényi distances of some diffusion processes*, Probab. Theory and Math. Statistics: Proc. of the Fifth Vilnius Conference (Vilnius, June 25–July 1, 1989) (B. Gregelionis et al., eds.), vol. 2, Mokslas, Vilnius; VSP Utrecht, 1990, pp. 529–534.
241. Yu. V. Linnik, *Statistical problems with nuisance parameters*, "Nauka", Moscow, 1966; English transl., Amer. Math. Soc., Providence, RI, 1968.
242. Yu. N. Lin'kov and L. A. Gabriel, *Large deviation theorems in Bayes testing problems*, Vestnik Volgograd. Univ. Ser. 1. Mathematics. Physics (1998), no. 3, 62–68. (Russian)
243. I. Csiszár, *On an extremum problem of information theory*, Studia Sci. Math. Hung. **9** (1974), 57–71.
244. A. Rényi, *Statistics and information theory*, Studia Sci. Math. Hung. **2** (1967), no. 1–2, 249–256.
245. ———, *On some basic problems of statistics from the point of view of information theory*, Proc. of the Fifth Berkeley Symposium on Math. Statist. and Probability (California, June 21–July 18, 1965; December 27, 1965–January 7, 1966) (L. Le Cam and J. Neyman, eds.), vol. 1, Univ. California Press, Berkeley–Los Angeles, 1967, pp. 531–543.
246. I. A. Ibragimov and R. Z. Has'minskii, *Amount of information about a parameter contained in a sample*, Vestnik Leningrad. Univ., Mathematics. Mechanics. Astronomy **19** (1973), no. 4, 42–48. (Russian)

Index

ε-entropy of a conditional distribution, 186
σ-algebra
  completion, 10
  optional, 2
  predictable, 1
ε-entropy of a conditional distribution, 186
p-quantile, 54

asymptotic expansion of the likelihood ratio, 65

canonical representation of a semimartingale, 8
central limit theorem for local martingales, 21
  uniform, 25
compensator
  of a measure, 6
  of a process, 3, 19
  of a renewal process, 124
complete asymptotic distinguishability of hypotheses, 36
complete asymptotic indistinguishability of hypotheses, 40
condition
  Cramér's, 106, 155, 156
  for the absolute continuity of measures, 11
  for the equivalence of measures, 15
  for the local absolute continuity of measures, 11
  for the local equivalence of measures, 11
  Lindeberg's, 25
  local asymptotic at a point (LANC), 139
  Lyapunov's, 25
  uniform local asymptotic normality in a set, 139
conditional entropy
  of a distribution, 168
contiguity of hypotheses, 42
Cramér
  condition, 129
  series, 106, 129
Cramér–Wold method, 23, 195
cumulant of a process, 121

decomposition
  asymptotic, 89
  Lebesgue, 33
  multiplicative, 80
Dellacherie conditions, 10
distance
  in variation, 36
  Kakutani–Hellinger, 37
distribution
  a posteriori, 169
  a priori, 36
  degenerate, 22
  one-vertex infinitely divisible, 122
distribution density
  a posteriori, 138
  a priori, 137
  conditional, 168
  of a parameter, 161
distribution function of a stable law, 110, 130
Doléans stochastic exponent, 80
dual projection, 3

entropy
  average conditional of a distribution, 169
  conditional differential, 168
  differential, 161
  of a distribution, 168
  of a measure with respect to another measure, 46, 175
  Shannon, 52
equivalence of measures, 11
  local, 11
estimate
  asymptotically efficient, 138
  asymptotically minimax, 138
  Bayes, 137
  maximum likelihood, 137
  statistical, 137
  uniformly consistent, 152

family
  dominated by a measure, 9
  locally asymptotically normal at a point (LAN), 138
  of binary statistical experiments, 15, 18, 21, 79
  of parametric statistical experiments, 16, 19, 21, 137, 161

of statistical experiments, 15
uniformly locally asymptotically normal in a set, 139
family of functions
tight with respect to another family of measures, 43
uniformly integrable with respect to a family of measures, 43
family of probability measures, 10, 36
completely asymptotically inseparable, 40
completely asymptotically separable, 36
contiguous to another family of measures, 42
mutually contiguous, 42
mutually noncontiguous, 42
noncontiguous to another family of measures, 42
not completely asymptotically separable, 36
not mutually contiguous, 42
parametric, 88
relatively compact, 58
tight, 58
family of statistical hypotheses, 36
completely asymptotically distinguishable, 36
completely asymptotically indistinguishable, 40
contiguous to another family of hypotheses, 42
mutually contiguous, 42
mutually noncontiguous, 42
not completely asymptotically distinguishable, 36
not mutually contiguous, 42
filtration, 1
satisfying ordinary conditions, 1
formula
for the change of variables for semimartingales, 8
asymptotic for the Shannon information, 162
conditional information, 168, 170
Itô's, 8
frequency modulation, 103
function
mean-square differentiable, 151
slowly varying in the Karamata sense, 130

Girsanov theorem, 11

indistinguishable processes, 1
inequality
Cramér–Rao's, 186, 187, 191
Hájek's, 186, 192, 193
information, 186, 187
Lenglart's, 3
information
conditional, 168
Kullback–Leibler's, 175
information sufficiency, 194
asymptotical, 194
integral
Hellinger, 37
of order ε, 37, 79
Lebesgue–Stieltjes, 3
Itô formula, 8

Kullback–Leibler
distance, 46
divergence, 46
information, 46

large deviations, 48, 84
law of large numbers, 46, 83, 112, 126
strong, 126
lemma
fundamental Neyman–Pearson, 34
Stein's, 48
level of a test, 33
likelihood ratio, 33, 139
normalizing, 139
local absolute continuity of measures, 10
local class of processes, 2
local density of a measure, 11, 18, 20
localizing sequence for a process, 2
lower bounds for the information on a parameter
asymptotic, 179, 184
nonasymptotic, 176
lower bounds for the risk function
asymptotically minimax, 192
Hájek's, 193
Hájek's type, 192
integral Cramér–Rao's, 187
nonintegral of Cramér–Rao's type, 191

martingale, 2
continuous, 2
continuous Gaussian, 119
local, 2
purely discontinuous, 2
square-integrable local, 2
measure
σ-finite invariant, 109
$\widetilde{\mathcal{O}}$-σ-finite, 6
$\widetilde{\mathcal{P}}$-σ-finite, 6
compensator, 6
Dirac's, 6
Doléans, 6
dominating, 9
dual predictable projection, 6
finite invariant, 105, 154
Lebesgue invariant, 111
locally absolutely continuous with respect to another measure, 10
locally equivalent to another measure, 11
probability, 9

mixture of distribution laws, 60
mixture of normal laws, 60

observation, 9
  of a counting process, 21, 112
  of a diffusion-type process, 18, 19, 98, 151
  of a martingale, 140
  of a semimartingale, 15, 79, 194

power of the test, 33
probabilities of errors of a test, 33
  of the first kind, 33
  of the second kind, 33
probability space, 9
  with a filtration satisfying ordinary Dellacherie conditions, 10
  with filtration, 1
problem
  of distinguishing between hypotheses, 79
  of distinguishing between two simple hypotheses, 33
  of estimating parameter, 10
  of testing hypotheses, reduction, 70
  of testing two simple hypotheses, 10
process
  F-adapted, 1
  F-optional, 2
  F-predictable, 2
  adapted (to a filtration), 1
  compensator, 3
  counting, 19
  counting with determinate compensators, 116, 156
  diffusion-type, 17, 98, 151
  dual predictable projection, 3
  Hellinger of order ε, 80, 144
  homogeneous diffusion, 105, 154
    null recurrent, 109, 111
    positively recurrent, 105
    with exact growth at infinity, 110
  nondecreasing, 2
  of the local density of a measure, 11, 89, 112, 115
  optional, 2, 6
  Poisson, 116
  Poisson homogeneous, 116
  Poisson inhomogeneous, 116
  positive recurrent, 154
  predictable, 2, 6
  quadratic variation, 4
  renewal, 124, 126, 159
  standard Wiener, 16
  stochastic, 1
  with independent increments, 8
property of (τ_n)-uniqueness, 11
quadratic characteristic, 4
  mutual, 4
quantile of order p, 54

Radon–Nikodym derivatives, 33, 81
random measure, 5
  integral-valued, 5
  jump, 6
  local martingale, 6
  optional, 6
  predictable, 6
reduction of the problem of testing hypotheses
  first, 71
  second, 75
return time
  first, 106
risk function for an estimate, 175

sample space, 10, 15
semimartingale, 2, 8
  left-quasicontinuous, 8
  special, 2
  with independent increments, 121
set
  F-optional, 2
  F-predictable, 2
  P-negligible, 1
  negligible, 1
  nonessential, 1
  optional, 2
  predictable, 2
  random, 1
Shannon
  entropy of the distribution, 168
  information contained in an estimate, 175
  information contained in an observation, 161
  information with respect to a parameter, 161
  lower bound for the ε-entropy, 190
signal in white noise, 151
state space of a process, 1
statistic, 137, 152
  asymptotically information sufficient, 194
  information sufficient, 194
statistical experiment, 9
  binary, 9, 15
  generated by an observation, 9
statistical experiments
  generated by counting processes, 19
  generated by diffusion-type processes, 16
  generated by semimartingales, 10
  parametric, 16
statistical hypotheses, 33
  close, 102
  close noncontiguous, 104
  simple, 33
stochastic basis, 1
  P-complete, 1
  complete, 1
  satisfying ordinary conditions, 1
stochastic differential equation, 17

stochastic integral, 3
over a local martingale, 5
over a local martingale measure, 7
over a random measure, 5
over a semimartingale, 5, 18
over a square integrable local martingale, 4
stochastic interval, 2
stochastic process
P-modification, 1
regular **P**-modification, 3
stochastic processes
P-equivalent, 1
P-indistinguishable, 1
stopping time, 2
predictable, 2

test for distinguishing between hypotheses, 33
Bayes, 36
Neyman–Pearson, 34
nonrandomized, 35
theorem
characterization of complete asymptotic distinguishability of hypotheses, 38
characterization of complete asymptotic indistinguishability of hypotheses, 40
characterization of contiguity of hypotheses, 43
Girsanov's for semimartingales, 11
Krafft and Plachky, 52
on large deviations, 50, 84, 129
on the weak convergence, 86, 130
on the weak convergence of semimartingales, 31
Prokhorov's, 58
total variation, 36
trajectory of a stochastic process, 1
triple of predictable characteristics of a semimartingale, 8
type of asymptotic distinguishability, 45

Wald identity, 133, 135
weak convergence
of distribution laws, 53
of the likelihood ratio, 59
Wiener process, 112
self-similarity property, 112